DATE DUE

JUL 29			

Industrial Control Handbook
Volume 2: Techniques

Industrial Control Handbook

Volume 2: Techniques

E.A. Parr
BSc, CEng, MIEE

BSP PROFESSIONAL BOOKS
OXFORD LONDON EDINBURGH
BOSTON PALO ALTO MELBOURNE

First published 1987

British Library
Cataloguing in Publication Data
Parr, E.A.
Industrial control handbook
Vol. 2: Techniques
1. Process control
I. Title
670.42′7 TS156.8

ISBN 0-632-01835-6

BSP Professional Books
Editorial offices:
Osney Mead, Oxford OX2 0EL
(*Orders:* Tel. 0865 240201)
8 John Street, London WC1N 2ES
23 Ainslie Place, Edinburgh EH3 6AJ
52 Beacon Street, Boston
Massachusetts 02108, USA
667 Lytton Avenue, Palo Alto
California 94301, USA
107 Barry Street, Carlton
Victoria 3053, Australia

Typeset by Scribe Design
Gillingham, Kent
Printed and bound in Great Britain by
Billing & Sons of Worcester

For Alison

Technology has brought meaning to the life of many technicians.

Ed Bluestone

Contents

Preface

This book is the second in a three-volume series on the topic of industrial process control. The aim of the series is to present the subject in a practical manner that reflects the way industry works. Volume 1 covers the vital sensors and transducers that gather information from the plant. Volume 3 gives the background theory and mathematics normally associated with process control, and provides the link between theory and practice. The series should be of value to both the newly qualified engineer (for whom the bridge from theory to practice is often difficult to cross) and the practising engineer who wishes to widen his base of knowledge.

The process control engineer has to be competent in a wide range of technologies–power electronics, pneumatics, hydraulics and computing to name a few. This volume is concerned with these and other techniques, and gives sufficient detail for the engineer to use and apply the equipment described.

Many people have helped in the preparation of this book, particularly my fellow engineers at the Sheerness Steel Company who have given helpful suggestions and advice. Special thanks are due to my long-suffering family for the ever increasing backlog of work that remains unfinished about the house. Apologies are due to a very patient Bernard Watson of Collins who tolerated a deadline that seemed to follow the pattern of suppliers' delivery dates and slip further and further back! Many companies have provided information and photographs. These are acknowledged at the relevant places in the text.

Andrew Parr
Isle of Sheppey,
Kent

June 1987

Chapter 1
DC amplifiers

1.1. Introduction

1.1.1. DC amplifier requirements

Signals in instrumentation and process control are generally represented digitally, pneumatically or as an analog voltage or current. Digital signals are covered in chapter 3, and pneumatics in chapter 6. This chapter is concerned with the manipulation of signals represented as an electrical voltage or current.

Instrumentation signals are essentially static for long periods, and as such require amplifiers with predictable characteristics down to 0 Hz, i.e. DC amplifiers. A typical AC audio amplifier, for comparison, would have little gain below about 20 Hz, and would 'droop' on low-frequency signals as fig.1.1a. The lower frequency limit in AC amplifiers is usually determined by the impedance of coupling capacitors between stages and emitter decoupling capacitors. In fig.1.1b, the impedances of C1 to C5 will all increase with decreasing frequency, causing the gain to fall. DC amplifiers therefore use direct coupling between stages.

Direct coupling, however, brings its own problems. Figure 1.2a shows a possible design for a simple direct coupled amplifier. Unfortunately almost all transistor parameters vary with temperature and from device to device. V_{be}, for example, changes by 2 mV per °C, and collector/emitter leakage current doubles every 10°C. These, and similar effects, will cause the output of fig.1.2a to vary in a manner which is indistinguishable from changes caused by the signal itself.

Most DC amplifiers are based on the so-called long tailed pair circuit of fig.1.2b. TR1 and TR2 are identical transistors maintained at the same temperature (both conditions being ensured by constructing the circuit on a single integrated

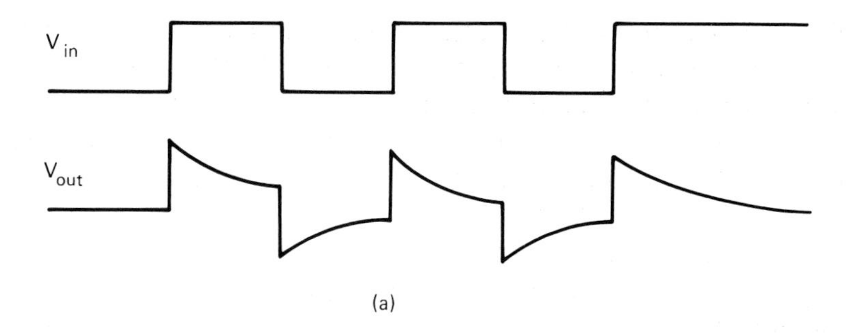

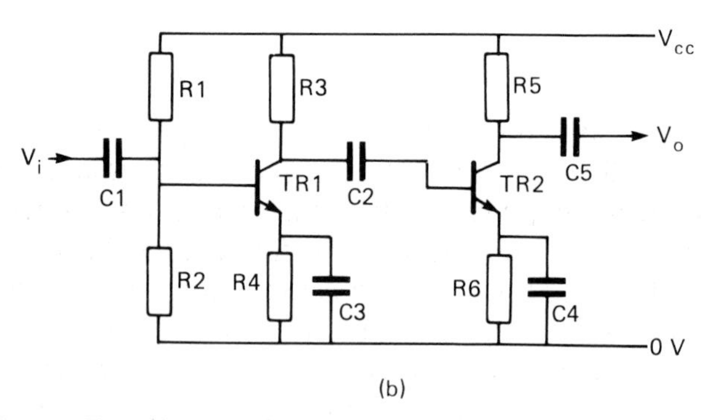

Fig. 1.1 The effect of AC amplifiers on low frequency signals. (a) The effect of poor low frequency response on a signal. (b) A simple AC amplifier. The low frequency response is determined by capacitors C1–C5.

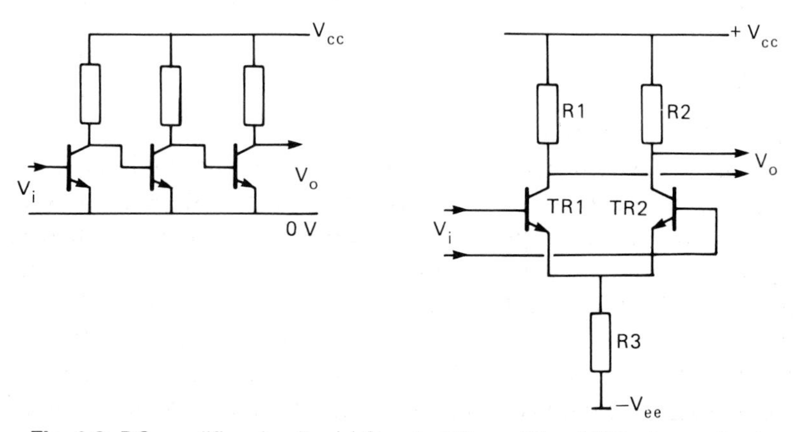

Fig. 1.2 DC amplifier circuits. (a) Simple DC amplifier. (b) The long tail pair.

circuit–IC–silicon wafer). Resistor R3 acts as a constant current sink, with the current being split between TR1/TR2. Because the two transistors are identical and at the same temperature, leakage currents and offsets will cancel and the current split between TR1, TR2 will depend solely on V_{in}. The output voltage is therefore an amplified version of V_{in}, and is unaffected by temperature changes. Note that fig.1.2b is, in effect, a *differential* amplifier because it amplifies the voltage difference between its two inputs.

1.1.2. Integrated circuit DC amplifiers

It is exceedingly rare nowadays for DC amplifiers to be constructed of individual transistors. The requirements of fig.1.2b are identical components and a uniform circuit temperature. These conditions are best met by fabricating the entire circuit as an IC. There are many IC DC amplifiers available with different characteristics, but by far the commonest is the ubiquitous 741, arguably the most successful IC ever designed. Other DC amplifiers are similar in principle, differing only in, say, greater or lesser gain or perhaps frequency response. DC amplifier specifications are described in section 1.2.

A DC amplifier can be represented by fig.1.3a; this has two inputs, two power supply connections (usually symmetrical 15 volts positive and negative) and the output. Note that all voltages (including the power supplies) are referred to the 0V rail, even though the DC amplifier itself does not need an 0V connection. The output voltage is given by:

$$V_o = A\,(V1 - V2) \qquad (1.1)$$

where A is the amplifier gain.

Input 1 is usually called the non-inverting input and input 2 the inverting input, for reasons that can be seen by linking each input to 0V in turn. With input 2 linked to 0V, and a signal applied to input 1, $V_o = AV1$; the output moves in the same sense as the input signal. With input 1 linked to 0V, and a signal applied to input 2, $V_o = -AV2$; the output moves in the opposite sense to the input signal. Often a + sign is used for the non-inverting input and a − sign for the inverting input; these should not be confused with the power supply connections.

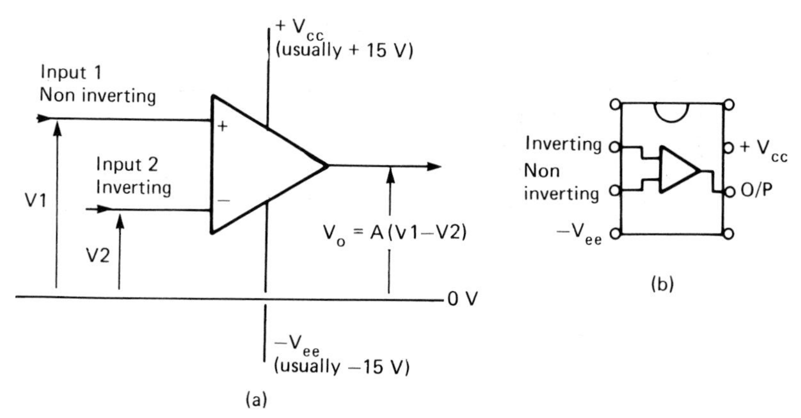

Fig. 1.3 Practical DC amplifiers. (a) Representation of a DC amplifier. (b) Integrated circuit DC amplifier (741).

DC amplifiers are usually encapsulated in an eight-pin dual in line (DIL) IC. The arrangement of fig.1.3b is widely used, although there are slight variations in the use of pins 1, 5 and 8.

High-gain DC amplifiers were originally used in analog computers where the term operational amplifier, shortened to op amps, was used. This description is usually given to IC DC amplifiers and will be used in the rest of this chapter.

1.2. Op amp specifications

1.2.1. Introduction

There are probably several hundred different op amp ICs. The process control engineer needs to be able to select a device for a specific application. This section describes the relevant items found on a specification sheet. Where typical values are given, the 741 specification has been used.

1.2.2. DC gain (A_{VD})

This is defined as the change in output volts divided by the change in input volts. Referring to fig.1.4a:

$$A_{VD} = \frac{\Delta V_o}{\Delta(V1 - V2)} \tag{1.2}$$

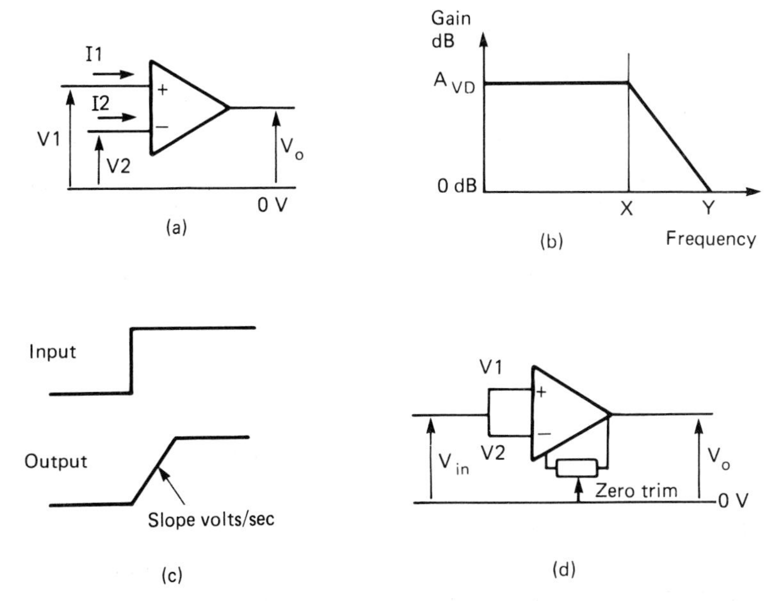

Fig.1.4 Definition of DC amplifier terms. (a) DC gain and input bias currents. (b) Unity gain bandwidth. (c) Slew rate. (d) Common mode gain.

A typical value for a 741 is around 100 000. Often the gain is given as V/mV, so a gain of 100 000 would be given as 100 V/mV. Sometimes the gain is given in dB:

$$A_{VD} = 20 \log_{10}(\Delta V_o/(\Delta(V1 - V2))) \text{ decibels} \qquad (1.3)$$

The output of an op amp must, obviously, stay within the supply voltages, so a differential input voltage of less than 1 mV is required to drive the amplifier into saturation. In practice, however, feedback is used to define the gain as described in section 1.3.

1.2.3. Unity gain bandwidth (BW)

All amplifiers have a frequency response similar to fig.1.4b. In audio amplifier specifications it is usual to define point X: the 3 dB point. With op amps this point is of little interest because the closed loop gain is usually far lower than A_{VD}. The bandwidth is commonly specified as the frequency at which the gain drops to unity (0 dB), point Y on fig.1.4b. Many op amps (including the 741) have a deliberate low-frequency roll of a few hertz to assist stability when feedback is applied. Point X for a 741 is actually 10 Hz, whereas the unity gain bandwidth is 1 MHz. Choosing an op amp with a better frequency response than an application requires can lead to stability problems. (see section 1.3.5).

1.2.4. Slew rate (SR)

In fig.1.4c a step input has been applied to an amplifier input, with the amplitude of the step such that it is just sufficient to drive the amplifier into saturation. The output of the amplifier will not be a step, but a ramp as shown. The ramp slope is a measure of the useful frequency range for the amplifier. A 741 has a slew rate of 1 V/μS, but amplifiers are available with slew rates of over 100 V/μS.

1.2.5. Input offset voltage (V_{IO})

If V1, V2 in fig.1.4a are both tied to 0 V, V_{out} should be zero. In practice, V_o will be either positive or negative by a significant amount. The offset voltage is the differential input voltage that is needed to make V_o zero. The figure for a 741 is 2 mV maximum. In itself, V_{IO} is not particularly important as it can be removed by the zeroing circuits of section 1.3.4. What is usually more important is how V_{IO} changes with temperature.

1.2.6. Offset voltage temperature coefficient (αV_{IO})

This specification states how V_{IO} may change with temperature and is typically 5 μV/°C. This error cannot be zeroed out and can only be controlled by choosing an amplifier with a sufficiently low value of αV_{IO} for the application. Values as low as 0.5 μV/°C are available (at a price).

1.2.7. Input bias current (I_{IB})

With V1 = V2 = 0 V in fig.1.4a input currents I1, I2 will be non-zero (being the base currents for the input long tail pair). A typical value is 0.1 μA. This will not normally cause any problems provided the source impedances of V1, V2 are equal.

1.2.8. Input offset current (I_{IO})

This is the difference between I1 and I2, and will cause an offset voltage even if the source impedances are equal. A typical value for a 741 is 20 nA. Errors due to I_{IO} can be removed by a zeroing potentiometer.

1.2.9. Common mode gain (A_{CM})

In fig.1.4d, V1 and V2 have been linked and V_{in} set to 0 V. The zeroing potentiometer has been adjusted to set V_o to 0 V. If V_{in} is varied, V_o should stay at zero volts, but in practice will vary slightly. The common mode gain is defined as:

$$A_{CM} = \frac{\text{change in } V_o}{\text{change in } V_{in}} \qquad (1.4)$$

The common mode gain is an indication of how well an amplifier will reject common mode noise. Usually, an amplifier specification does not give A_{CM}, but uses the common mode rejection ratio defined below.

1.2.10. Common mode rejection ratio (CMRR)

The DC gain A_{VD} was defined in section 1.2.2, and the common mode gain A_{CM} above. The CMRR is defined as:

$$\text{CMRR} = \frac{A_{VD}}{A_{CM}} \qquad (1.5)$$

The CMRR is large, and is usually given in dB. The 741 has a CMRR of 90 dB.

1.3. Basic circuits

1.3.1. Inverting amplifiers

The commonest DC amplifier circuit is the inverting amplifier of fig.1.5. The output voltage V_o will be in the range ±15 V, so the junction of R1/R2 will be in the range $\pm 15/A_{VD}$; typically less

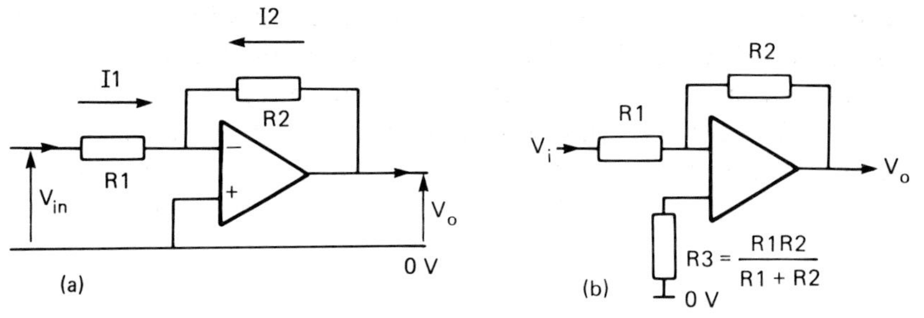

Fig.1.5 The inverting amplifier. (a) Theoretical circuit. (b) Practical circuit to eliminate offset from input bias currents.

than a mV. For all practical purposes, we can assume that both inputs of the amplifier remain at 0 V (the junction of R1, R2 being called a virtual earth).

If the input bias current is small compared with I1, I2, we can say:

$$I1 + I2 = 0 \tag{1.6}$$

$$\frac{V_{in}}{R1} + \frac{V_o}{R2} = 0 \tag{1.7}$$

or

$$V_o = -\frac{R2}{R1} V_{in} \tag{1.8}$$

The minus sign denotes inversion, i.e. positive V_{in} gives negative V_o. The closed loop gain (−R2/R1) is independent of the amplifier gain.

If R1 and R2 are large (> 100K) resistor R3 should be added as fig.1.5b to equalise the offset due to I_{IB}. The value of R3 is chosen to equal the values of R1, R2 in parallel (e.g. if R1 = R2 = 330K, R3 would be chosen to be 150K, the nearest preferred value to the theoretical 165K).

Equation 1.8 assumes the amplifier has infinite input impedance, zero output impedance and an open loop gain (A_{VD}) much larger than the closed loop gain (R2/R1). This will be true for almost all circuits using IC op amps. If the open loop gain is included, equation 1.8 becomes:

$$V_o = -\frac{R2}{R1 + \dfrac{(R1 + R2)}{A_{VD}}} \cdot V_{in} \tag{1.9}$$

For a typical circuit, the difference between equations 1.8 and 1.9 is less than 0.05%. The effect of finite input impedance (typically 2MΩ) and non-zero output impedance (about 50Ω) is even less.

The closed loop input impedance of fig.1.5 is simply the value of R1. This can be unacceptably low if high gain is required. The non-inverting amplifier below has very high input impedance.

1.3.2. Non-inverting amplifiers

The non-inverting amplifier circuit is shown in fig.1.6. Resistor R3 is included purely to minimise the input bias current offset and should equal the value of R1 and R2 in parallel. R3 plays no part in the determination of the gain.

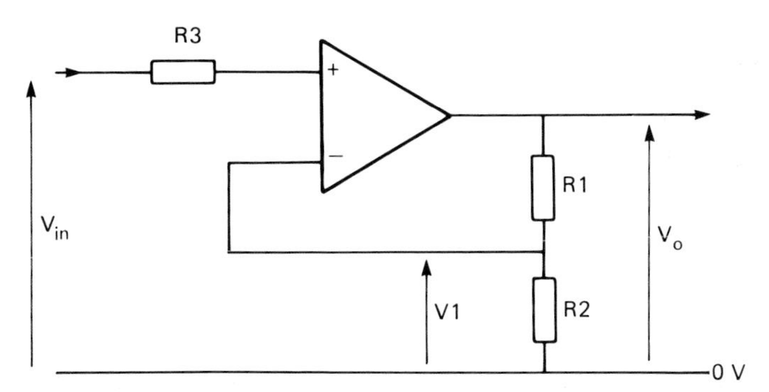

Fig.1.6 Non-inverting amplifier

The voltage at the junction of R1 and R2 is given by:

$$V1 = \frac{R2}{R1 + R2} V_o \tag{1.10}$$

As explained previously, the voltages at the two amplifier inputs can be considered equal, so:

$$V1 = V_{in} \tag{1.11}$$

$$\frac{R2}{R1 + R2} V_o = V_{in} \tag{1.12}$$

$$V_o = \frac{(R1 + R2)}{R2} V_{in} \tag{1.13}$$

The closed loop gain is determined solely by R1 and R2 and is independent of the actual amplifier gain (provided the closed loop gain is small compared with the open loop gain).

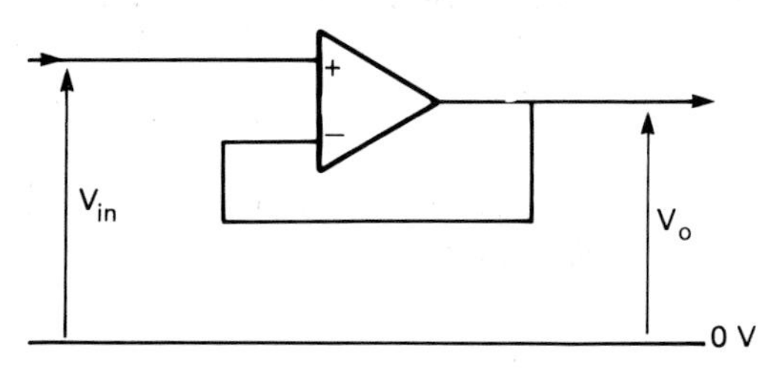

Fig.1.7 Unity gain buffer

The input impedance of the non-inverting amplifier is very high, and can be considered infinite for most purposes. If a high input impedance unity gain amplifier is required, the buffer circuit of fig.1.7 may be used. Obviously $V_o = V_{in}$.

1.3.3. Differential amplifiers

The circuits of figs. 1.5 to 1.8 amplify a signal which is referred to the 0 V line. In many applications a true differential input is

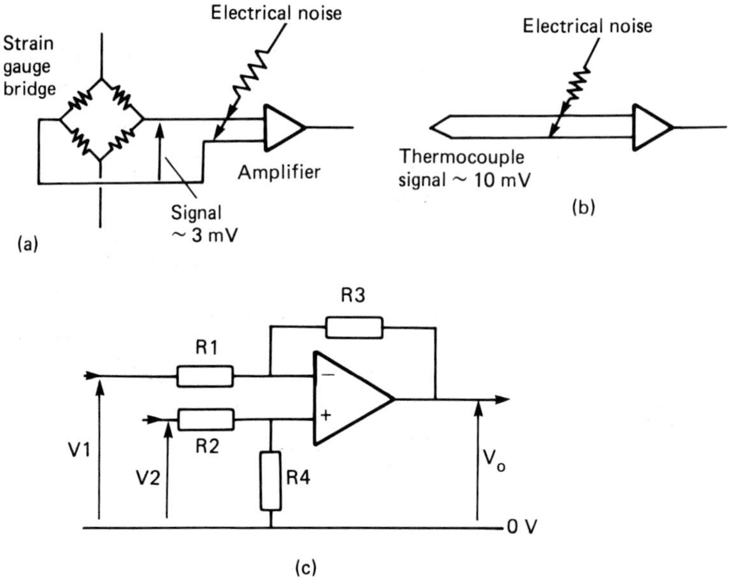

Fig.1.8 The differential amplifier. (a) Strain gauge circuit. (b) Thermocouple circuit. (c) Practical circuit.

required. The output from the strain gauge bridge of fig.1.8a, for example, is a voltage which is not referred to 0 V. Another common requirement is shown on fig.1.8b where a thermocouple signal is to be amplified despite common mode noise of several volts being superimposed on the signal lines. In both cases, V_o is to be an amplified version of V_{in} and superimposed input voltages with respect to 0 V are to be ignored.

Figure 1.8c shows a differential output amplifier. If R1 = R2 = R_a and R3 = R4 = R_b, then:

$$V_o = -\frac{R_b}{R_a}(V1 - V2) \tag{1.14}$$

The circuit's common mode rejection depends on how close the resistors match. Where a mV signal is to be amplified, precision resistors should be used.

The impedance seen by V2 is R2 + R_4. The impedance seen by V1 varies with the signal, but is of the order of R1. The circuit can handle input voltages outside the supply voltages, provided the actual amplifier inputs stay within the supply range. With unity gain (all resistors equal), the circuit will remove 30 V common mode voltage from a signal of a few mV.

1.3.4. Zeroing circuits

The input voltage offset V_{IO} is typically a few mV. In many circuits, where the closed loop gain is small and the signal levels are larger than the offset, the effect of V_{IO} can be ignored. When the signal levels are similar to the offset, a zeroing (or nulling) circuit must be included in the design.

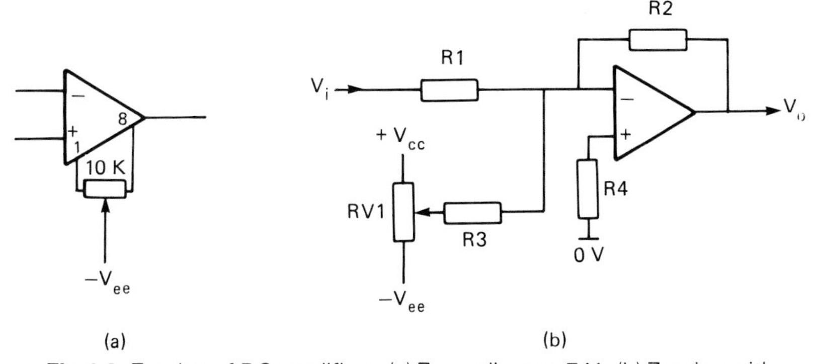

Fig.1.9 Zeroing of DC amplifiers. (a) Zero adjust on 741. (b) Zeroing with inverting amplifier.

Many ICs, including the 741, have provision for a single zeroing potentiometer to be connected direct to the amplifier, as in fig.1.9a. An alternative method for inverting amplifiers is shown in fig.1.9b. RV1 effectively adds an offset current to the virtual earth to produce a voltage output which cancels V_{IO}. The circuit of fig.1.9b can also be used to generate an Ax + B function where R2/R1 determines A, and the setting of RV1 determines B.

1.3.5. Stability

Almost all op amp circuits use feedback and, as shown in Volume 3 of this series, feedback can cause instability. The analysis techniques given in Volume 3 can be used to analyse op amp circuits, but in most applications are not necessary.

In general, stability problems only occur where an op amp has high gain at high frequencies, and stray capacitance and lead inductance become significant. Most signal processing circuits deal with frequencies of at most a few kHz. Op amps such as the 741 have a deliberately low high-frequency gain to prevent high-frequency oscillation. Such ICs are said to be unconditionally stable.

If high-frequency operation is required, the gain/frequency response of many op amps can be tailored by the use of external components as shown for the 709 and 308 ICs in fig.1.10. Op amps with useful gain up to 500 kHz are available, but care in the

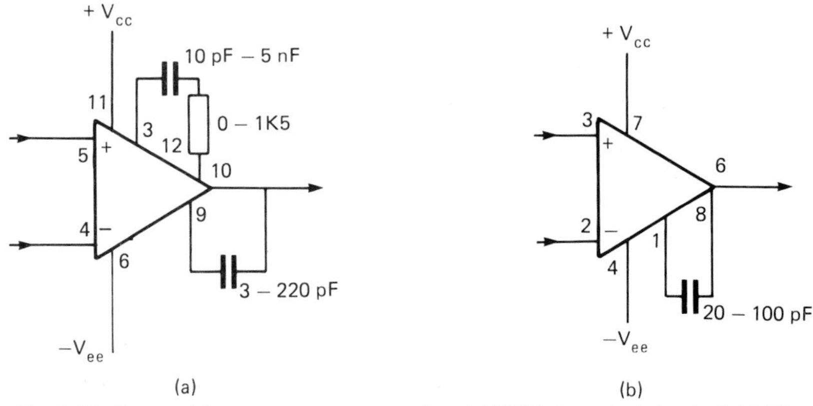

Fig.1.10 External frequency compensation. (a) 709 integrated circuit. (b) 308 integrated circuit.

circuit layout is required to avoid oscillation. The cardinal rule is choose a frequency response that is just good enough for the application.

1.4. Computing circuits

1.4.1. Addition

Figure 1.11 is used where two or more voltages are to be added. By the reasoning of section 1.3.1.:

$$I1 + I2 + I3 + I4 = 0 \tag{1.15}$$

$$\frac{V1}{R1} + \frac{V2}{R2} + \frac{V3}{R3} + \frac{V_o}{R4} = 0 \tag{1.16}$$

or

$$V_o = -\left(\frac{R4}{R1}V1 + \frac{R4}{R2}V2 + \frac{R4}{R3}V3\right) \tag{1.17}$$

if R1 = R2 = R3 = R

$$V_o = -\frac{R4}{R}(V1 + V2 + V3) \tag{1.18}$$

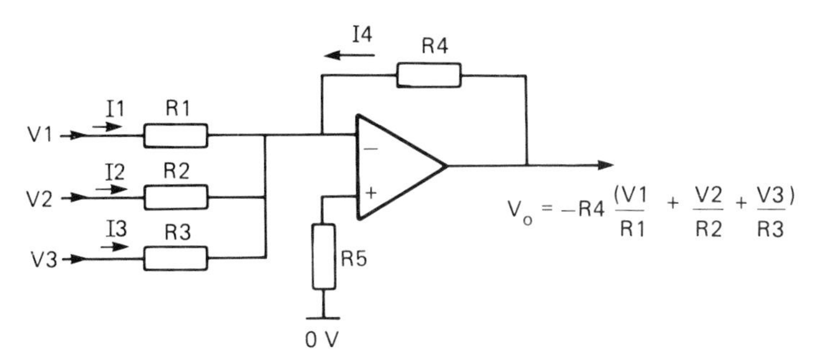

Fig.1.11 Summing amplifier

Obviously the circuit can be expanded to any number of inputs. Resistor R5 is included to remove the effect of I_{IB}. The value of R5 should equal the parallel combination of the other resistors.

1.4.2. Subtraction

Where one voltage is to be subtracted from another, the differential amplifier of fig.1.8c may be used with all resistors equal in value. Where a combination of addition and subtraction (e.g. V1 + V2 − V3 + V4) or a subtraction involving multiplier constants (e.g. AV1 − BV2 where A and B are constants) is needed, unity gain inverters may be used as shown on fig.1.12.

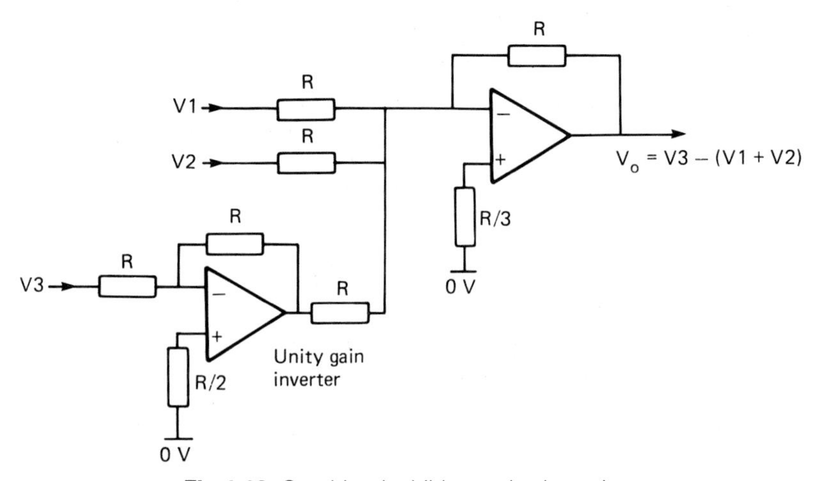

Fig.1.12 Combined addition and subtraction

1.4.3. Multiplication by a constant

If the constant is less than unity, a simple voltage divider as in fig.1.13 should be used. The output voltage should go to an impedance much higher than the parallel combination of R1 and R2. If this is not possible, a unity gain amplifier (as shown in fig.1.7) should follow the divider.

If the multiplication constant is greater than unity, a fixed gain inverting amplifier (fig.1.5) or non-inverting amplifier (fig.1.6) should be used.

1.4.4. Multiplication and division of two voltages

Multiplication of two voltages is not widely used because the circuits are complex and not very accurate. In general, input

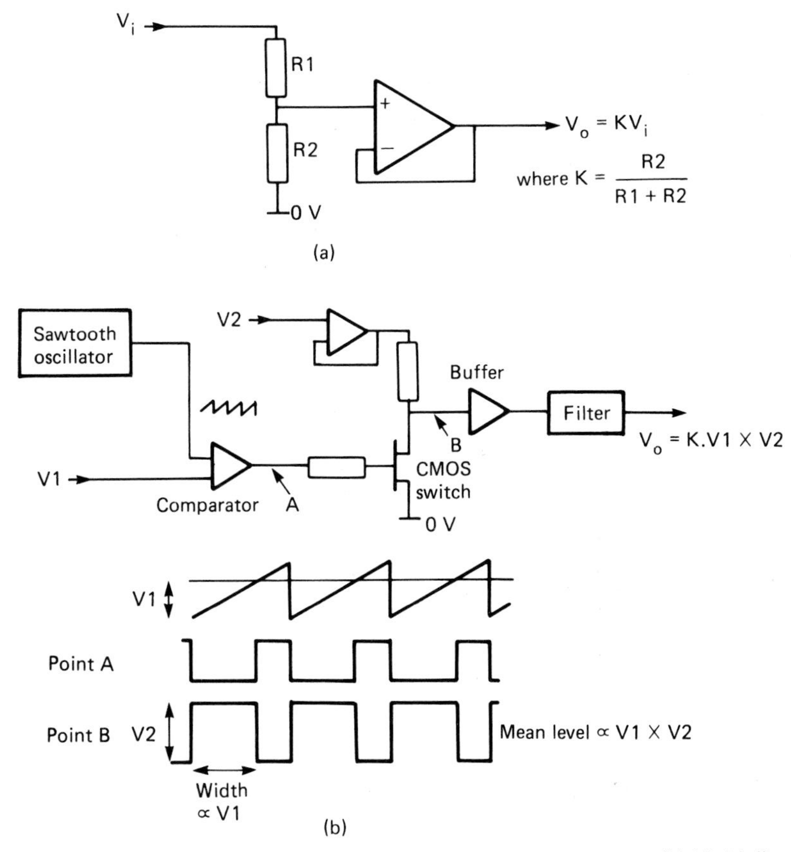

Fig.1.13 Multiplication circuits. (a) Multiplication by a constant. (b) Multiplier circuit.

voltages are scaled 0 to 10 volts, and the output is given by:

$$V_o = 0.1 \times V1 \times V2 \tag{1.19}$$

With V1 = 1.4 volts and V2 = 7.9 volts, V_o would be 1.106 volts. This keeps the output voltage at reasonable levels.

One possible multiplier circuit, shown in fig.1.13, uses a pulse with one voltage controlling the pulse width and the other the pulse height.

A sawtooth voltage is fed to a comparator along with the first voltage V1. The output of the comparator at point A is a square wave of constant frequency, but pulse width proportional to V1. This is used to control an electronic CMOS switch connected to input voltage V2. The pulses at point B have a width proportional to V1 and height proportional to V2. If this is passed through a

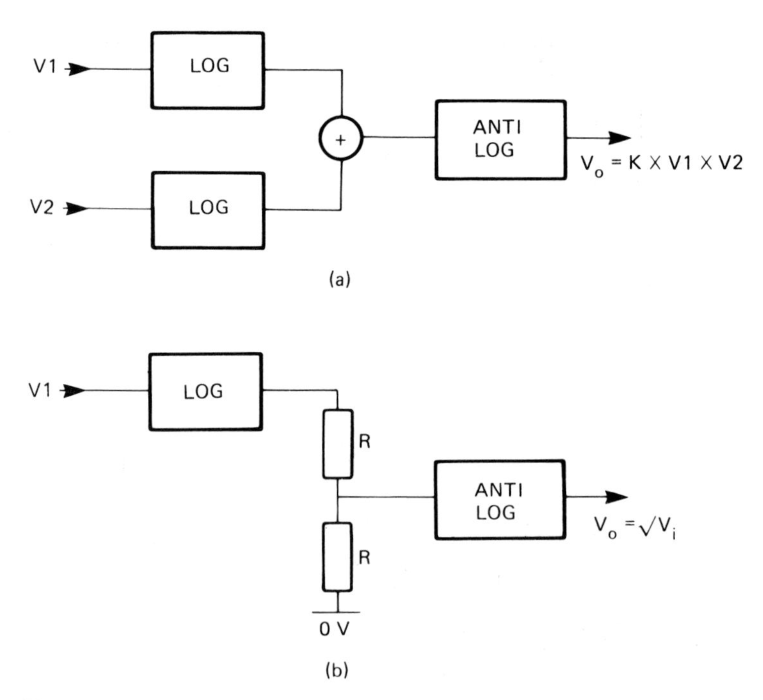

Fig.1.14 Use of logarithmic elements. (a) Multiplication using log/antilog amplifiers. (b) Square root extractor.

low pass filter, the output voltage is proportional to the product of V1 and V2.

The voltage drop across a semiconductor diode is related to the current through it by the exponential relationship:

$$I = Ae^{BV} \tag{1.20}$$

where A and B are constants. This allows the construction of logarithmic and antilogarithmic amplifiers (by the use of voltage to current and current to voltage amplifiers).

Multiplication of numbers can be performed by the addition of the numbers' logarithms, so a multiplier can be constructed as shown in fig.1.14. In practice, ICs such as the AD534 are available which perform multiplication using the log/antilog principle.

Figures 1.13 and 1.14 can only perform multiplication of two positive numbers (called single quadrant multiplication). Correctly signed multiplication of positive and negative numbers (called four quadrant multiplication) is, of course, far more complex but is available in IC form.

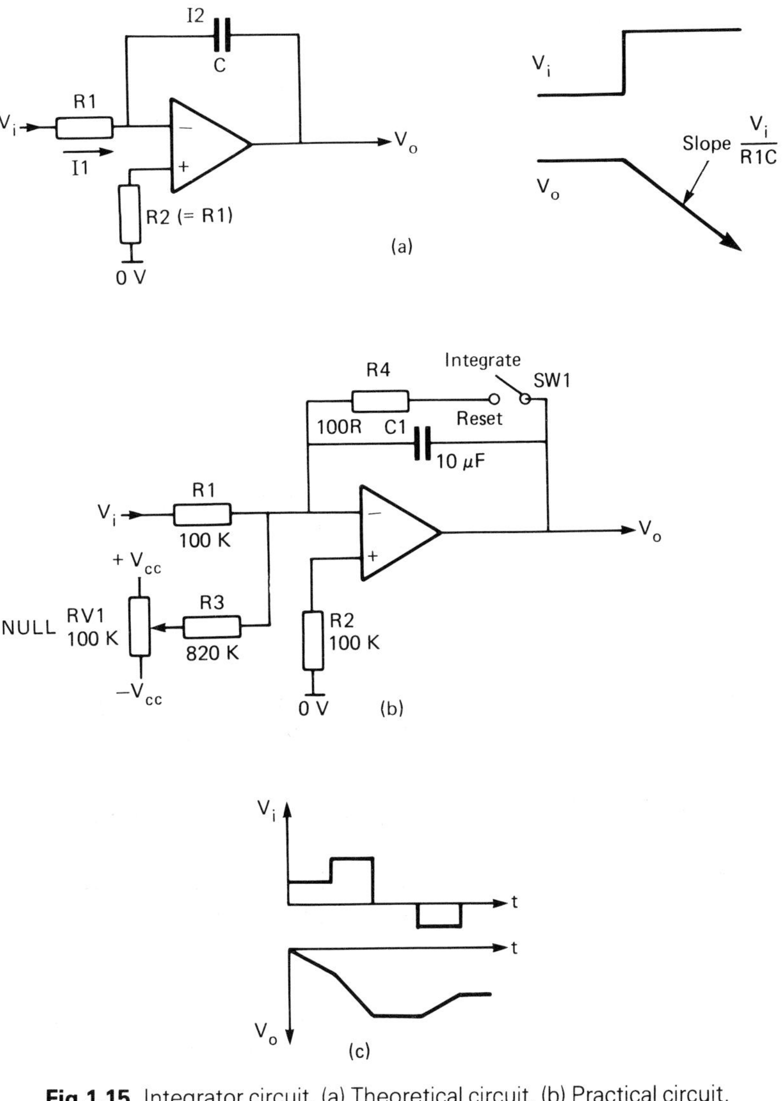

Fig.1.15 Integrator circuit. (a) Theoretical circuit. (b) Practical circuit. (c) Typical action.

Division can also be performed with log/antilog amplifiers by subtracting logarithms. If the adder in fig.1.14 is replaced by a subtractor, V_o will become V1/V2. Division is rarely performed in practice.

Square root extraction can also be performed with log/antilog amplifiers by taking the log of the input voltage, dividing by 2 and taking the antilog as shown in fig.1.14b. Square root circuits are often needed in flow measurement.

1.4.5. Integration

The theoretical circuit for an integrator is shown in fig.1.15a; as before, if the amplifier has infinite input impedance, I1 and I2 are equal so:

$$\frac{V_i}{R} + C\frac{dV_o}{dt} = 0 \tag{1.21}$$

$$\frac{dV_o}{dt} = -RCV_i \tag{1.22}$$

$$\text{or } V_o = -\frac{1}{RC}\int V_i\,dt + A \tag{1.23}$$

where A is an initialising constant.

A practical circuit is shown in fig.1.15b. Difficulties are usually caused by the input bias current of the amplifier (which is integrated by C) and leakage current through the capacitor. Low leakage (non-electrolytic) capacitors and FET op amps should therefore be used.

RV1 is more than a simple zero control. Because V_o is the integral of V_i, the output will not be zero for zero input voltage, but should stay at its current value. RV1 should be adjusted so that V_o neither rises nor falls with zero input voltage.

Switch SW1 (which can be a CMOS switch) is used to short out C1 when the integrator is not in use (to stop integral wind-up when a PID controller is in manual, for example), or to initialise the circuit. R4 is purely to limit the current through the capacitor when SW1 is first closed.

Figure 1.15c shows the action of an integration circuit.

1.4.6. Differentiation

Figure 1.16a shows the theoretical circuit for a differentiator. By usual analysis:

$$V_o = -RC\frac{dV_i}{dt} \tag{1.24}$$

A differentiator implies a continually rising gain, which is impossible because of factors such as stray capacitance and the

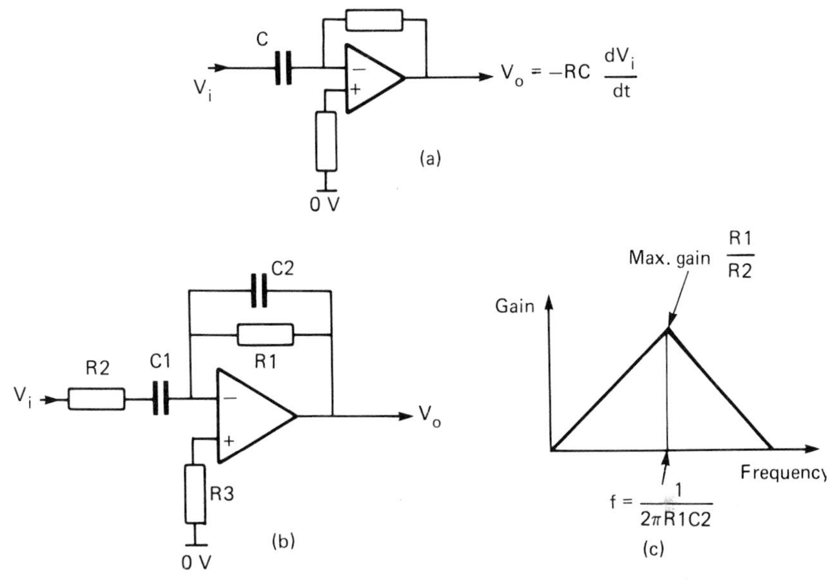

Fig.1.16 Differentiator circuit. (a) Theoretical circuit. (b) Practical circuit. (c) Frequency response.

frequency response of the amplifier itself. To give a predictable high-frequency response, the circuit of fig.1.16b is usually used. R1 and C1 form the differentiator, and R2, C2 are chosen such that:

$$R1\,C1 = R2\,C2 \tag{1.25}$$

The circuit has the response of fig.1.16c, and acts as a differentiator up to a frequency given by:

$$f = \frac{1}{2\pi\ R1\ C2} \tag{1.26}$$

at which point the gain is R1/R2.

It should be noted that the increasing gain with frequency characteristic makes differentiation prone to noise pickup. In designing a circuit based on fig.1.16b, the frequency at which the gain peaks should be set as low as the application allows.

1.4.7. Analog computers and differential equations

Using the circuits in sections 1.4.1. to 1.4.6, DC amplifiers can be used to perform complex arithmetical operations and solve simultaneous and differential equations. An analog computer is a collection of DC amplifier circuits which can be interconnected by jumper leads to perform a particular calculation. They are particularly useful for real time simulation (for testing the instrumentation and control strategy for a plant, for example).

Suppose we wish to simulate the equation:

$$5 = \frac{d^2x}{dt^2} + 3\frac{dx}{dt} + 2x \tag{1.27}$$

Figure 1.17a shows that given d^2x/dt^2 we can obtain $-dx/dt$ and x by two integrators. Equation 1.27 can be rearranged:

$$\frac{d^2x}{dt^2} = 5 - 3\frac{dx}{dt} - 2x \tag{1.28}$$

The terms on the right-hand side of the equation are obtained from the two integrators by suitable scaling amplifiers and added to form d^2x/dt^2 as shown in fig.1.17b. Switch SW1 allows the step response of the circuit to be investigated by switching the constant term in and out. Switch SW2 is used to short out the integrating capacitors to prevent drift when the circuit is not in use.

Circuits similar to fig.1.17 can be used to simulate most differential equations, and non-linear effects which are difficult to

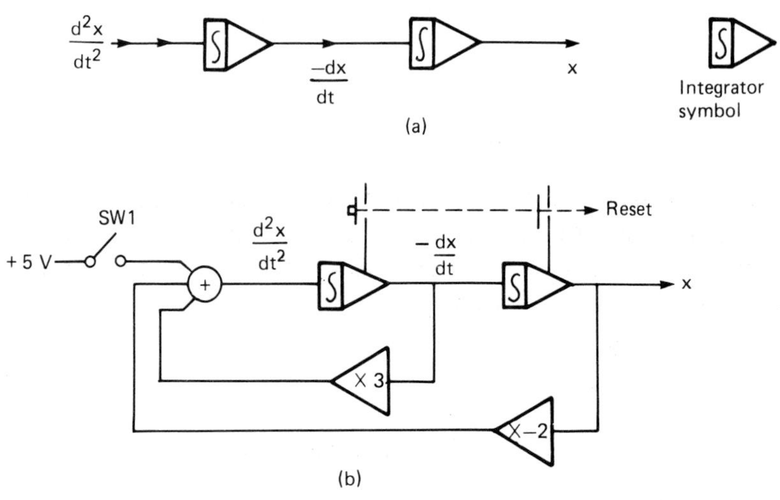

Fig.1.17 Analog computing circuits. (a) Basic integrator action. (b) Circuit for solution of a different equation.

analyse mathematically (e.g. saturation) can be included by means of the circuits of section 1.6. Where a process has long time constants (e.g. blast furnaces), the equations can be rearranged to give a scaled time (e.g. 1 second represents 1 minute).

1.5. Filter circuits

1.5.1. Low pass filters

The circuit of fig.1.18 simulates a simple exponential rise, and as such is a simple low pass filter. The DC gain is, obviously, −R2/R1 and the 3 dB cut-off frequency is given by:

$$f = \frac{1}{2\pi CR2} \tag{1.29}$$

The circuit has a roll off of 6 dB/octave above this frequency.

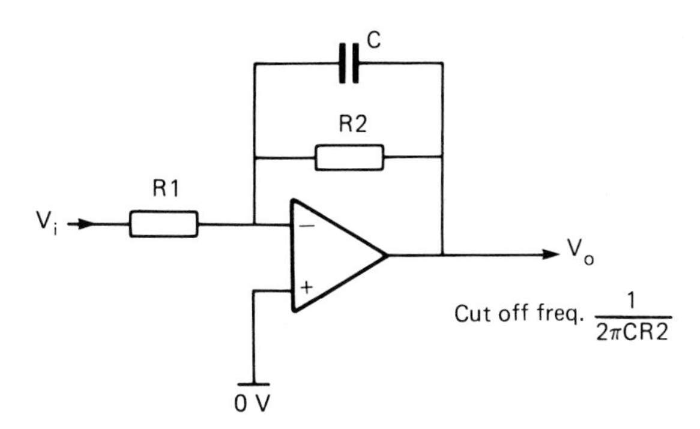

Fig.1.18 Simple low pass filter.

The classical low pass filter is shown in fig.1.19a. The circuit has unity gain, and a roll off of 12 dB/octave, and should be designed with R1 = R2 = R and C2 = 0.5C1 = C (i.e. C1 = 2C). The cut-off frequency is then given by:

$$f = \frac{\sqrt{2}}{4\pi RC} \tag{1.30}$$

If C1 is not 2C2, the cut-off frequency is given by:

$$f = \frac{1}{2\pi R\sqrt{C1.C2}} \tag{1.31}$$

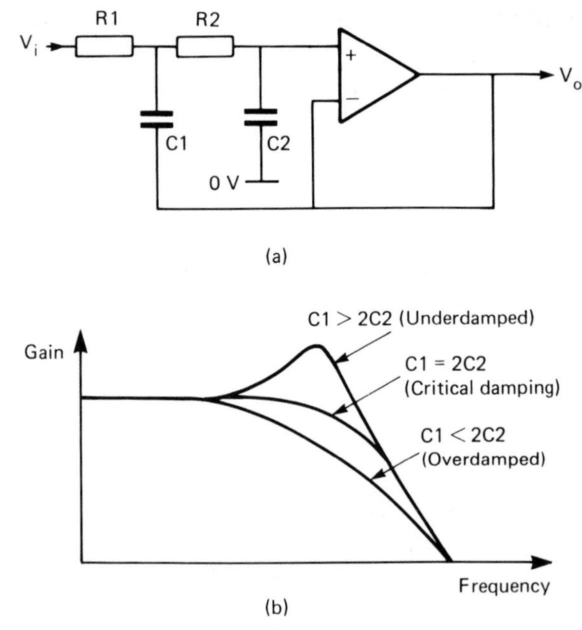

Fig.1.19 Two stage low pass filter. (a) Classical low pass filter. (b) Frequency response.

and the damping of the circuit depends on the ratio C1/C2 as shown in fig.1.19b.

1.5.2. High pass filters

A perfect high pass filter is not feasible because the frequency response of the amplifier itself will ultimately cause high frequency attenuation. The amplifier frequency response is therefore a critical part of the design.

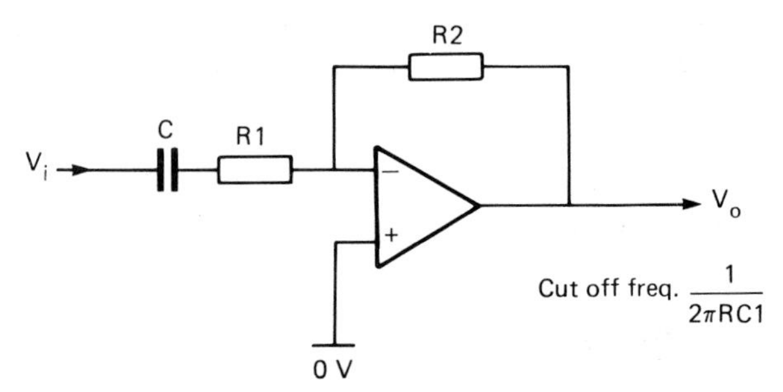

Fig.1.20 Simple high pass filter.

The simplest high pass filter is shown in fig.1.20. This has a high frequency gain of R2/R1 (below the point at which the fall off in amplifier gain becomes significant). The cut off frequency is given by:

$$f = \frac{1}{2\pi RC1} \tag{1.32}$$

An alternative high pass filter is shown in fig.1.21a. If the components are chosen such that R1 = 0.5R2 = R and C1 = C2 = C, the cut off frequency is given by:

$$f = \frac{\sqrt{2}}{4\pi RC} \tag{1.33}$$

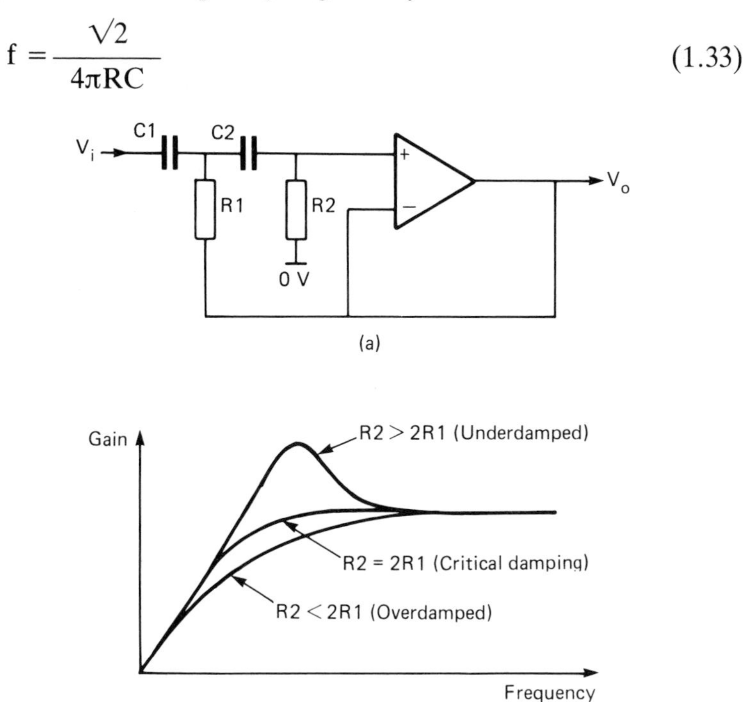

Fig.1.21 Two stage high pass filter. (a) Classical high pass filter. (b) Frequency response. Gain will fall again at high frequencies because of amplifier limitations.

The response of the circuit, like that of fig.1.19, can be adjusted by the ratio of R1 to R2. The cut-off frequency is then given by:

$$f = \frac{1}{2\pi C\sqrt{R1.R2}} \tag{1.34}$$

The circuit has unity gain at high frequencies.

1.5.3. Bandpass filters

A bandpass filter passes a specific band of frequencies, and is specified by the centre frequency and the width of the frequency band at the −3 dB point. The latter characteristic is usually expressed as the Q of the circuit, being the ratio of the centre frequency to the bandwidth. The higher the value of Q, the sharper the peak.

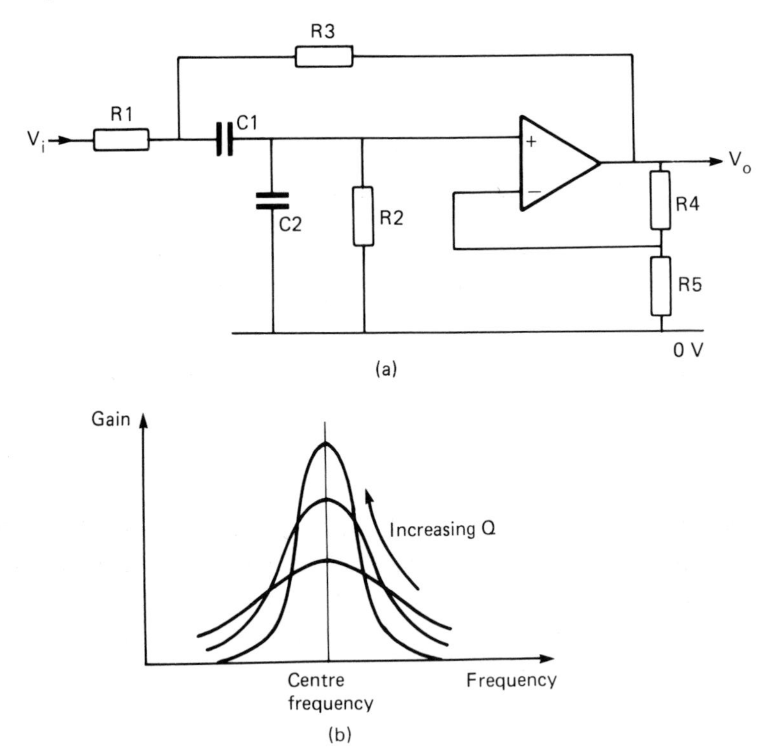

Fig.1.22 Bandpass filter. (a) Bandpass filter circuit. (b) Frequency response.

A simple circuit is shown in fig.1.22. The components should be chosen so R1 = R2 = R3 = R and C1 = C2 = C. The centre frequency is then given by:

$$f = \frac{\sqrt{2}}{2\pi RC} \tag{1.35}$$

and the Q of the circuit by:

$$Q = \frac{R5\ \sqrt{2}}{4R5 - R4} \tag{1.36}$$

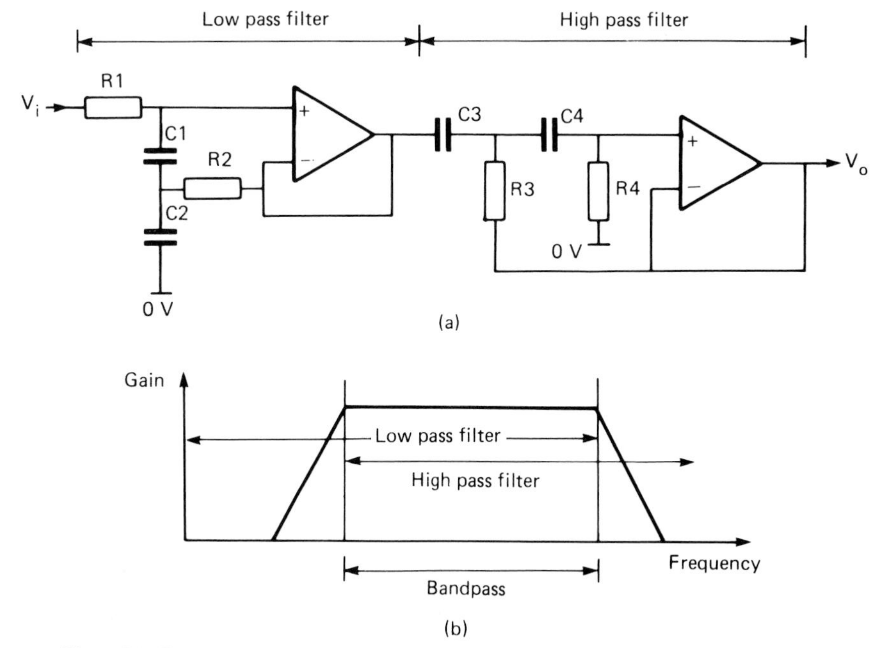

Fig.1.23 Bandpass filter. (a) Circuit diagram. (b) Frequency response.

The circuit of fig. 1.22a is useful where a small bandwidth is required; decreasing Q gives a poorly defined cut-off for a larger bandwidth, as shown in fig.1.22b. Where a wide bandwidth with sharply defined cut-off is required, a high pass and low pass filter can be combined as in fig.1.23a, which is effectively figs. 1.19 and 1.21 combined. The upper and lower cut-off frequencies can then be determined separately as in fig.1.23b.

1.5.4. Notch filters

The notch filter is used to reject a band of frequencies, usually to suppress 50 Hz (or 60 Hz) mains noise induced onto signal lines. A narrow notch is usually required, which implies a high Q.

The circuit of fig.1.24 is commonly used. The values should be chosen such that:

$$R1 = R2 = R$$
$$R3 = 0.5R$$
$$C1 = C2 = C$$
$$C3 = 2C$$

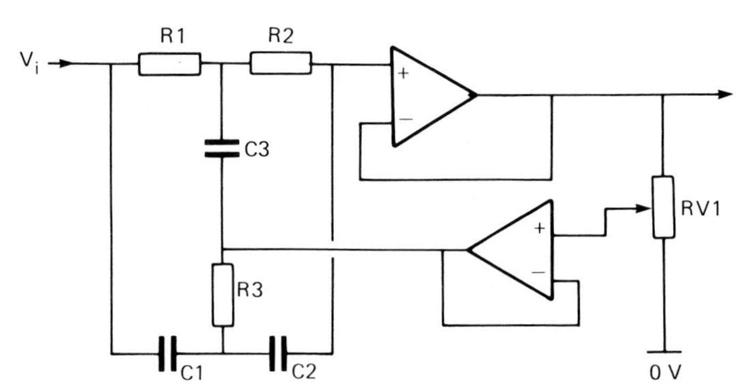

Fig.1.24 Notch filter circuit.

The centre frequency is then given by:

$$f = \frac{1}{2\pi RC} \tag{1.37}$$

The value of Q is determined by RV1, the maximum value being obtained with the wiper connected to the output.

1.6. Miscellaneous circuits

1.6.1. Voltage-to-current and current-to-voltage conversion

Signals in instrumentation are often transmitted as an electrical current (4–20 mA, for example) because of the high common mode rejection of a current loop. This requires circuits to give voltage-to-current and current-to-voltage conversion.

Voltage-to-current conversion is shown in fig.1.25a. The current loop goes through the load, then through the current setting resistor R. The voltage at both amplifier inputs must be equal, so:

$$IR = V_{in} \tag{1.38}$$

$$\text{or} \quad I = \frac{V_{in}}{R} \tag{1.39}$$

which is independent of load provided the amplifier output does not saturate.

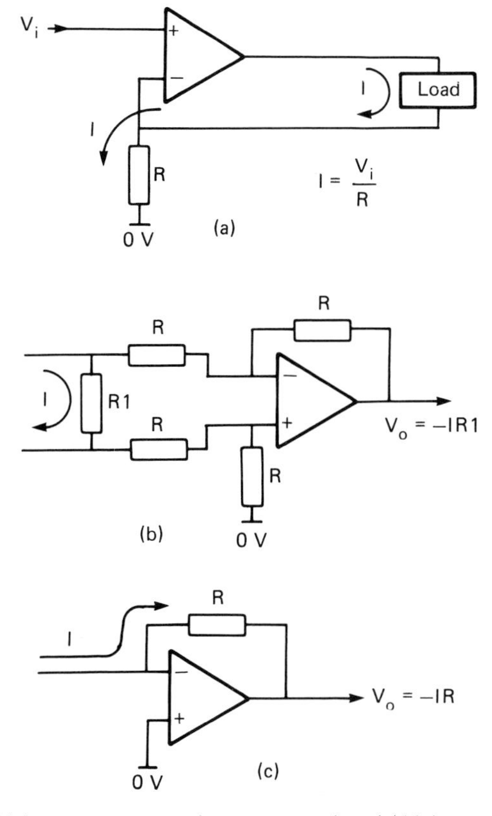

Fig.1.25 Voltage–current–voltage conversion. (a) Voltage to current conversion. (b) Floating current to voltage conversion. (c) Non floating current–voltage conversion.

Current-to-voltage conversion is achieved by connecting a differential amplifier across the load resistor. The output voltage is then given by:

$$V_o = -IR_L \tag{1.40}$$

if the amplifier is set for unity gain (all resistors equal). The value of the load resistor should be small compared with the amplifier resistors R1 to R4.

If the current is not from a current loop, with the return route being via the 0 V or supply rails, the circuit of fig.1.25c may be used. The inverting input acts as a virtual earth, and

$$V_o = -IR \tag{1.41}$$

A common application of this circuit is in photometry where the current from a photodiode is to be converted to a voltage.

1.6.2. Ramp circuit

Motor drive circuits require acceleration and deceleration to be limited to prevent mechanical damage to couplings and gearboxes. This can be achieved by the low pass filter of fig.1.18, but the exponential rise causes the drive to take an excessive time to reach speed.

The circuit of fig.1.26a gives a constant ramp change as shown in fig.1.26b. IC1 acts as a comparator (it has no feedback, so the

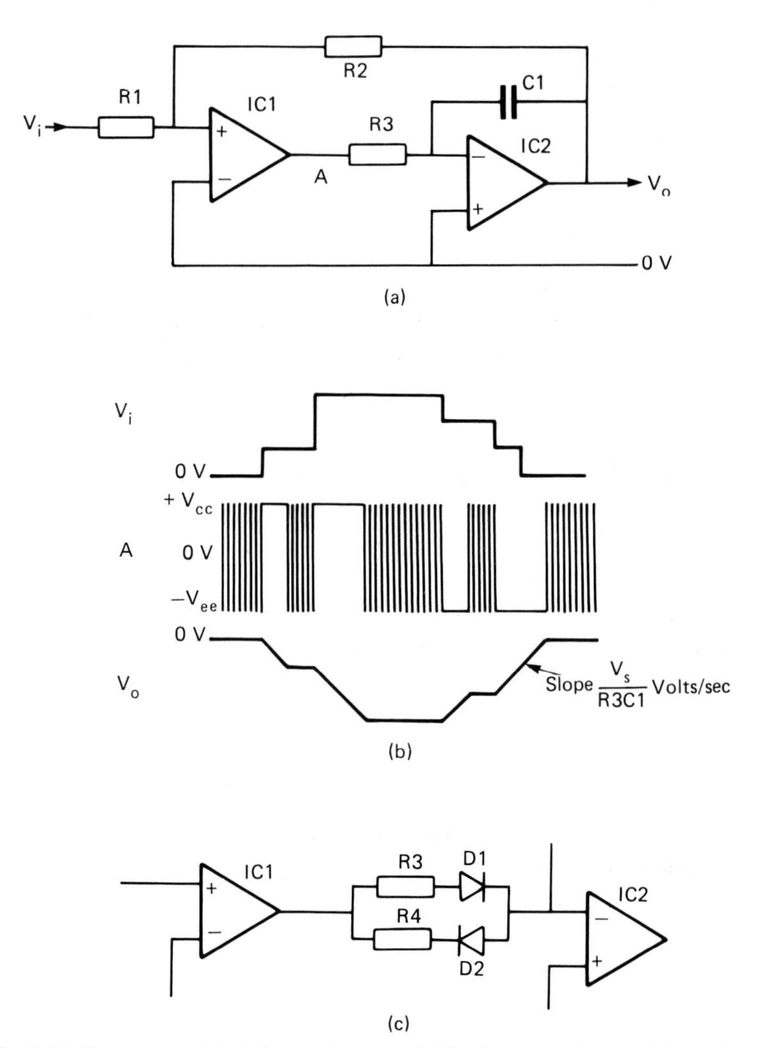

Fig.1.26 Ramp circuit. (a) Circuit diagram. (b) Typical waveform. (c) Modification to give different positive and negative ramp slopes.

full amplifier gain is used) and compares V_i with V_o. At balance $V_o = -V_i$. Any deviation will cause the output of IC1 to saturate positive or negative.

IC2 is an integrator, whose output will ramp negative at a constant rate when IC1 is saturated positive, and ramp positive when IC1 is saturated negative. The circuit waveforms are shown in fig.1.26b. When $V_o = -V_i$ the output of IC1 is theoretically zero, but in practice small random variations will cause it to dither as shown.

The gain of the circuit is given by:

$$A = -\frac{R2}{R1} \tag{1.42}$$

and the ramp slope by:

$$\frac{dV}{dt} = -\frac{V_s}{R3C1} \tag{1.43}$$

where V_s is the saturation voltage of IC1.

If different acceleration and deceleration rates are required, R3 can be replaced by the two-diode circuit of fig.1.26c.

1.6.3. Inverting/non-inverting amplifier

The circuit of fig.1.27 has either unity positive gain or unity negative gain dependent on the state of SW1. With SW1 closed, the circuit acts as a conventional inverting amplifier with unity negative gain. With SW1 open the circuit becomes a unity gain buffer, with $V_o = V_i$.

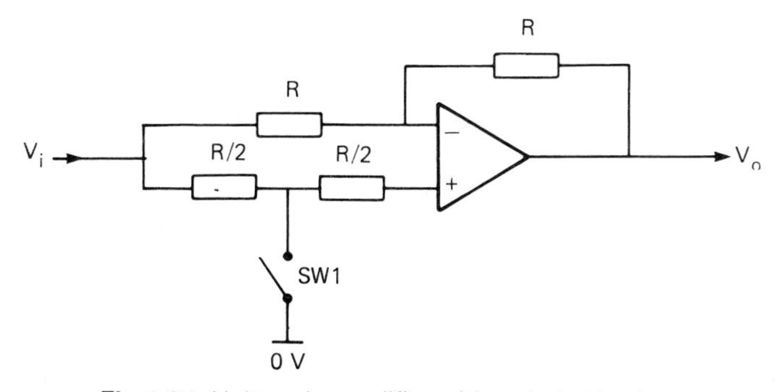

Fig.1.27 Unity gain amplifier with switchable sign.

SW1 can be a mechanical contact (e.g. a relay or a switch) or a CMOS analog switch (e.g. the 4016). If V_i is only positive, an NPN transistor may be used in place of SW1.

1.6.4. Phase advance

The circuit of fig.1.28a delays the feedback signal by the inclusion of capacitor C. This causes the amplifier output to saturate on

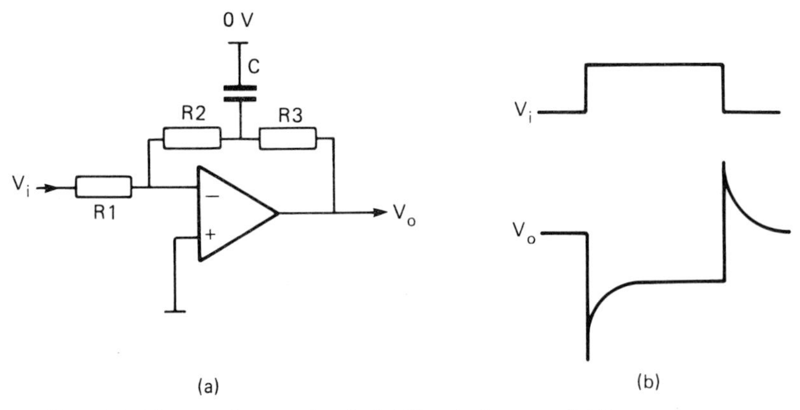

Fig.1.28 Phase advance circuit. (a) Circuit diagram. (b) Step response.

change of input signal as shown in fig.1.28b. The circuit is commonly used with hydraulic proportional valves to give an initial kick to overcome stiction.

1.6.5. Peak picker

A peak picker circuit is used to hold the maximum value of a variable for subsequent measurement. The circuit, and its action, is shown in fig.1.29. IC1 acts as a comparator; if V_o is less than V_i, IC1 output goes positive charging the capacitor via the diode until V_o equals V_i. IC2 simply acts as a buffer to prevent the output load discharging C. Ideally IC2 should be an FET amplifier.

The output voltage therefore follows the maximum values of V_i as shown. Resistor R is included to give a controlled discharge of C if required. If the diode is reversed the circuit follows the minimum value of V_i.

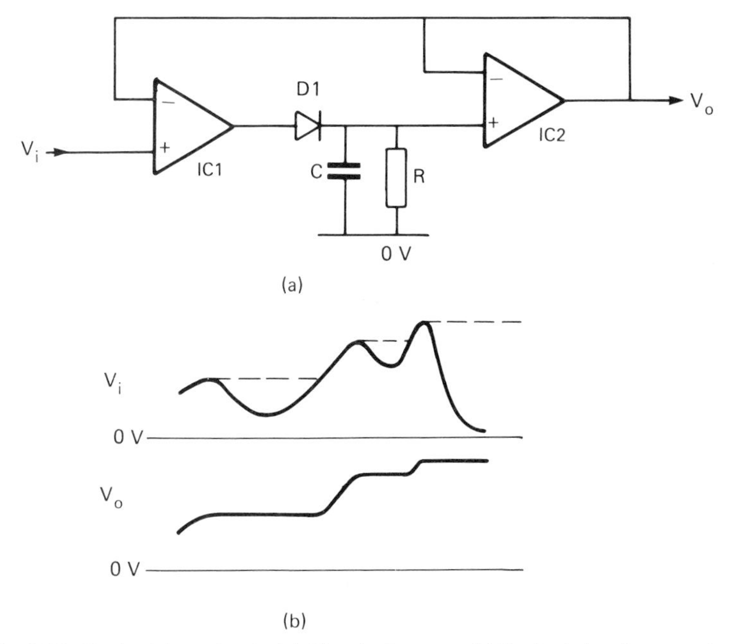

Fig.1.29 Peak picker circuit. (a) Circuit diagram. (b) Typical circuit operation.

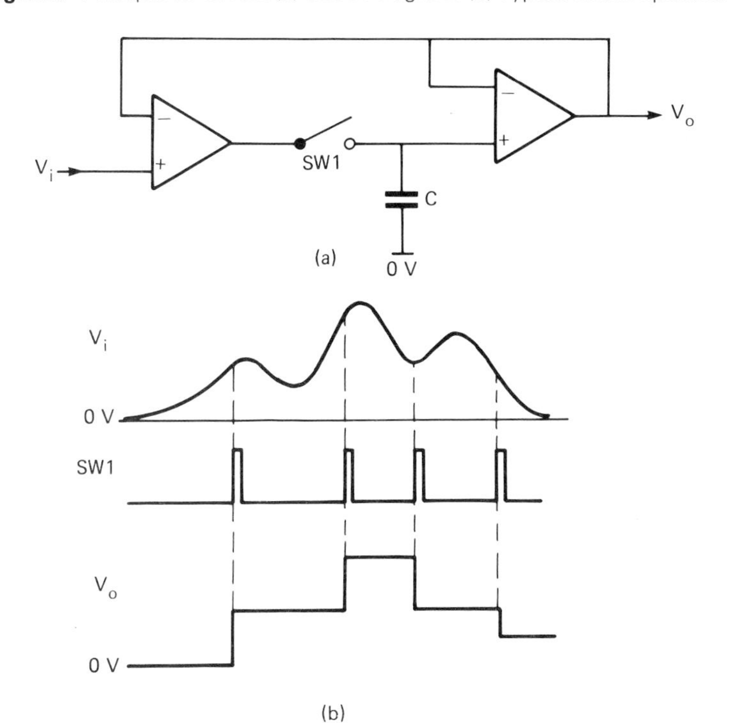

Fig.1.30 Sample and hold circuit. (a) Circuit diagram. (b) Typical circuit operation.

1.6.6. Sample and hold

If the diode in fig.1.29 is replaced by a 'switch' SW1 as fig.1.30a, the circuit can be used to 'freeze' a varying signal for measurement. With SW1 closed, V_o will follow V_i. With SW1 open, V_o will hold the value of V_i at the instant that the switch was opened as shown in fig.1.30b.

Sample and hold circuits are an essential part of analog-to-digital conversion, a topic covered in section 3.9.2.

1.6.7. Schmitt triggers

Voltage comparison is often required in control systems. This is usually provided by a circuit with built in 'backlash', or 'hysteresis' as it is more commonly called. This is summarised in figs. 1.31a and b. The circuit is characterised by two voltages, the upper and lower trigger points. The output changes when the input goes above the UTP and below the LTP. The differential between the upper and lower trigger points gives protection against output jitter which can occur when a simple single point comparison circuit is used and the input voltage is near the switching point.

A comparison circuit with hysteresis is usually called a Schmitt trigger, and is shown in its simplest form in fig.1.31c. Suppose V_{in} is above the UTP; the output will be saturated negative, and the voltage at the junction of R1 and R2 will be:

$$V1 = -V_{sat} \cdot \frac{R2}{R1 + R2} \qquad (1.44)$$

This is the lower trigger point LTP because V_{in} will need to go more negative than V1 before the circuit will switch. The output will now go positive, and the voltage at the junction of R1 and R2 will be:

$$V2 = +V_{sat} \cdot \frac{R2}{R1 + R2} \qquad (1.45)$$

This is the upper trigger point UTP because the input must go more positive than V2 before the circuit will switch. The hysteresis of the circuit is therefore the difference between V1 and V2 (remembering that V1 is negative).

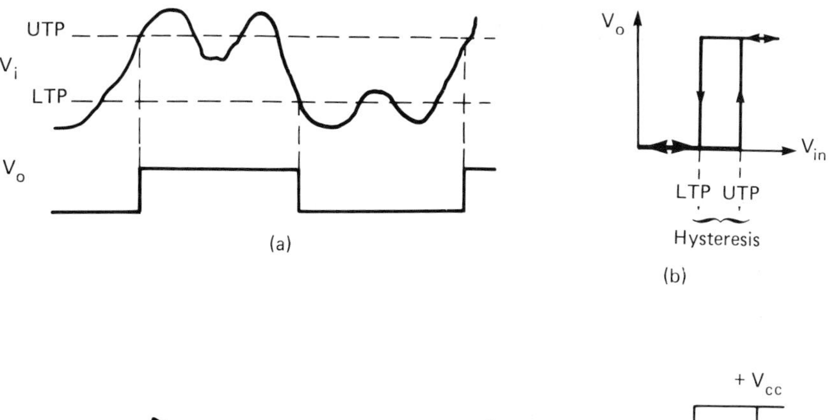

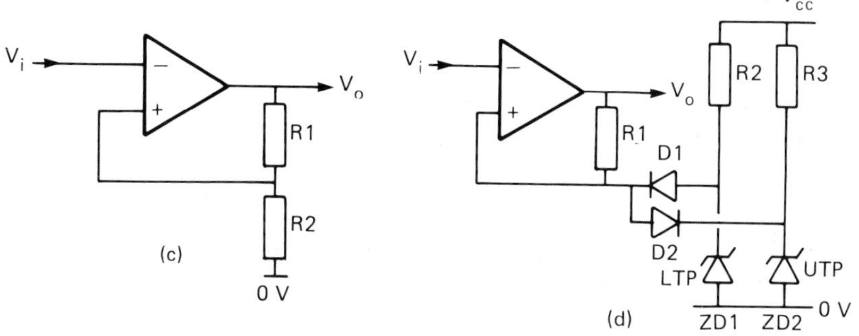

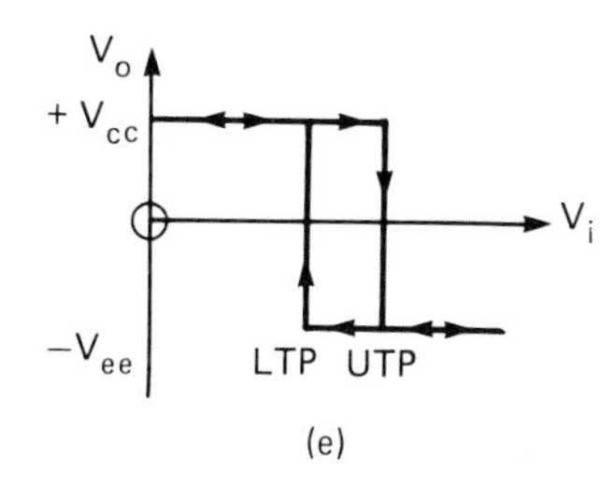

Fig.1.31 The Schmitt trigger. (a) Schmitt trigger operation. (b) Circuit hysteresis. (c) Simple Schmitt trigger with UTP and LTP symmetrical about zero. (d) Schmitt trigger with UTP and LTP having same sign. (e) Operation of circuit d.

The circuit of fig.1.31c has a symmetrical response, with positive UTP and negative LTP. The circuit of fig.1.31d has positive UTP and LTP determined by zener diodes. R1 should be an order of magnitude larger than R2 and R3 to avoid the output switching significantly affecting the zener voltages. The circuit has the response of fig.1.31e.

1.6.8. Limiting circuits

A limiting circuit is a linear amplifier provided the input signal stays within predefined levels. If these levels are exceeded, the output signal stays constant as shown in fig.1.32a. A typical example is error limiting in speed control circuits to prevent excessive motor currents.

Figure 1.32b is a non-inverting amplifier, with input limiting by D1 and D2. RV1 sets the upper limit and RV2 the lower limit. The value of R1 should be at least an order of magnitude greater than RV1 and RV2.

The inverting circuit is shown in fig.1.32c. If R1 and R2 are equal, the limiting input voltages will be twice the voltage set on RV1 and RV2. Again, the value of R1 should be at least an order of magnitude greater than RV1 and RV2.

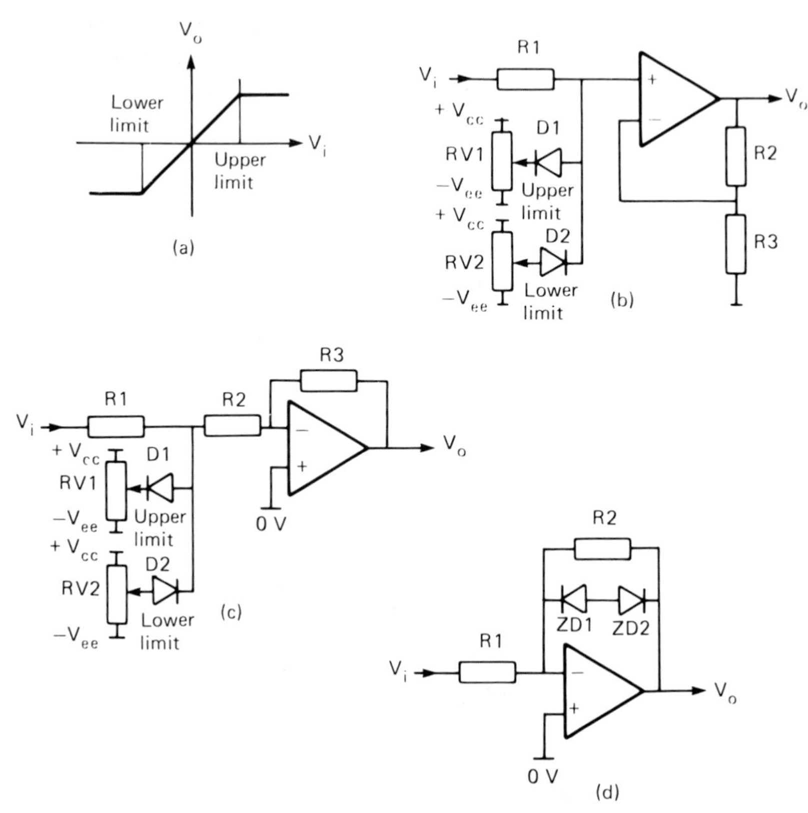

Fig.1.32 Limiting circuits. (a) Circuit operation. (b) Non inverting circuit. (c) Inverting circuit. (d) Inverting circuit with output limits set by zener diodes.

Shunt limiting can be provided by zener diodes as shown in fig.1.32d. ZD2 directly sets the output limiting positive voltage, and ZD1 the negative voltage.

Figures 1.32b and c should be used for variable gain amplifiers when the limiting is to be performed on the *input* signal level and fig.1.32d where the limiting is to be performed on the *output* signal level.

1.6.9. Deadband amplifier

A deadband amplifier has the response of fig.1.33a, and gives a region of zero sensitivity. The circuit is often used on the error signal in position control systems to prevent 'dither' about the home positions.

With V_{in} at zero, V1 and V2 can be set by RV1 and RV2. The input voltage has to rise above V1 or fall below V2 before the output starts to change. RV1 therefore sets the positive deadband and RV2 the negative deadband. Note that R1 and R2 allow different slopes to be set for the positive and negative sections.

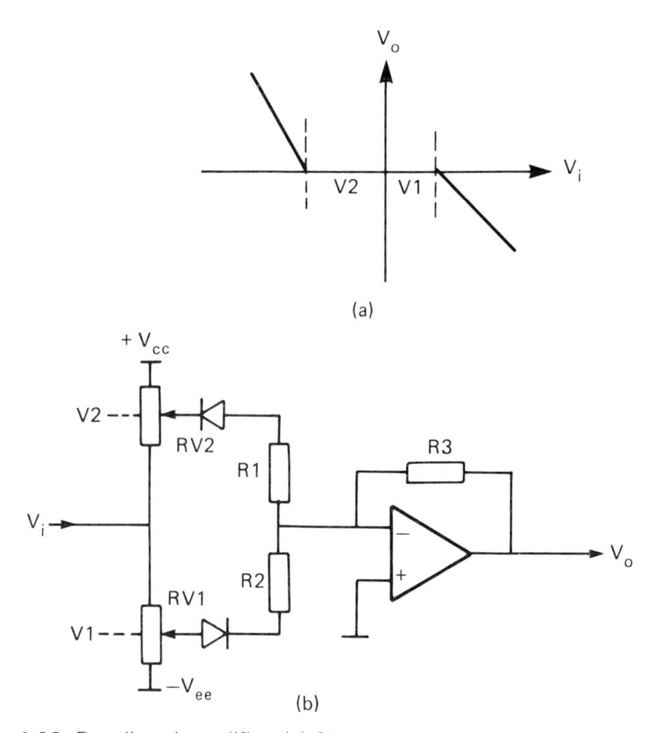

Fig.1.33 Deadband amplifier. (a) Circuit operation. (b) Circuit diagram.

1.6.10. Linearisation circuits

Many instrumentation circuits require linearisation of a signal; common examples are the square root function needed for many flow transducers and the multiterm series needed for thermocouples and PTRs. This linearisation can be performed by operational amplifiers.

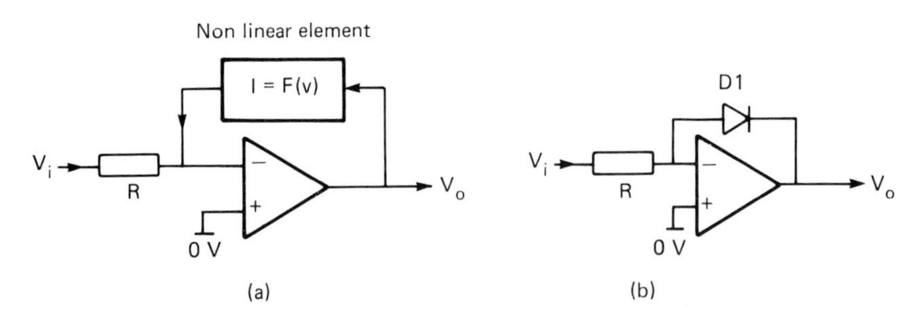

Fig.1.34 Linearisation circuit. (a) General principle. (b) Logarithmic circuit.

The generalised circuit is shown in fig.1.34a, where a non-linear element is placed on the feedback of an op amp. The output voltage is related to the input by:

$$V_o = G\left(\frac{V_{in}}{R}\right) \tag{1.46}$$

where $G(V_{in}/R)$ is the inverse function of $F(V_o)$.

In fig.1.34b, for example, a diode is used in the feedback, which has as exponential current/voltage relationship, i.e.:

$$F(V_o) = A\exp(BV_o) \tag{1.47}$$

where A and B are constants.

The inverse function is a logarithm, so:

$$V_o = C\log(DV_i) + E \tag{1.48}$$

where C,D and E are constants. This circuit is the basis of a DC amplifier based multiplier or divider (see section 1.4.4.).

An alternative approach is to approximate the required curves by straight lines. The curve in fig.1.35a, for example, is approximated by three straight lines. This can be achieved by the circuit of fig.1.35b, with RV1 to 3 setting the break points and RV4 to 6 the slopes of the individual sections. Practically any

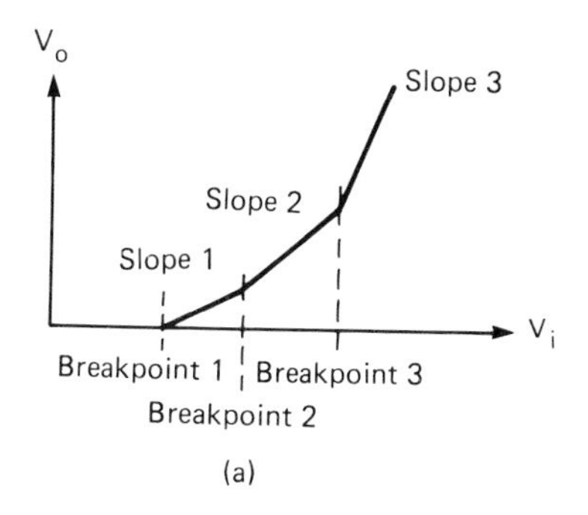

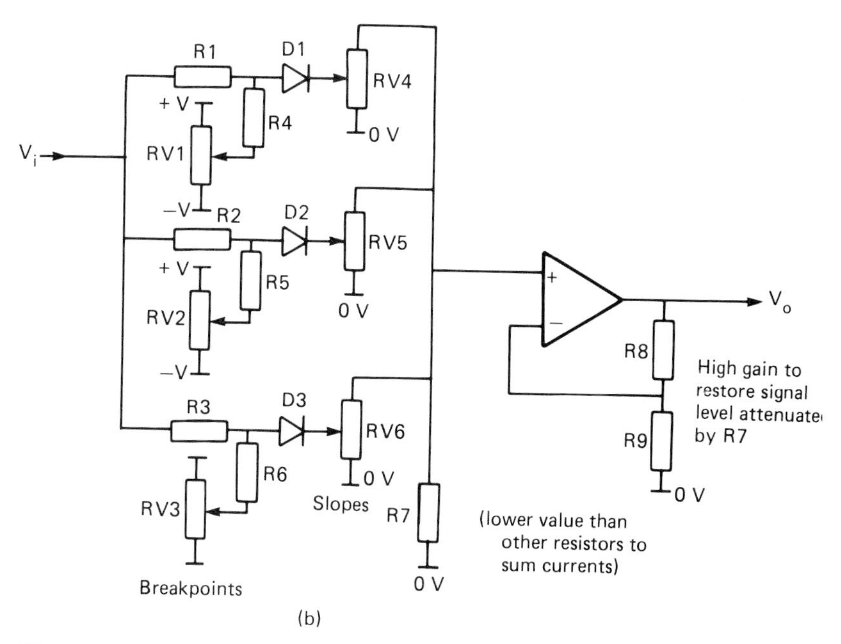

Fig.1.35 Linearisation by straight line approximation. (a) Circuit operation. (b) Practical circuit.

function can be achieved by combining amplifiers with various Ax + B characteristics, deadband amplifiers and limiters and summing the outputs.

1.6.11. P + I + D controllers

The classic three-term control algorithm uses three components: error, time integral of error, and time derivative of error. This can be achieved, in a rather sledgehammer way, by using four

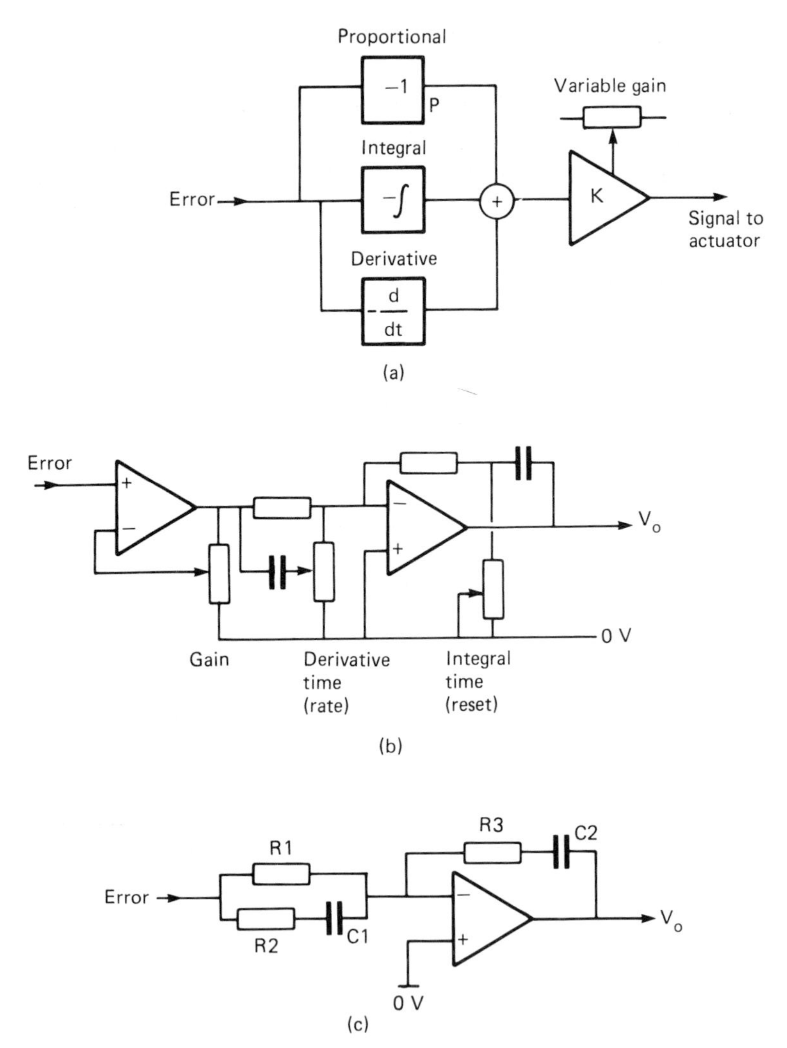

Fig.1.36 Three term controller circuits. (a) Block diagram of PID controller. (b) Two amplifier PID controller. (c) Simple interacting PID controller.

amplifiers as in fig.1.36a, with the integral term being provided by a circuit similar to fig.1.15 and the derivative term by one similar to fig.1.16.

More elegant circuits are shown in figs.1.36b and c. In the latter circuit the proportional gain is determined by R3/R1, the integral action mainly by R1, C2 and the derivative action by R2, R3, C1. If the resistances are made adjustable, it follows that there will be considerable interaction between the terms. Figures 1.36a and b allow relatively independent adjustments.

1.6.12. Precision rectifiers

Instrumentation signals are often obtained as an AC voltage (LVDT position transducers, for example). These signals must be rectified to give a linear DC signal. Voltage drops across the diodes in a bridge rectifier introduce errors, and make the circuit unusable with voltage below a few volts. Precision error-free rectifiers can be constructed using op amps.

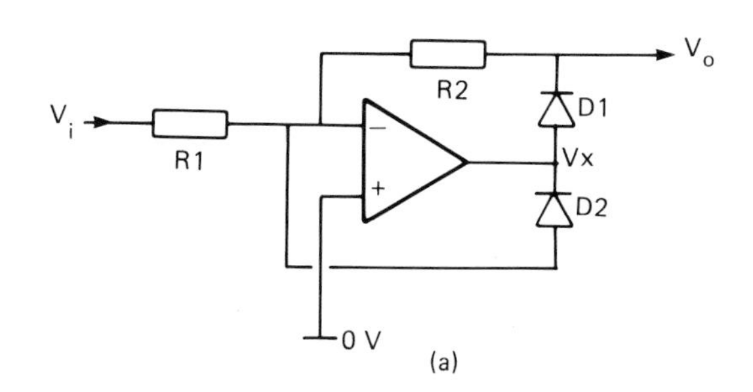

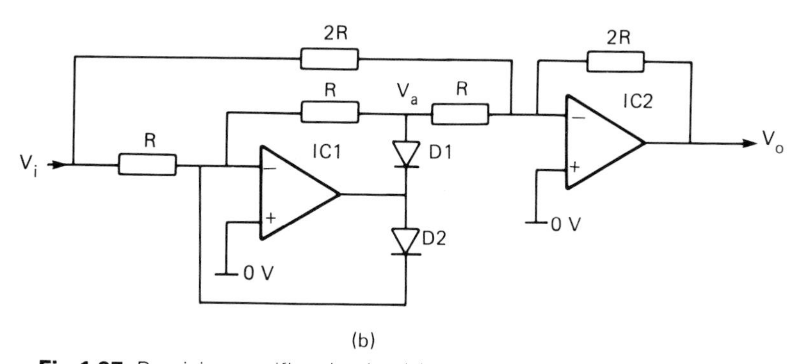

(b)

Fig.1.37 Precision rectifier circuits. (a) Half wave rectifier. (b) Full wave rectifier.

A half wave circuit is shown in fig.1.37a. On the negative half cycle, $V_{out} = -V_{in}$ because R1 and R2 are equal (i.e. V_{out} is a positive half cycle). Note that the amplifier output, V_x, will be equal to V_{out} plus the voltage drop of D1. On the positive half cycle D2 will conduct, limiting V_x to -0.8 V. The junction of R1 and R2 will still be a virtual earth, so provided negligible current is drawn through R2, V_{out} will be zero. If D1 and D2 are reversed, a negative half wave rectifier can be constructed.

A full wave rectifier is shown in fig.1.37b. IC1 is a negative half wave rectifier with output V_a. IC2 sums V_{in} and 2 V_a (because of the choice of resistor values). When V_{in} is positive, $V_a = -V_{in}$:

$$\begin{aligned} V_o &= -(V_{in} - 2V_{in}) \\ &= V_{in} \text{ (a positive signal)} \end{aligned} \tag{1.49}$$

When V_{in} is negative, V_a is zero, so:

$$\begin{aligned} V_o &= -(V_{in} + 2 \times 0) \\ &= -V_{in} \text{ (a positive signal again)} \end{aligned} \tag{1.50}$$

Whatever the polarity of V_{in}, the output signal is positive but equal in magnitude to V_{in}.

The circuits will work at fairly high frequencies (several tens of kHz). The limiting factor is the slew rate of the half wave rectifier which has to slew its output from +0.8 V (one forward diode drop) to −0.8 V as the input signal goes through zero.

1.6.13. Isolation amplifiers

Instrumentation signals are often low level and consequently prone to noise. Instrumentation amplifiers therefore need very high common mode rejection. The signals will also probably originate at some distance from the control cubicle, and will probably share cable ducts with high voltage signals. Under fault conditions it is possible for inter-cable shorts to occur, causing high voltages to appear on instrumentation lines. If no precautions are taken, the high voltage will go through all the instrumentation and destroy an entire control system. Both these problems are overcome by the use of isolation amplifiers, shown diagrammatically in fig.1.38.

In fig.1.38a, the input signal is chopped into AC by a CMOS switch. The resulting AC signal is amplified and passed to the primary of a transformer. The chopper circuit and the amplifier are powered by their own floating power supply derived from a DC to DC inverter. The AC signal on the transformer secondary is rectified and smoothed (see section 1.6.12) to give the original signal. The use of transformer coupling and the floating power supply gives almost perfect common mode noise rejection and protection against high voltages on the input. An inter-cable fault will still probably damage the transducer and the input stage of the isolation amplifier, but will not spread into the rest of the system.

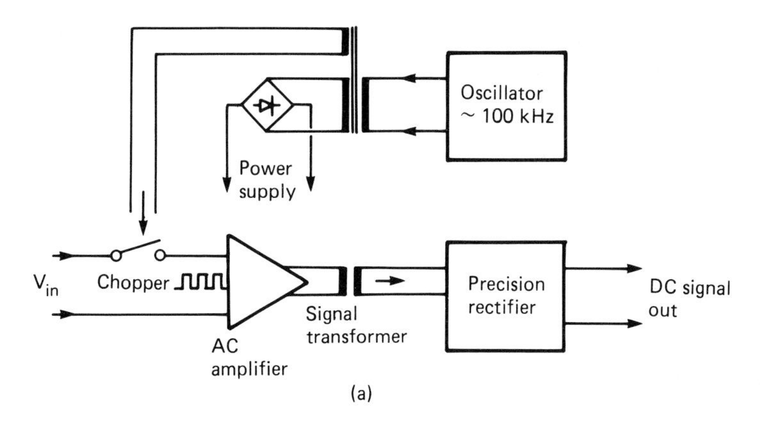

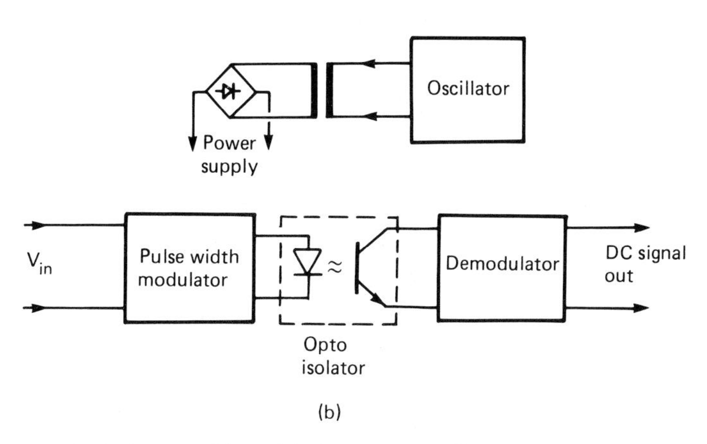

Fig.1.38 Isolation amplifiers. (a) Chopper amplifier transformer coupled. (b) PWM amplifier with opto isolation.

An alternative approach, shown in fig.1.38b, uses optical isolation. The input signal is converted to a pulse width modulated signal which drives the light emitting diode in an opto isolator. The input circuit and pulse width modulator are again driven from a floating supply derived from a DC to DC inverter. The pulse width modulated signal from the opto isolator is restored to DC form for use by the rest of the system.

Isolation amplifiers are usually purchased in encapsulated form, and are available with any required gain. A typical unit will give isolation with input voltages up to 1 kV.

Chapter 2
Rotating machines and power electronics

2.1. Introduction

The electric motor is the commonest industrial prime mover, and is often used as the final control in a process control scheme. This chapter describes various types of electric motor. Many applications require the speed of the motor to be accurately controlled. Thyristors and their uses in speed control are also described.

The hydraulic motor, which is an alternative form of rotating machine, is covered in chapter 5.

2.2. Basic principles

Electric motors and generators are closely related devices, and have a common background theory. Faraday found experimentally that an electrical potential is produced when a conductor is moved relative to a magnetic field. (It does not matter whether the conductor moves relative to a fixed field, or the field moves with respect to a fixed conductor; the resulting potential is the same.)

In fig.2.1a, a conductor of length L metres is moving at a uniform velocity v metres per second in a uniform magnetic field B webers per square metre. Faraday's experimental work showed that the induced voltage e is given by:

$$e = \text{flux linked per unit time} \tag{2.1}$$

$$= B.L.v \text{ volts} \tag{2.2}$$

Equation 2.2 is the basic equation for an electrical generator.

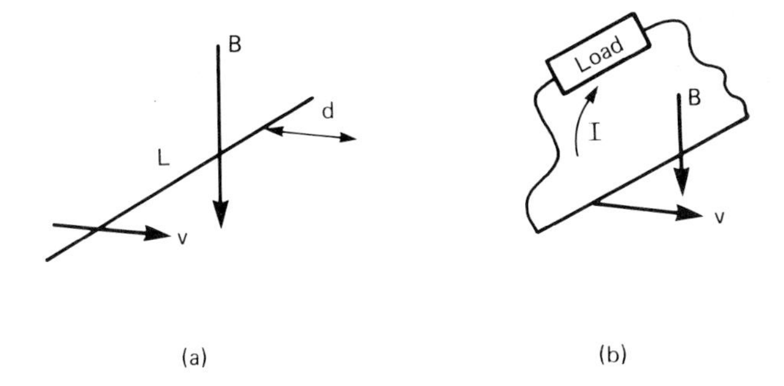

Fig.2.1 Principle of electrical motor and generator. (a) Conductor moving in a magnetic field. (b) Conductor connected to a load.

By conservation of energy, work must be done to generate the electrical potential. In fig.2.1b, a conductor moving in a magnetic field is connected to an electrical load. When the conductor moves, a current i will flow.

$$\text{power} = e.i \text{ watts} \tag{2.3}$$

If the conductor moves at uniform speed for time T secs:

$$\begin{aligned} \text{work done} &= \text{power} \times \text{time} \\ &= e.i.T \\ &= B.L.v.i.T \text{ joules} \end{aligned} \tag{2.4}$$

The generation of electrical energy produces an opposing force on the conductor. Mechanical work is done moving the conductor against this force. This work is given by:

$$\begin{aligned} \text{work done} &= \text{force} \times \text{distance} \\ &= \text{force}.v.T \end{aligned} \tag{2.5}$$

Combining equations 2.4 and 2.5 gives:

$$\text{force}.v.T = B.L.v.i.T$$

or

$$\text{force} = B.L.i \text{ newtons} \tag{2.6}$$

Equation 2.6 gives the force on a conductor in an electrical generator. It also follows that if a current is driven through a conductor in a magnetic field by an external power source, the conductor will experience a force. This force is also given by equation 2.6, and is the basis of electric motors.

If the conductor is not restrained, the force will cause it to accelerate. Once it starts to move, an emf will be generated. This will oppose the applied voltage (and hence is called the back emf), causing the current to fall. Eventually the conductor will reach a steady speed where the force given by equation 2.6 balances mechanical forces such as friction.

If the mechanical load decreases, the conductor will accelerate because the electrical force no longer balances the reduced mechanical force. The increased speed raises the back emf, reducing the current, and the conductor settles at a higher speed where the electrical force again balances the mechanical force.

2.3. Simple generators

2.3.1. Slip ring generator

In fig.2.2a a single turn coil is rotated in a fixed magnetic field. An electrical potential is generated in the coil by equation 2.2, with the instantaneous velocity relative to the field being a sinusoid. The generated voltage is therefore also sinusoidal, as in fig.2.2b. The frequency is exactly the same as the rotational

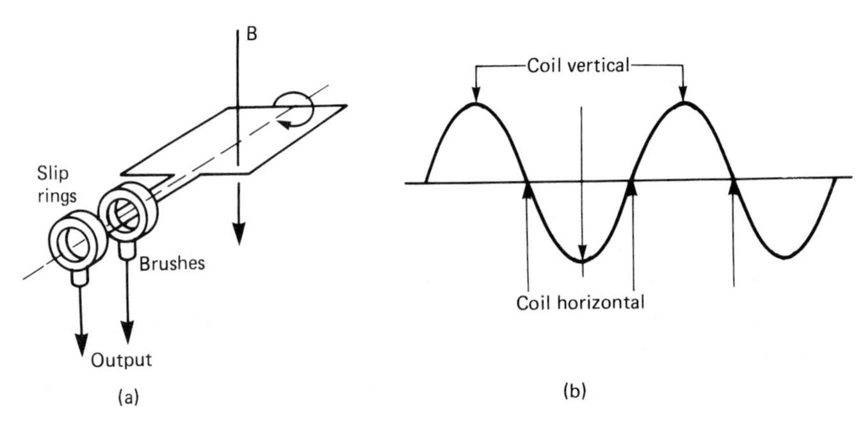

Fig.2.2 Slip ring AC generator. (a) Mechanical arrangement. (b) Output voltage.

frequency, and the amplitude depends on the size of coil and the magnetic field strength. Note that speeding up the rotational speed will cause the output voltage frequency and amplitude to increase.

The voltage induced in the coil is taken out by stationary carbon brushes rubbing on slip rings on the coil. In a practical generator, the coil will consist of many turns to increase the output voltage. The fixed magnetic field will be derived either from permanent magnets (in small machines) or via electromagnets (in larger machines).

2.3.2. DC generator

Figure 2.3a is similar to the slip ring generator, but the coil is brought out via a split copper cylinder called a commutator. This causes the coil to brush connections to reverse every 180 degrees of rotation.

The voltage generated in the rotating coil is, again, a sinusoid, but the commutator reverses the coil connections as the voltage passes through zero. The output waveform is therefore pulsating DC, as in fig.2.3c.

2.4. DC motors

2.4.1. Simple motor

Figure 2.3a can also be used as a simple DC motor. If a voltage is applied to the brush connections with the coil stationary and not horizontal (where the brushes would short the coil), a current i will flow which is determined by the voltage and the coil resistance, by Ohm's law.

By equation 2.6, a force will be produced on the coil which will start it turning. Inertia will take the coil through the commutator changeover point, and it will start to rotate and accelerate. As the coil speeds up, it also starts to act as a generator and a back emf is produced, as described in section 2.2.

It is common practice to use the symbol V for the applied voltage, and E for the back emf as shown in fig.2.4. At any speed these are related by:

$$V = E + iR \qquad (2.7)$$

E is directly proportional to motor speed, and for a fixed magnetic field the torque will be proportional to i. As the speed increases, the increased E causes a lower torque. The motor will

Fig.2.3 Commutator DC generator. (a) Mechanical arrangement.
(b) Commutator end view.
(c) Output voltage.
(d) Assembling the rotor on a DC machine (e) A typical commutator and brushgear on a DC machine. Photos courtesy of ASEA.

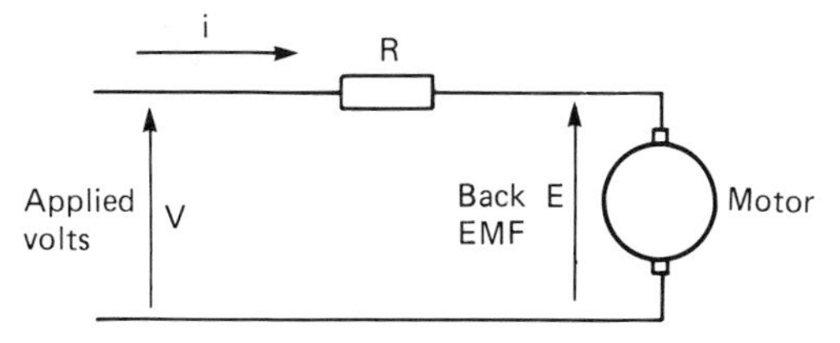

Fig.2.4 Basic motor equations.

settle at a speed where the torque exactly balances the load torque.

The motor based on fig.2.3a has many shortcomings; it gives a pulsating torque and is non-self-starting if the coil comes to rest in a position where it is shorted by the brushes. The next three subsections describe practical DC motor construction.

2.4.2. Ring wound machines

In fig.2.5a, the rotor consists of a laminated core wound with twelve coils. These are connected to twelve commutator sections as shown. The brushes are shown on the inside of the commutator for ease of illustration; in practice, of course, they bear on the outside surface.

A voltage is applied to the brushes, and this causes an equal current i to flow through each half of the coil as shown in fig.2.5b. Note that the total current at each brush is 2i. The current passing through the coils produces a torque on the rotor, causing it to rotate. As it rotates, however, the commutator switches the current to new coil sections to give an almost uniform torque.

The action of the commutator is critical, and is more involved than may at first be thought. Consider fig.2.5c; the coil is moving to the right. Coils A and C both have current i flowing through them, but the current directions are opposite. As the coil A moves to the right past the brush, its current must first be reduced to zero, then reversed. The coils will inevitably have large inductance, which inhibits fast changes of current. A failure to reduce the current in the coil to zero whilst it is shorted by the brush as in fig.2.5c leads to a large induced voltage in the coil, which in turn causes sparking and damage to the brush and commutator. The problem gets worse as speeds and armature current rise, and in large machines interpoles are used (see section 2.4.5 below).

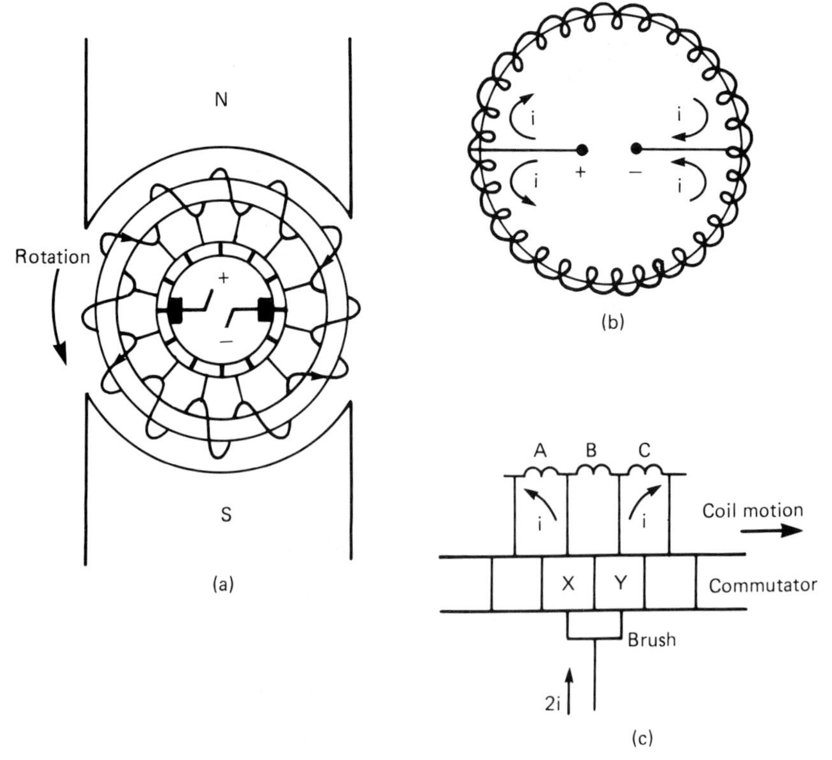

Fig.2.5 The ring wound DC motor. (a) Mechanical arrangement. (b) Current flow. (c) Commutator action.

The ring winding is difficult to construct, and DC machines are more commonly based on the lap or wave winding described below, which lend themselves to more automated assemblies.

2.4.3. Lap wound machines

The lap wound machine is based on prefabricated coils, as in fig.2.6a. The ends of the coil are connected to adjacent commutator segments as shown, and the coil laid into slots in a laminated armature as shown in fig.2.6b. Coils are overlaid as shown in fig.2.6c, with the finish of coil 1 connected (at the commutator) to the start of coil 2, and so on.

To give smooth torque, multiple magnetic poles are used. Figure 2.7a shows a typical four pole arrangement. In a lap wound machine, the number of brushes is equal to the number of poles, and brushes under like poles are paralleled as shown.

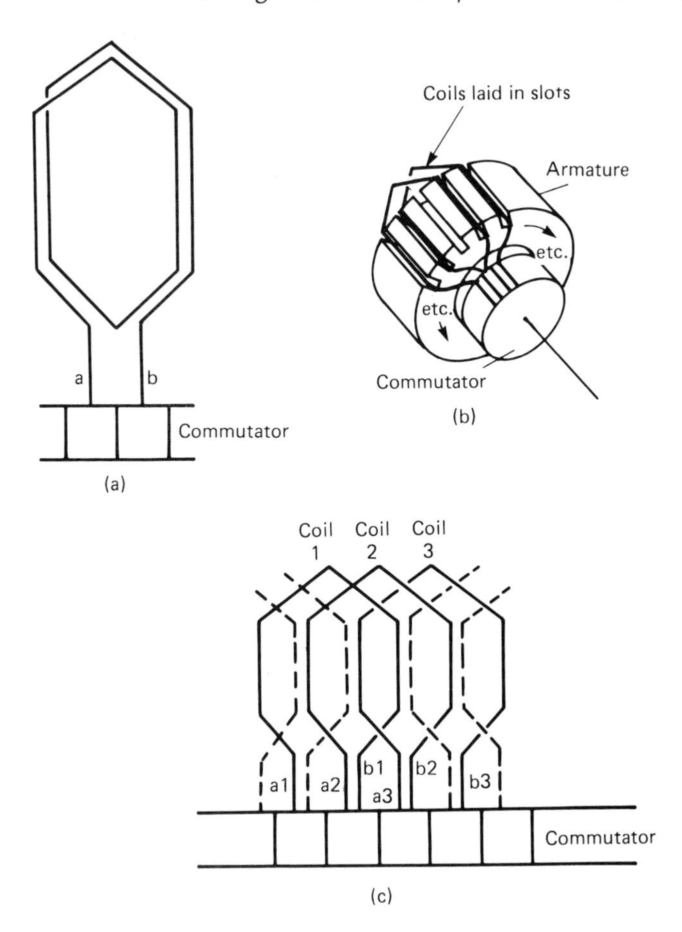

Fig.2.6 The lap wound machine. (a) Lap winding. (b) Machine construction. (c) Commutator connections.

Figure 2.7b shows what is happening under two adjacent pole faces for a machine with four poles and twelve coils (and hence twelve commutator segments). Current I1 enters via brush X, and proceeds via commutator segment a into the start of coil 1. It then flows through coils 2 and 3 (linked at segments b, c) to exit at segment d and brush Y. Because the coil widths and the pole-to-pole distances are equal, the forces on each coil are equal and in the same sense.

Figure 2.7b is not, however, a complete picture. Coil portions 10f, 11f, 12f will also be under the north pole, and an equal current I2 will also go via brush X, through segment a, coils 12, 11, 10 to brush W. Similarly coil portions 4s, 5s, 6s lie under the

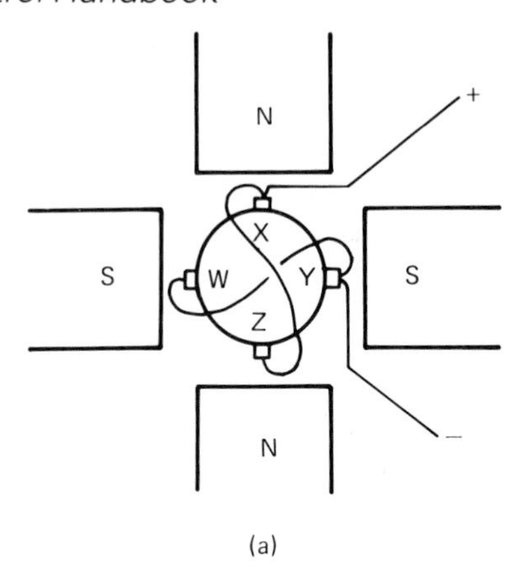

(a)

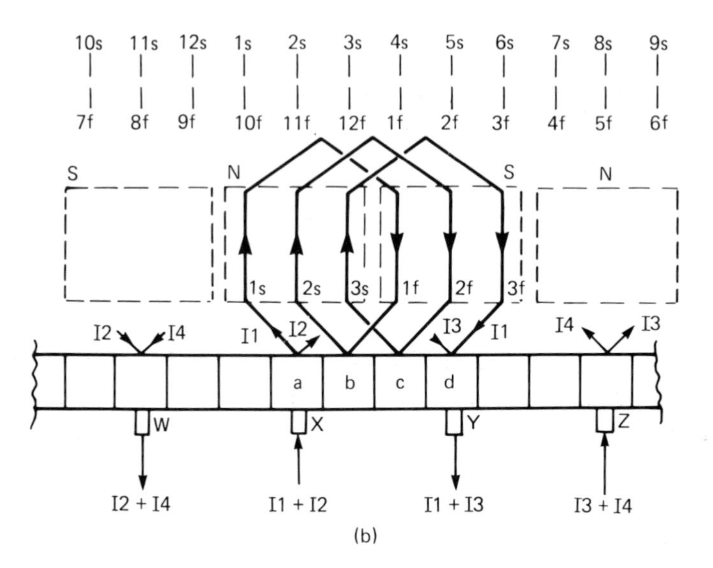

(b)

Fig.2.7 Coil connections in a lap wound machine. (a) Four pole arrangement. (b) Current flow in a four pole machine.

south pole, and a current I3 flows from brush Z to brush Y via coils 4, 5, 6. At any instant there are four parallel currents flowing: X to Y, X to W, Z to Y, Z to W.

2.4.4. Wave wound machines

The coil for a wave wound machine is shown in fig.2.8a. A wave wound machine has an odd number of coils and commutator

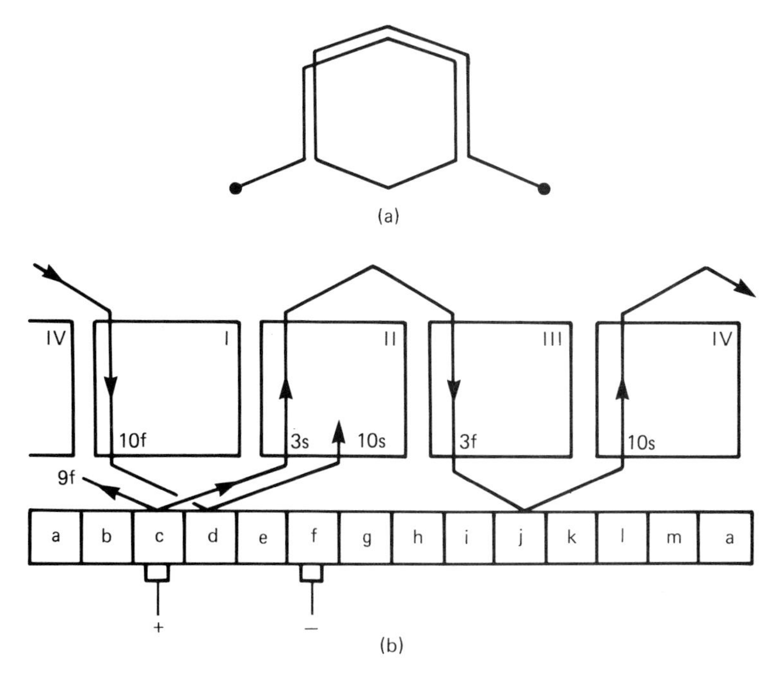

Fig.2.8 The wave wound machine. (a) The wave winding. (b) Coil connections.

segments; fig.2.8b has thirteen, for example. Each coil is connected to segments spaced just over 180° apart. In fig.2.8b, coil 3 is connected between segments c and j. This arrangement causes the coils to progress in waves and form a continuous series chain of coils. A wave wound machine has two brushes, spaced 90° apart, regardless of the number of poles.

Suppose the positive brush is at segment c; the negative brush is spaced 90° away on segment f. Current enters and passes through coil 3 to segment j, coil 10 to segment d, coil 4 to segment k, coil 11 to segment e, coil 5 to segment l and coil 12 to segment f and the negative brush. A parallel path can also be traced via coils 9, 2, 8, 1, 7, 13, 6.

As before, the current direction in front of each pole piece is such that a constant reinforcing torque is produced on the coils, causing the armature to rotate.

2.4.5. Compensating windings and interpoles

As the brushes move from one segment to the next, the current in an armature coil must be reduced to zero and then reversed (see

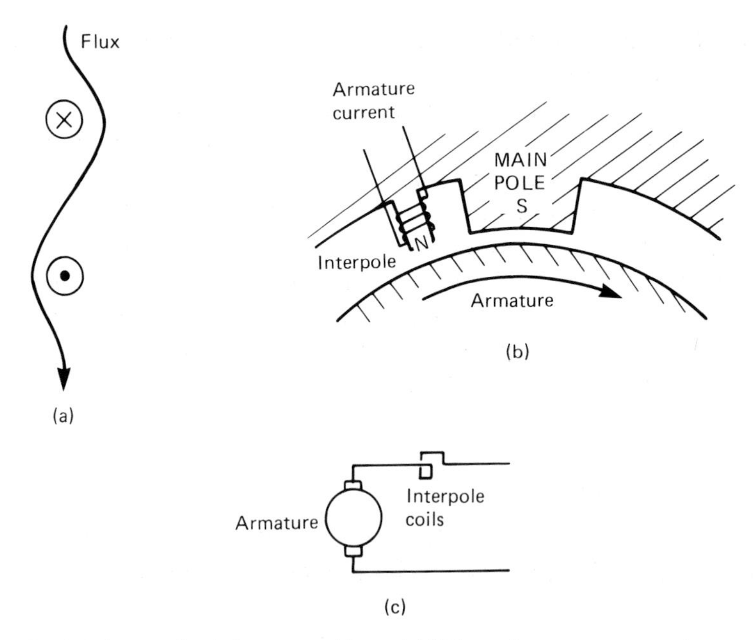

Fig.2.9 Interpoles in large machines. (a) Effect of armature current on magnetic field. (b) Interpole arrangement. (c) Interpole connection.

fig.2.5c above). This action is called commutation. If the changeover is not completed, excessive sparking occurs at the trailing edges of the brushes. The problem increases as speed or armature current rises.

The problem can be relieved, to some extent, by the use of high resistance brushes which span several commutator segments, thereby increasing the time for commutation. Such techniques are only suitable for small machines, however.

In large machines, another effect becomes apparent. The armature current will itself produce a magnetic field. This has the effect of 'twisting' the main field as shown in fig.2.9a. If the brushes were correctly aligned for zero current, they will be misaligned at full load. In old machines a lever for manual brush movement was provided, the operator adjusting the brush position for minimum sparking.

In more modern motors, an additional small pole called an interpole is added before each main pole, as shown in fig.2.9b. In motors this opposes the next pole, and is energised by armature current so its field strength depends on load current. The interpoles assist commutation, as their field assists the reversal of current in the coils undergoing commutation. As their field strength increases with current, they automatically compensate for changes in load.

A related problem can arise when rapid load changes occur. With a fast di/dt, transformer action in the armature coils can produce large voltages between adjacent commutator segments. Ultimately a flashover can occur, which will wreck the machine. To prevent this, a compensating winding is added. This consists of conductors in the surface of the pole faces, again carrying armature current. Their field cancels that of the armature conductors. The compensating winding also assists commutation at high currents as it reduces the armature reaction effect of fig.2.9a.

2.4.6. Motor equations

The first requirement is to derive equations for power and torque. In fig.2.10a a mass M kg is being raised at a constant

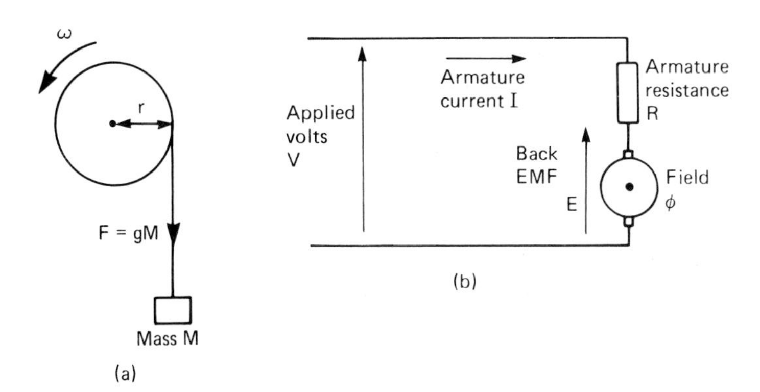

Fig.2.10 Funadamental motor principles. (a) Torque, work and power. (b) Motor equations.

speed by a rope being wound into a drum. The force being exerted on the mass is given by:

$$F = Mg \text{ newtons} \tag{2.8}$$

The torque at the drum periphery is:

$$T = Fr \text{ metre newtons} \tag{2.9}$$

In one revolution, the mass is raised by $2\pi r$ metres. The work done is given by:

$$w = 2\pi Fr \tag{2.10}$$

$$= 2\pi T \text{ joules (newton metres)} \tag{2.11}$$

Power is work done per second. In one second the drum rotates $\omega/2\pi$ revs, therefore:

$$\text{power} = 2\pi T \times \frac{\omega}{2\pi} \qquad (2.12)$$

$$= \omega T \text{ watts} \qquad (2.13)$$

The basic motor equation was given earlier in equation 2.7. Figure 2.10b shows the representation of a simple motor. V denotes the applied voltage, E the back emf, I the armature current, R the armature resistance and ϕ the field strength. We have, by simple circuit theory:

$$V = E + IR \qquad (2.14)$$

The back emf is given by:

$$E = A\phi\omega \qquad (2.15)$$

where A is a constant for the specific machine, and is related to the motor constructional details such as the number of poles, coils, turns on coils, etc.

The useful electrical power is given by:

$$P = EI \text{ watts} \qquad (2.16)$$

and by substitution:

$$P = A\phi\omega I \text{ watts} \qquad (2.17)$$

The motor torque is given by:

$$T = A\phi I \text{ metre newtons} \qquad (2.18)$$

Note that for a fixed ϕ, $T \propto I$ and $P \propto \omega I$.

The power equations above relate useful electrical power. The total power balance is:

$$VI = EI + I^2R \qquad (2.19)$$

where VI is the applied power, and I^2R the armature loss which appears as heat. Not all the motor power EI is delivered to the load, however, as frictional losses and losses through windage, etc., must be subtracted. The efficiency of the machines is therefore lower than the equation above would imply.

The above equations indicate that the speed of a motor can be controlled in two ways: by control of armature volts (and hence armature current) or by control of the field strength.

2.4.7. Field circuits

So far it has been assumed that the magnetic field has been produced by permanent magnets. Such an arrangement is only suitable for small motors, an electromagnetically produced field being used on most machines.

There are essentially five different field connections, shown in fig.2.11. The first, and most versatile, has a totally separately controlled field. This is the commonest arrangement for applications such as steel rolling mills, mine winders and paper manufacture. The field will have its own control circuit.

The parallel (shunt) field of fig.2.11b is used where the armature voltage is not likely to vary widely. The series field of fig.2.11c gives a large starting and accelerating torque, as the field strength is directly proportional to armature current. The equations in section 2.4.6, however, indicate that the speed of a motor rises for a reduced field. Unlike the connections in fig.2.11a, b, a series connected motor does not have a clearly defined maximum speed (for the separate and shunt field, E cannot exceed V). If a series connected machine is deprived of a load, its speed can rise to a point where the commutator is destroyed by centrifugal force. It follows that series connected motors should not be used for chain or belt drives.

The final circuits of fig.2.11d, e combine the advantages of series and shunt machines and are known as compound machines. The arrangement of fig.2.11d is known as a short shunt, and fig.2.11e a long shunt.

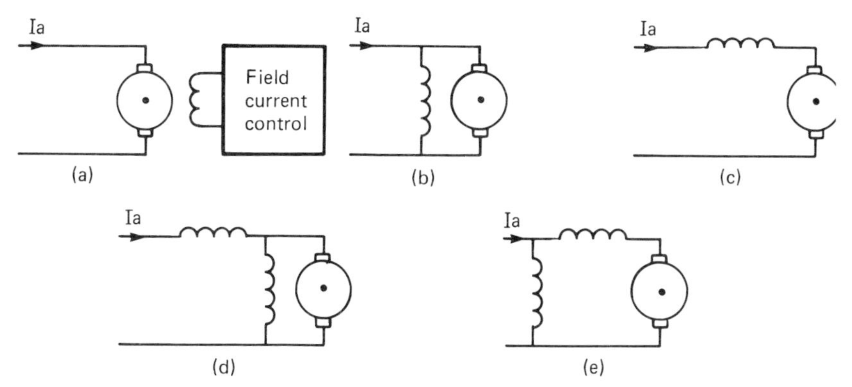

Fig.2.11 Field circuits. (a) Separately excited field. (b) Shunt field. (c) Series field. (d) Compound field short shunt. (e) Compound field long shunt.

Field windings, by their nature, inherently have a large inductance. This can cause problems as large voltages can be induced by changes of field current. Later sections deal with the characteristics of circuits containing significant inductance.

2.5. Power semiconductors

2.5.1. Rectifier diodes

Power diodes are used where alternating voltage (from the AC mains supply) is to be converted to DC. Typical examples are electrolysis processes, battery charging, electromagnets and motor field circuits.

Rectifiers based on copper oxide or selenium plates were once common, but these devices are physically large and modern devices are almost entirely constructed from semiconductor materials with a pn junction, as in fig.2.12a. Silicon and germanium diodes are used, with silicon being more common. The mechanism of rectification need not concern us, but a typical power diode will have a voltage/current response similar to fig.2.12c.

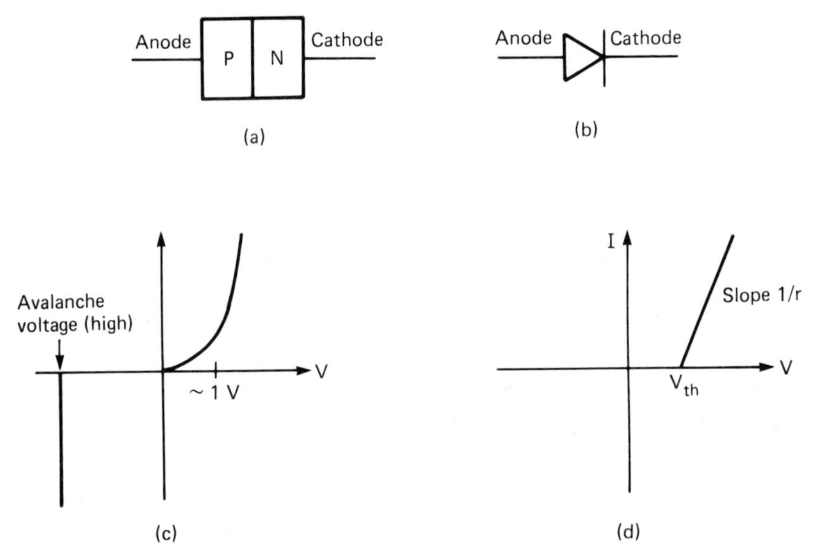

Fig.2.12 The power diode. (a) Diode construction. (b) Diode symbol. (c) V/I relationship. Note different voltage scales for positive and negative voltages. (d) Straight line approximation.

In the reverse direction, a negligible leakage current flows (typically 1 mA for every 5 A of rated forward current). This remains fairly constant until an avalanche voltage is reached, where the current increases dramatically (and the diode is irreversibly damaged).

In the forward direction, the current-to-voltage relationship is approximately exponential, but can be considered as a straight line from a threshold voltage V_{th} and slope 1/r where r is the slope resistance. Typically V_{th} is 0.8 V and r is 0.001 ohms. If I_m is the mean current, and I_r the rms current (which may not be the same as I_m), the heat dissipated in the diode is:

$$P = I_m V_{th} + I_r^2 r \tag{2.20}$$

To calculate I_r from I_m, the form factor k for the application must be known, since:

$$I_r = k I_m \tag{2.21}$$

k varies from 1.0 for a pure resistive load to 2.5 for a capacitive load.

As a rough rule of thumb, power diode losses can be approximated as:

0.5 W per A DC per germanium diode
1.2 W per A DC per silicon diode

This is obviously a lot of heat for, say, a 500 A diode, so the construction of the diode encapsulation is made to encourage dissipation via an external heat sink. Typical devices are shown in fig.2.13.

Diodes usually fail, however, not from overcurrent or overtemperature, but from overvoltage in the reverse region of fig.2.12c. Figure 2.14a shows a simple half wave diode rectifier, and fig.2.14b the voltage appearing across the diode. This is small whilst the diode is conducting, but follows the AC whilst the diode is blocking.

There are three voltages of interest. These are, as shown, the peak working voltage, the maximum recurrent voltage and the peak transient voltage. The peak transient voltage must never be exceeded, even briefly, or device failure will result. Transients are difficult to predict, and can arise internally from switching of inductive loads, or 'ringing' of transformer cores at switch-on. Externally generated transients can come from other users (such as electric arc furnaces) or from naturally occuring sources such

Fig.2.13 Typical power diodes.

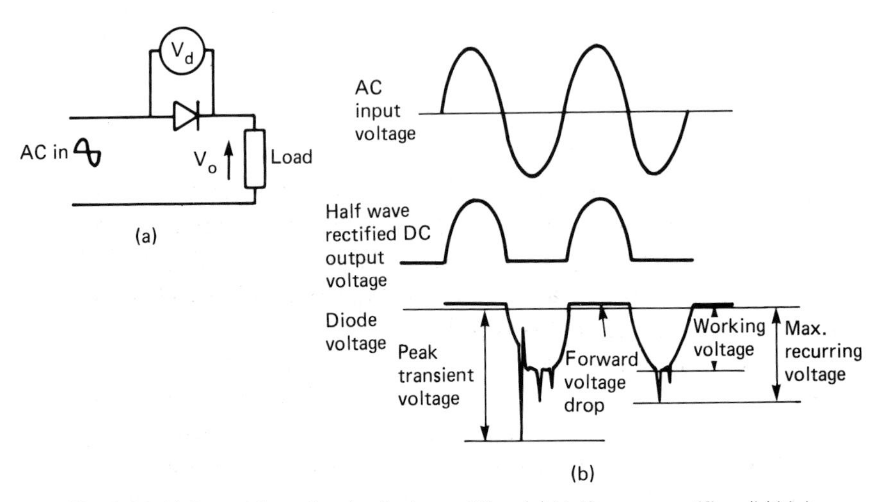

Fig.2.14 Voltages in a simple diode rectifier. (a) Half wave rectifier. (b) Voltages in circuit. Transients are not shown on AC in and V_o waveforms for simplicity.

as lightning strikes on the grid. A voltage margin, defined as PTV/PWV, of 2.5 is usually found to be adequate. If in doubt, protection snubber circuits (resistor and capacitor in series) can be placed across the diodes. Semiconductor manufacturers' data sheets give design criteria for these snubber circuits.

2.5.2. Thyristors

The workhorse of power electronics is the thyristor, also known as the silicon controlled rectifier or SCR. (This term arises from the device as controlled rectifier constructed from silicon not, as might be thought, a rectifier controlled by silicon.)

A thyristor is a four layer semiconductor device as shown in fig.2.15a. It is usually denoted by the circuit symbol of fig.2.15b. The operation of a thyristor is complex, but can be considered by rearranging fig.2.15a as fig.2.15c which approximates to the transistor pnp/npn pair of fig.2.15d.

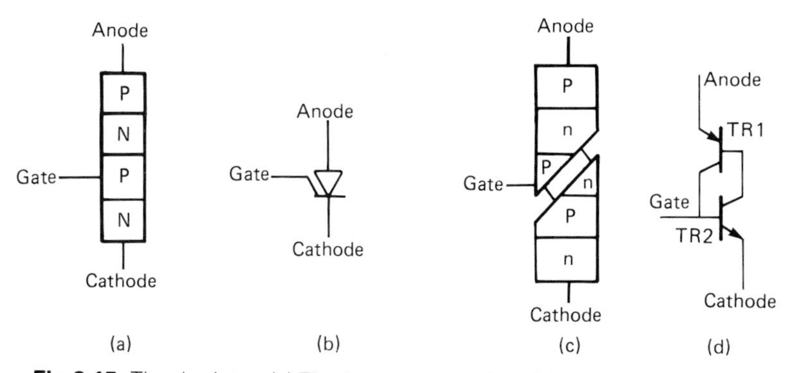

Fig.2.15 The thyristor. (a) Thyristor construction. (b) Thyristor symbol. (c) Transistor analogy. (d) Transistor equivalent.

Figure 2.16 shows typical V/I characteristics with no gate signal. In the reverse direction, a thyristor behaves like a normal diode, having a small leakage current (typically 1 mA per 5 A rated current) until the avalanche voltage is reached, at which point the current rises catastrophically.

In the forward direction, minimal leakage current occurs until the breakover voltage, V_{bo}, is reached. At this point the device suddenly switches to a low impedance state with low voltage drop. The conducting portion of fig.2.16 corresponds to a conventional diode with a slightly higher threshold V_{th} of about 1.5 V.

Thyristors are rarely, if ever, used without a gate signal, however. Suppose the thyristor is forward biased and sitting as point A on fig.2.16. Referring to the two transistor relationships of fig.2.15d, the emitter of TR1 will be biased positive, and the emitter of TR2 negative. Current is injected into the gate terminal; current will enter the base of TR2 causing a small collector current to flow. This acts as base current for TR1 which turns on, giving more base current to TR2. A regenerative action takes place until both transistors are turned hard on.

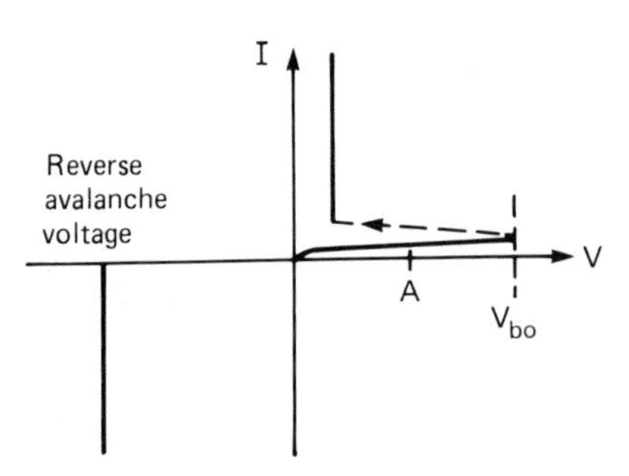

Fig.2.16 The thyristor V/I relationship.

The effect on fig.2.16 is that the device goes from point A to a point on the conduction region of the characteristic determined by the load and the rest of the circuit. The gate is therefore used to trigger the thyristor into conduction in the forward direction. A typical gate signal is 1.5 V at 100 mA for 100 μS.

Once conduction starts, it will continue until stopped by one of two conditions:

(1) The forward current is reduced below a specified level called the holding current (typically a few mA).
(2) The voltage across the device reverses, taking it into the reverse section of fig.2.16.

Once conduction has ceased, it will only restart by application of gate current in the forward section, or by taking the forward voltage above V_{bo}.

Figure 2.17 shows some simple thyristor circuits. In fig.2.17a, the operation of PB1 will inject gate current causing Th1 to turn on. The bulb will light, and stay on (as the current is above the holding current). Operation of PB2 breaks the current, causing Th1 to turn off, and stay off.

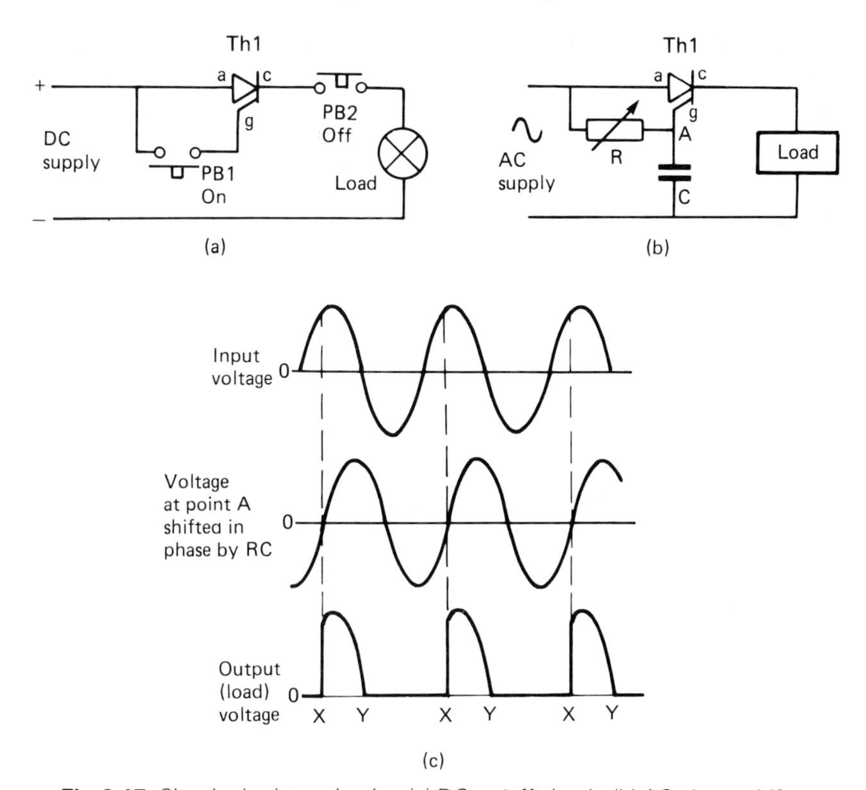

Fig.2.17 Simple thyristor circuits. (a) DC on/off circuit. (b) AC phase shift circuit. (c) Voltages in phase shift circuit.

Figure 2.17b is an AC circuit. The RC network delays the gate waveform as shown in fig.2.17c. At point X the gate goes positive causing Th1 to turn on. Current flows through the load until the AC supply reaches zero at point Y. Load current then falls to zero and the voltage across Th1 reverses. Conduction ceases until point X again in the next positive half cycle. Load voltage waveforms are shown in fig.2.17c; the point of conduction is determined by the value of R and C. Practical versions of this circuit are used for light dimmers and speed control for electric drills.

Thyristors, like diodes, can be considered to have a threshold voltage and slope resistance when conducting. These lead to a loss (heat dissipation) which is typically 1.8 W per A DC per thyristor. The construction of thyristors is again made to facilitate heat removal; typical devices are shown in fig.2.18. The gate connection is usually a short tag or lead.

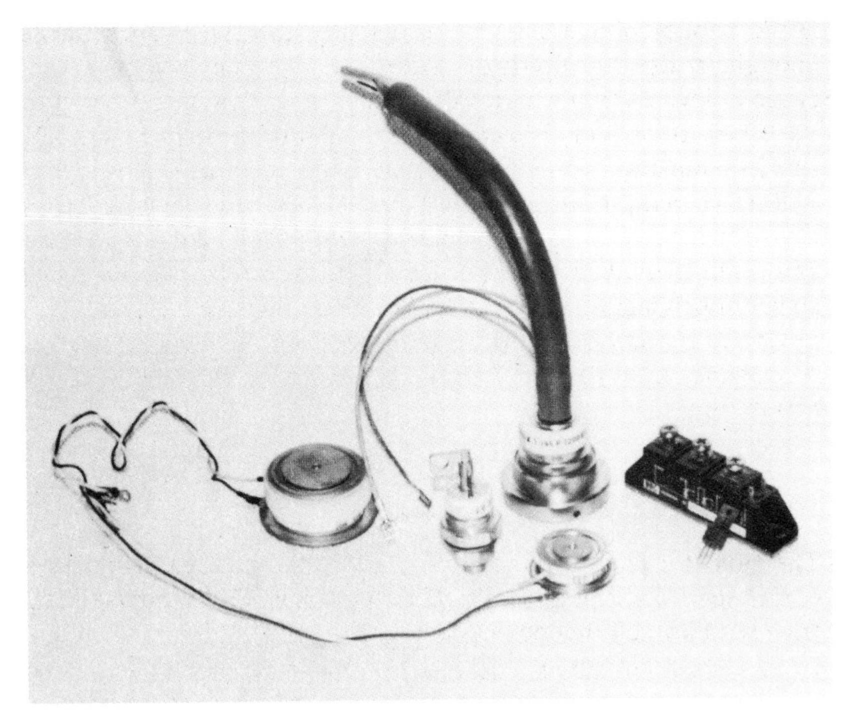

Fig.2.18 Typical thyristors.

Comments in section 2.5.1. on reverse voltage protection and current rating apply equally to thyristors. There are other constraints, however, that need to be considered.

The construction of a thyristor creates capacitance between anode and gate as shown in fig.2.19. When the anode voltage changes, a current I will flow given by:

$$I = C\frac{dV}{dt} \tag{2.22}$$

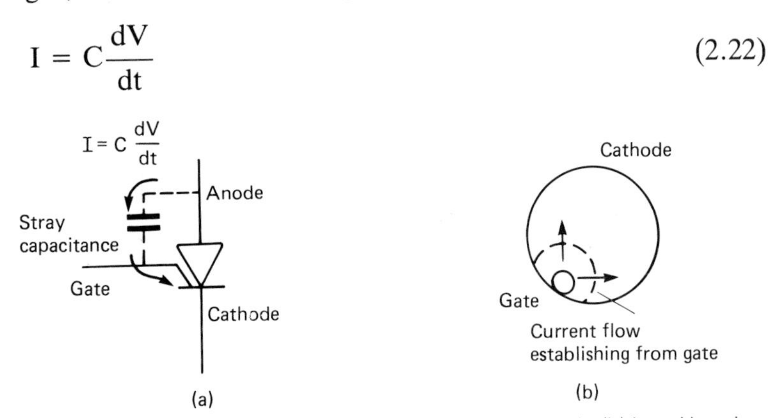

Fig.2.19 Thyristor failure modes. (a) False triggering by dV/dt. (b) Local heating by di/dt.

If the anode voltage changes sufficiently quickly, i can be large enough to fire the thyristor. Although this does not, in itself, cause a device failure, in most applications harm would result to the thyristor or other components. Manufacturers accordingly specify a maximum dV/dt on their data sheets (typically 2 kV/μS).

Rate of rise of current (di/dt) must also be limited. When a thyristor is fired, conduction initially starts in a small area by the gate, and spreads across the device as shown in fig.2.19b. The speed of propagation is typically 0.1 mm/μS, so for a typical 10 A device with 4 mm slice, some 40 μS must elapse for the thyristor to conduct completely. Too high a di/dt leads to local high dissipation.

2.5.3. Triacs

The thyristor is a unidirectional device, and as such produces a controlled (albeit pulsating) DC output from an AC supply. The triac is a bidirectional device that can be used to give a controlled AC output. In some respects it can be considered as two thyristors connected, as in fig.2.20 (although this arrangement is a somewhat incomplete analogy).

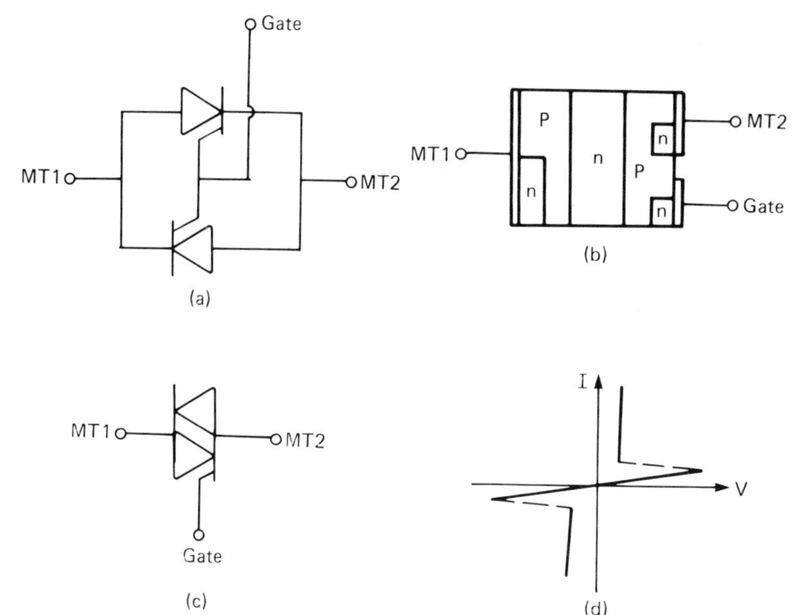

Fig.2.20 The triac. (a) Simplified representation. (b) Construction. (c) Symbol. (d) V/I relationship.

A triac is a three terminal device constructed as in fig.2.20b with the circuit symbol of fig.2.20c. Current flow is between MT1 and MT2 (for main terminal) and conduction is possible in either direction. Its characteristic is shown in fig.2.20d. Conduction is blocked in either direction until a gate pulse of *any* polarity is applied when the device goes to a low impedance state. Conduction continues until the current falls below a holding current (typically a few mA).

Figure 2.21a shows a controlled AC circuit using a triac (and is a full wave version of fig.2.17b). As before, R and C form a phase shifting circuit to delay the gate signal. D1 is a device called a diac (described in the following section) which is used as a bidirectional voltage switch, only passing the voltage at A to the gate when its amplitude (positive *or* negative) exceeds the diac's breakdown voltage.

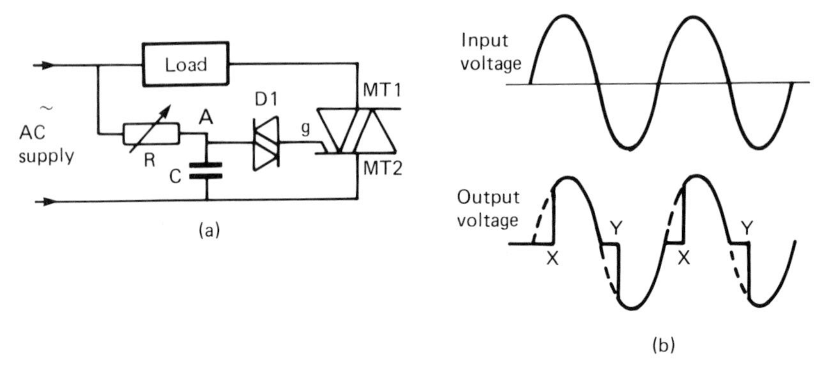

Fig.2.21 Simple triac circuit. (a) Circuit diagram. (b) Waveforms.

Waveforms are shown in fig.2.21b. The gate is triggered by the diac at X on the positive half cycle and Y on the negative half cycle. The position of X (and Y) is controlled by the values of R and C. The circuit of fig.2.21 gives full wave control of an AC load.

Triacs are subject to the same design constraints (transient voltage rating, di/dt, etc.) as thyristors.

2.5.4. Diacs

The diac is a two-terminal symmetrical device constructed as in fig.2.22a with circuit symbols as in fig.2.22b. The device is essentially two thyristors connected as in fig.2.20a but with no

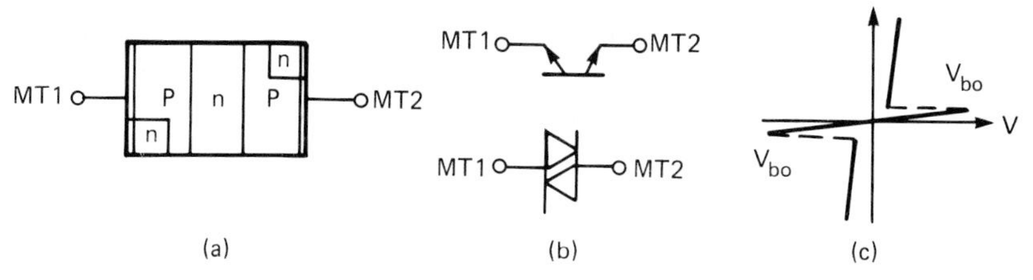

Fig.2.22 The diac. (a) Construction. (b) Symbols. (c) V/I relationship.

gate connection. The device is therefore characterised by the symmetrical curve of fig.2.22c, in particular the two breakover voltages; typically 30 V in each direction.

Diacs are used as a voltage switch, blocking a signal until it reaches some predetermined level. Diacs are typically used as part of the gate triggering circuits for triacs.

2.5.5. Gate turn off (GTO) thyristors

A conventional thyristor blocks in the reverse direction, and can be triggered into conduction by a positive gate pulse (or by exceeding the breakover voltage). Once conducting, it can only be returned to an off state by reducing the current below the holding value, or by reversing the voltage across the device.

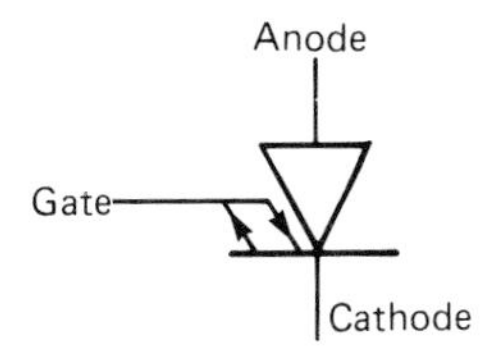

Fig.2.23 The gate turn off (GTO) thyristor.

A variant is the gate turn off thyristor, whose circuit symbol is shown in fig.2.23. This behaves as a conventional thyristor, with the additional feature that it can be taken out of conduction by a negative gate pulse. The gate pulse to turn on the device is similar to a conventional thyristor (typically 1.5 V at about 100 mA for about 100 μS). The turn off pulse is negative (typically −5 V) but the required current is of the same order of magnitude as the anode current, albeit for a few μS. The higher the gate current, the faster the turn off.

The ability to turn off a large circuit has its side effects, however, particularly if the load contains significant inductance when large transients can be generated. A conventional thyristor only turns off at zero current, and tends to suppress inductive transients by itself.

Gate turn off thyristors are commonly used in PWM inverters, described in section 2.10.4.

2.6. Rectifier circuits

Table 2.1 shows common arrangements of rectifier circuits, converting AC from the mains supply to DC (either directly, or via a suitable transformer). The table shows circuit diagrams, and the relationship between AC and DC current and voltages. It should be noted that the PIV quoted is the theoretical value, and in practice a considerable safety margin should be made to allow for transients, etc.

The table is for the most part self explanatory, but there are a few points that may need amplification. In the single phase bridges (a to c) the unsmoothed output is pulsating DC, whereas the three phase circuits have inherently a low ripple content (4% fundamental for the full wave bridge e).

It is also interesting to compare circuits b and c. The biphase circuit gives the same output as the full wave bridge, but is apparently a more economical circuit. Examination, however, shows that the PIV of the diodes for circuit b is twice that for circuit c, and in addition a probably more expensive transformer is required. Similar comparisons apply to circuits d, e and f with the additional observation that circuit e can be used with Star or Delta connected transformers. Note also the difference in ripple frequency between the various three phase circuits; a higher ripple frequency is easier to smooth.

Circuit f is used (albeit rarely) for heavy current low voltage applications, and gives low output ripple. The interphase reactor is used to improve transformer utilisation, and is called a hexaphase rectifier. If the reactor is omitted (and the two star points linked) the circuit is called a diametric connection.

Another consideration is the currents flowing in the AC legs of the rectifier. In general, non-bridge circuits will have a DC component which will reflect into the AC supply (for direct connected rectifiers) or require a larger than necessary transformer because of the resultant magnetising current. The zig-zag

Table 2.1 Common rectifier arrangements

Arrangement	Vdc	Diode current Peak	Diode current Mean	Diode PIV	Controlled rectifier Vdc continuous I
(a) Single phase half wave	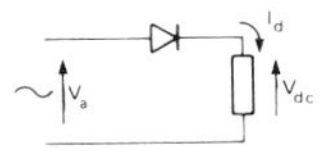 $\frac{Va\sqrt{2}}{\Pi}$ (0.45 Va) 0 with continuous current	Id	Id	3.14 Vdc	Zero
(b) Biphase half wave	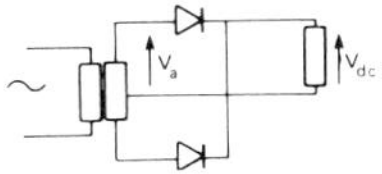 $\frac{Va2\sqrt{2}}{\Pi}$ (0.9 Va)	Id	$\frac{Id}{2}$	3.14 Vdc	$Va \frac{2\sqrt{2}}{\Pi} \cos\alpha$
(c) Single phase full wave bridge	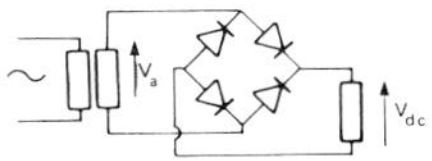 $\frac{Va2\sqrt{2}}{\Pi}$ (0.9 Va)	Id	$\frac{Id}{2}$	1.57 Vdc	$Va \frac{2\sqrt{2}}{\Pi} \cos\alpha$
(d) Three phase half wave	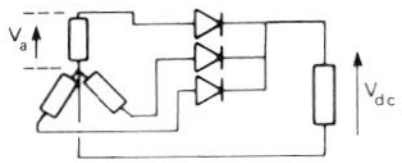 $Va \frac{3\sqrt{6}}{2\Pi}$ (1.17 Va)	Id	$\frac{Id}{3}$	2.09 Vdc	$Va \frac{3\sqrt{6}}{2\Pi} \cos\alpha$
(e) Three phase full wave	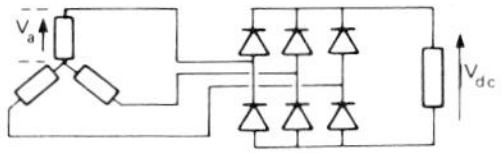 $\frac{Va3\sqrt{6}}{\Pi}$ (2.34 Va)	Id	$\frac{Id}{3}$	1.045 Vdc	$Va \frac{3\sqrt{6}}{\Pi} \cos\alpha$
(f) Double star	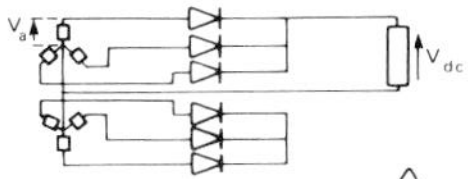 $Va \frac{3\sqrt{2}}{\Pi}$ (1.35 Va)	0.5 Id	$\frac{Id}{6}$	2.09 Vdc	$Va \frac{3\sqrt{2}}{\Pi} \cos\alpha$
(g) Zig zag	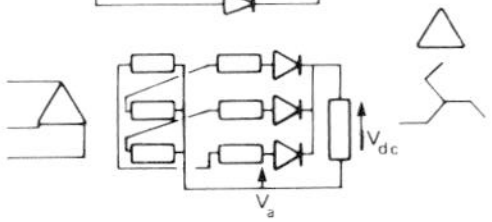 $\frac{Va3\sqrt{6}}{2\Pi}$ (1.17 Va)	Id	$\frac{Id}{3}$	2.09 Vdc	$Va \frac{3\sqrt{6}}{2\Pi} \cos\alpha$

No capacitive smoothing is assumed
Note that AC voltages are given with respect to neutral. For phase to phase AC voltages reduce Vdc by $\sqrt{3}$. For example, 240V phase to neutral, and 415V phase to phase give the same Vdc

connection of the circuit g gives zero DC component whilst minimising the number of rectifiers (albeit at the expense of transformer complexity and cost).

2.7. Burst fired circuits

The simplest thyristor and triac control circuits are those used to control heaters and similar devices. In these the triac (say) is simply used as a switch or relay to turn power on and off. The power on/off ratio controls the power fed to the load.

Figure 2.24a shows a block diagram of a burst fired circuit. A synchronising circuit produces the gate signal to fire the triac. The control signal enables the gate signals and could come from a thermostat (for temperature control with a heater load) or from an oscillator with adjustable mark/space ratio for proportional power control.

Waveforms are shown in fig.2.24b. The control signal is present from time A to time B. The synchronising circuit, however, delays firing the triac to the next voltage zero at time

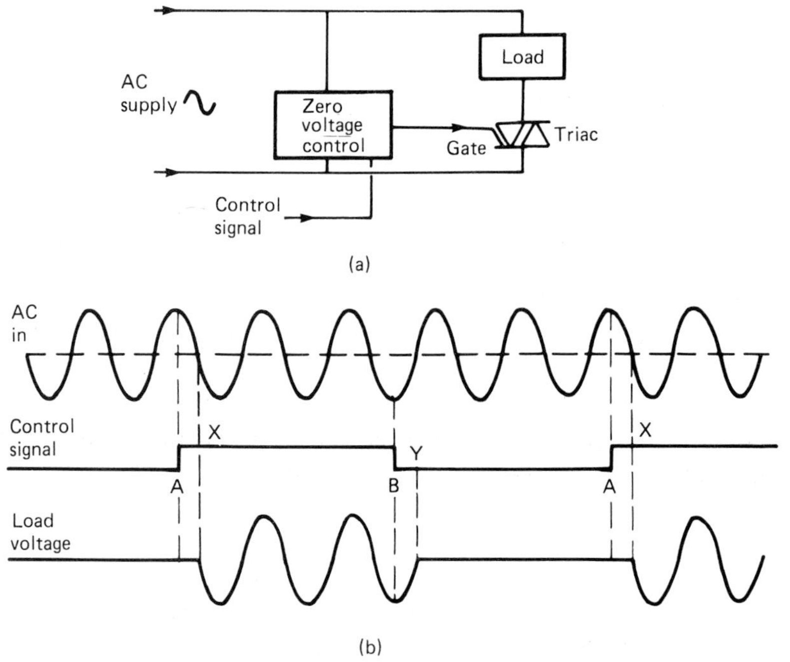

Fig.2.24 Burst firing of triacs. (a) Circuit diagram. (b) Waveforms.

X, and the natural action of the triac keeps current flowing after time B to point Y. This switch on at zero voltage, off at zero current, minimises transients fed back into the supply, and prevents interference with radio and television receivers.

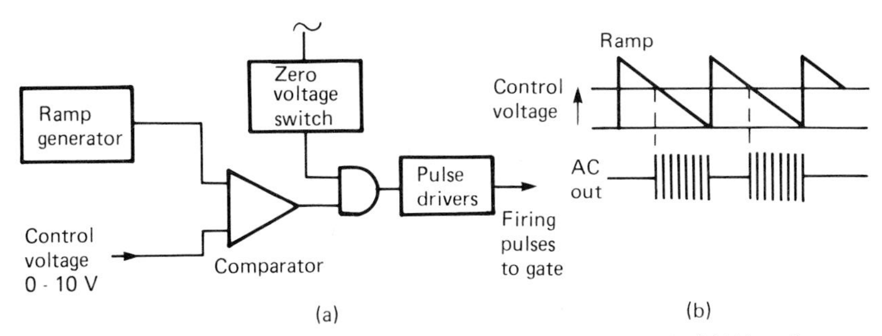

Fig.2.25 Proportional control circuit for burst firing. (a) Circuit. (b) Waveforms.

Synchronising circuits are, not surprisingly, available in IC form. Figure 2.25a shows the block diagram of a proportional device. The control voltage is compared with a relatively slow ramp (typical period 10 to 100 seconds). When the control voltage is above the ramp pulse (as detected by the comparator) the triac is enabled as shown in Fig.2.25b.

2.8. Phase shift control

2.8.1. Basic principles

Burst firing is simple, but is only suitable for loads with long time constants such as heaters. For speed control, the pulsating torque would be totally unacceptable. The commonest type of power control varies the power fed to the load by varying the percentage of each cycle of the AC supply.

Figure 2.26a shows a simple biphase rectifier similar to that in Table 2.1 except that the diodes have been replaced by thyristors. These are fired by a synchronising circuit, and the gate pulses can be adjusted over 180° of the cycle; thyristor Th1 over one half cycle, thyristor Th2 over the other. The details of the synchronising circuit need not concern us at present.

Waveforms for a resistance load are shown in fig.2.26b. During time X, the thyristors are being fired at 30° after the zero crossing point, each thyristor conducts for the rest of the half cycle once

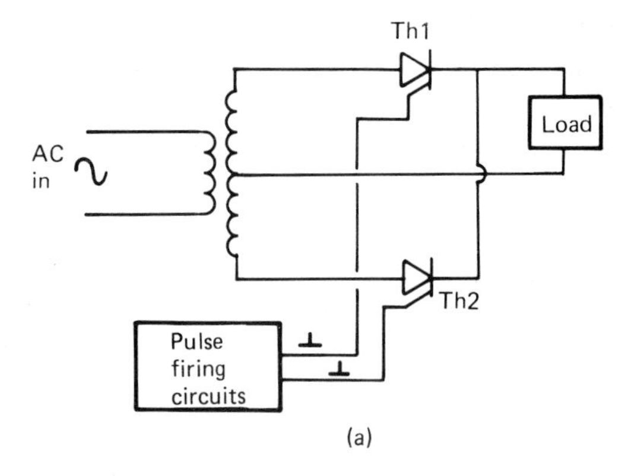

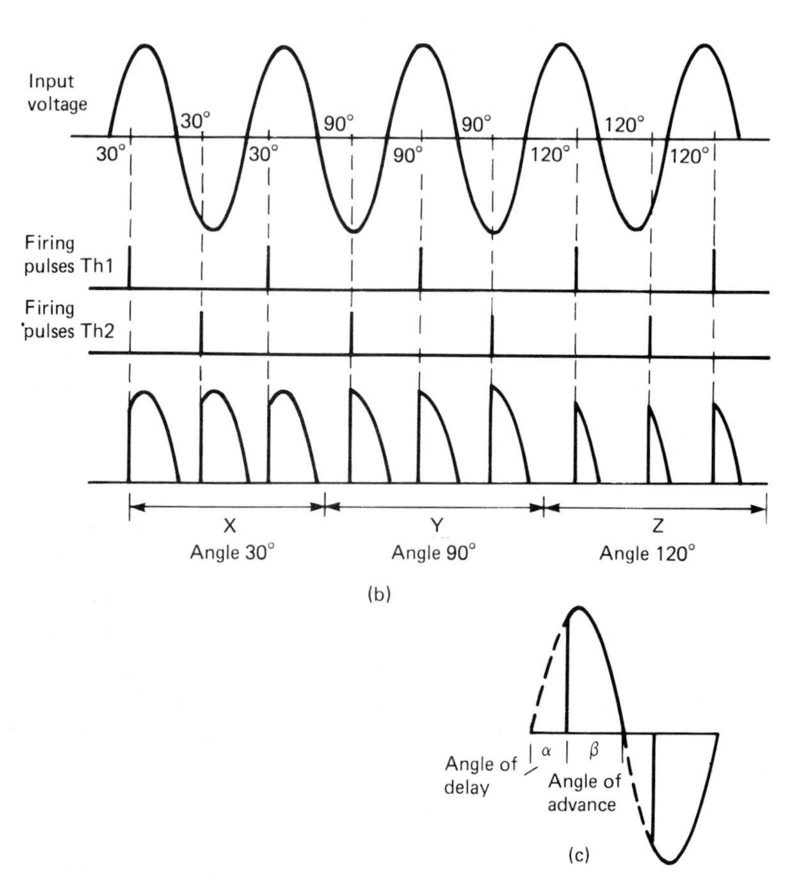

Fig.2.26 The phase shift controlled rectifier. (a) Phase controlled rectifier. (b) Waveforms for various angles. (c) Firing angles.

fired, and the power fed to the load is nearly the same as that obtained from a simple biphase rectifier.

During time Y, the firing of the thyristors is delayed to 90°, and the power fed to the load is significantly reduced. Finally, during time Z, the firing is delayed to 120°, and the power fed to the load is reduced even further.

The power fed to the load is therefore controlled by the point in the cycle at which the thyristors are fired. The firing angle as described above is called the angle of delay, denoted by α Maximum power is delivered for $\alpha = 0°$, and minimum power for $\alpha = 180°$ for a resistive load (inductive loads are described later). The firing point can also be described by the angle of advance, denoted by β, as shown in fig.2.26c. Obviously $\beta = 180° - \alpha$, and maximum power occurs for $\beta = 180°$.

The average output voltage of fig.2.26 is given by:

$$V = \frac{\sqrt{2}E}{\pi}(1 + \cos\alpha) \qquad (2.23)$$

2.8.2. *Inversion and the effects of inductance*

Figure 2.26 was drawn for a purely resistive load. Most industrial loads, particularly motor armatures and fields, have significant inductance and this significantly modifies the behaviour of controlled rectifier circuits.

Figure 2.27a is the simplest controlled rectifier arrangement: a single half wave bridge. In fig.2.27b this is fired at an α of 60° at time T1. At time T2 the transformer voltage passes through zero (and conduction would cease for a resistive load). The inductance, however, keeps current flowing until time T3, so the voltage follows the line voltage negative from T2 to T3.

During the time T1 to T2, power is being transferred from the supply to the load; during T2 to T3 power is being returned from the stored energy in the inductor to the supply.

When the controlled rectifier is returning power to the supply it is said to be acting as an inverter. When power is flowing from the supply to the load, the circuit is said to be acting as a rectifier.

Inversion occurs in fig.2.27 because of the load inductance. It will also occur in any situation where the load current is continuous. Later sections will describe dynamic braking of a

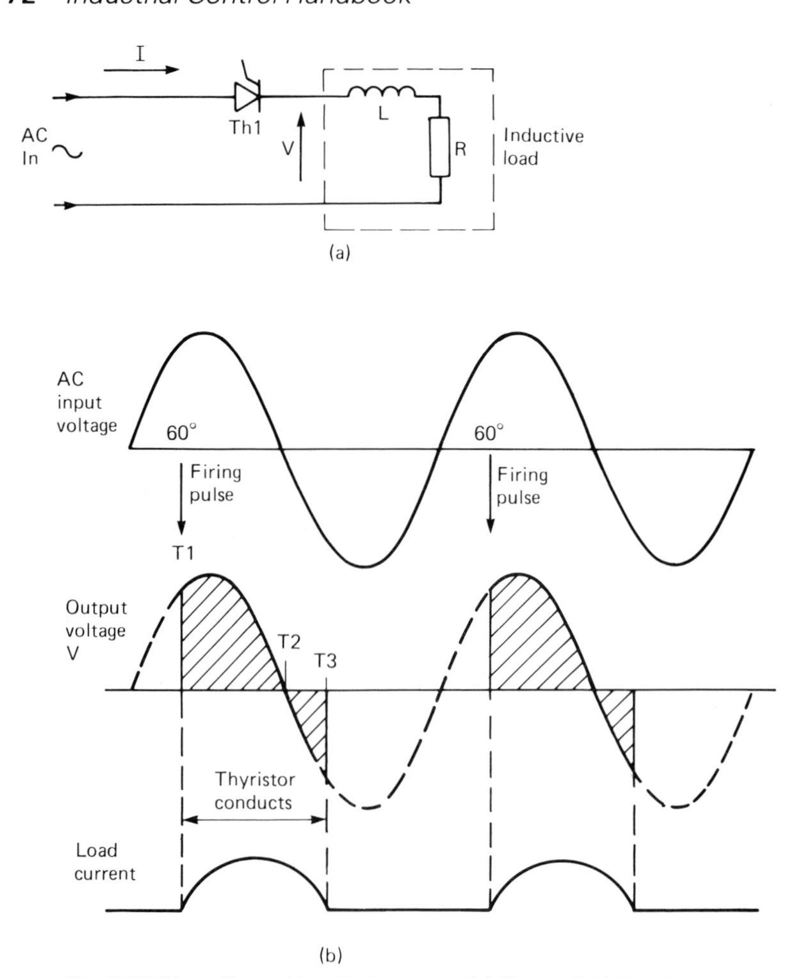

Fig.2.27 The effect of load inductance. (a) Circuit. (b) Waveforms.

motor by returning its kinetic energy to the supply via its own back emf and a controlled rectifier circuit acting as an inverter.

Figure 2.28a shows a controlled full wave bridge connected to a load which draws continuous current – a highly inductive load, or a motor acting as a generator, for example. The output voltage for various values of α are shown in fig.2.28b. Note that the continuous current causes the output voltage to follow the line voltage negative until the other thyristor pair is fired.

For $\alpha = 30°$ and $\alpha = 60°$ the average voltage is positive and power is, overall, transferred to the load, i.e. the circuit behaves as a rectifier. For $\alpha = 90°$ the positive and negative portions of the waveforms are equal and there is no net transfer of power. For α

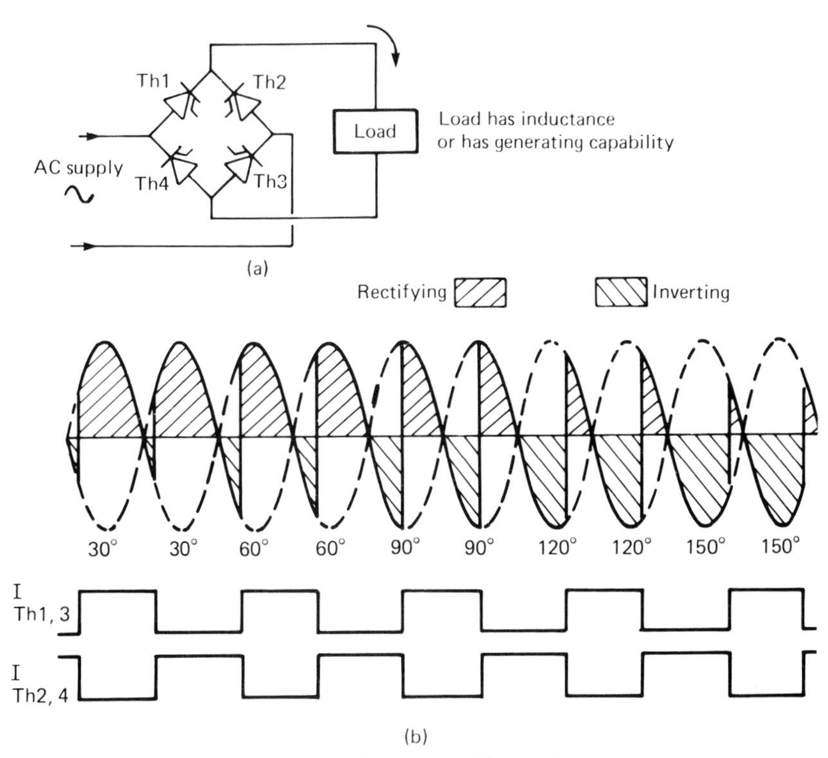

Fig.2.28 The fully controlled full wave rectifier with continuous current. (a) Circuit. (b) Waveforms.

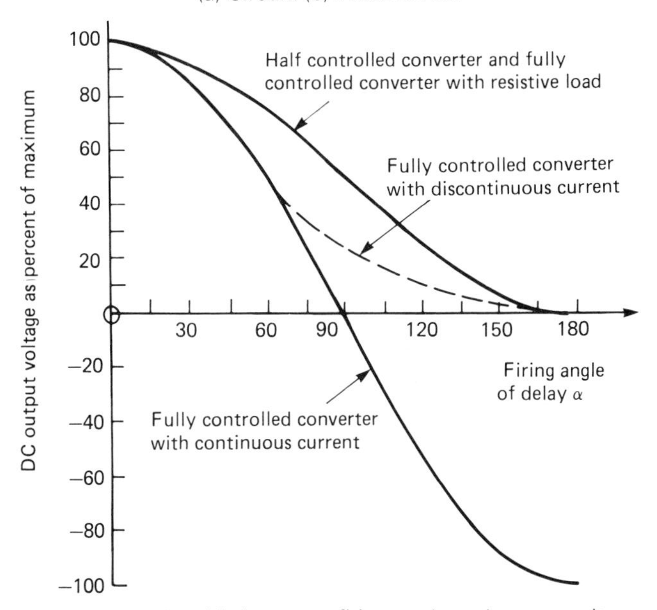

Fig.2.29 Relationship between firing angle and output voltage.

For $\alpha = 120°$ and $\alpha = 150°$ the average voltage is negative and power is being transferred from the load to the supply, i.e. the circuit acts as an inverter.

If the load current is continuous, therefore, the circuit acts as a controlled rectifier from $\alpha = 0°$ to $\alpha = 90°$, and as an inverter from $\alpha = 90°$ to $\alpha = 180°$. The output voltage is given by:

$$V = \frac{2\sqrt{2}E}{\pi} \cos\alpha \tag{2.24}$$

This should be compared with the resistive case of equation 2.23. Both are plotted in fig.2.29 for comparison. A circuit containing inductance, but not sustaining continuous current, will follow an intermediate curve similar to the dotted curve in fig.2.29.

It is conventional, but by no means universal, to define the firing angle in terms of α for rectification and β for inversion.

2.8.3. Three phase circuits

Most motor control circuits convert a three phase AC supply to DC. These are usually based on the half wave bridge d and the full wave bridge e of table 2.1.

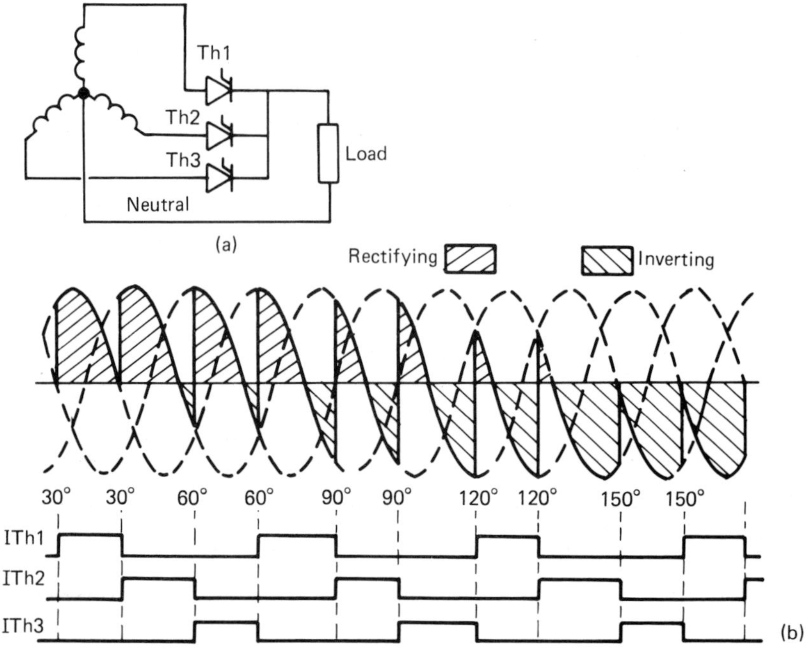

Fig.2.30 The three phase half wave controlled rectifier with continuous current. (a) Circuit diagram. (b) Waveforms.

The half wave bridge circuit is shown in fig.2.30a. As before, the production of the firing pulses need not concern us at this stage. The firing angle α is defined from the natural commutation point of the bridge, as shown in fig.2.30b which gives waveforms for continuous current from 30° to 150°. As we saw for the single

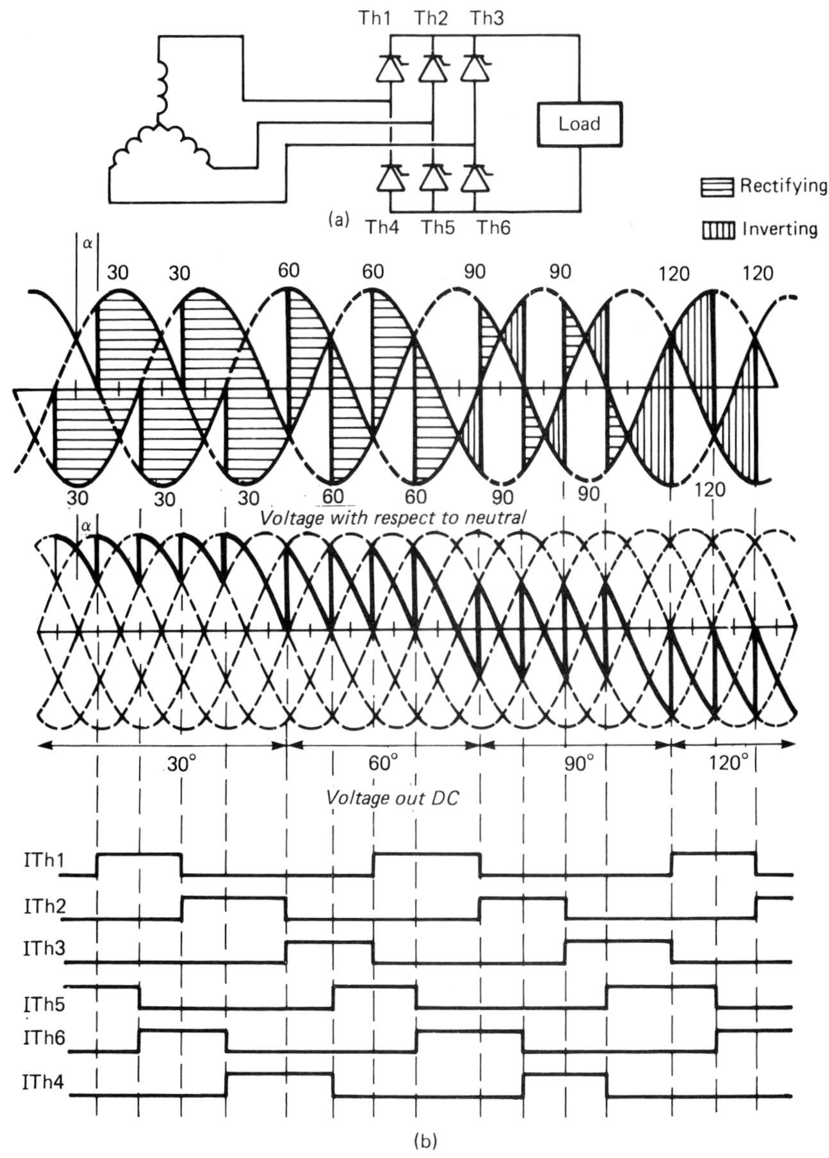

Fig.2.31 The three phase full wave controlled rectifier. (a) Circuit. (b) Waveforms.

phase circuits with continuous current, the circuit rectifies for $\alpha < 90°$, and inverts for $\alpha > 90°$.

Figure 2.31a shows the full wave circuit – probably the commonest controlled rectifier arrangement. Waveforms are again given in fig.2.31b with the firing angle defined from the natural commutation point. As before, the circuit rectifies for $\alpha < 90°$ and inverts for $\alpha > 90°$.

2.8.4. *Half controlled circuits*

Superficially, fig.2.32a is a full wave thyristor bridge, but closer inspection will show that devices D1 and D2 are simply rectifiers. A circuit with mixed thyristors and diodes is called a half controlled, or mixed, bridge.

Assuming continuous current, let us follow the circuit action from the point where Th1 fires. Current flows through Th1 and D1 until the end of the half cycle. At this point, current switches from D1 to D2, the continuous current circulating through Th1, the load and D2 with zero transformer current. This state continues until Th2 fires, when current flows from the transformer through Th2, the load and D2. At the end of the half cycle, the current switches from D2 back to D1 and current circulates through Th2 and D1 until Th1 fires again. Waveforms are shown in fig.2.32b.

The action of D1 and D2 prevents the bridge output voltage from going negative, and consequently the circuit can only operate as a rectifier, not an inverter. The output voltage is similar to that for a purely resistive load, and the average voltage given by equation 2.23.

Figure 2.32a is not the only half controlled circuit. Figure 2.33a shows an alternative single phase circuit, and fig.2.33b and c three phase equivalents. Figure 2.33c is commonly used where inversion is not required from a three phase motor control circuit.

2.8.5. *Commutation and overlap*

In any practical controlled or uncontrolled rectifier, the AC source will have significant inductance in each phase from the transformers, supply lines and any suppression circuits on the AC

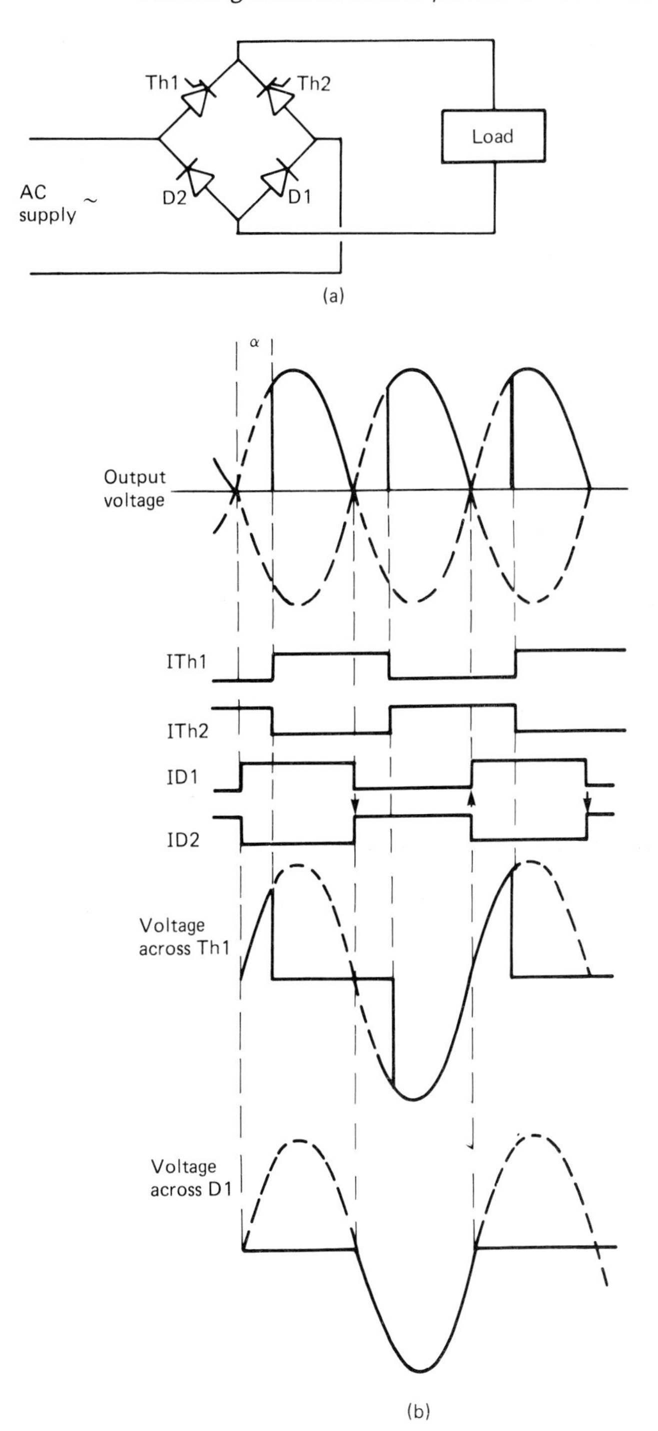

Fig.2.32 The half controlled rectifier. (a) Circuit. (b) Waveforms.

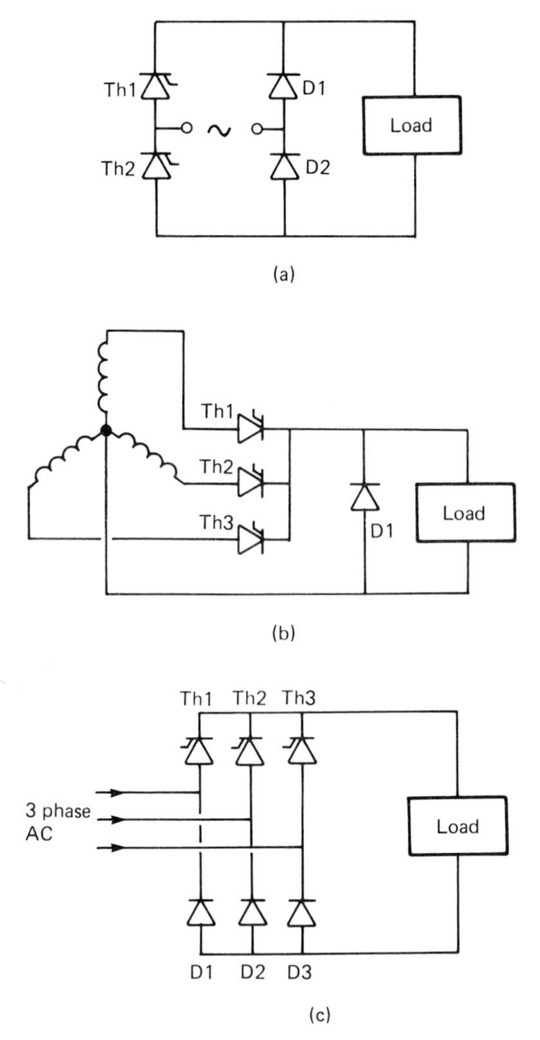

Fig.2.33 Common half controlled rectifier circuits. (a) Circulating current flows through D1, D2. (b) Three phase half wave rectifier. Circulating current flows through D1. (c) Three phase full wave rectifier. This operates in a similar manner to Fig.2.32.

supply. This inductance will prevent the instantaneous switch of current from one rectifier to another.

A three-phase half wave rectifier is shown in fig.2.34a. Current theoretically switches instantaneously from D1 to D2 to D3 and back to D1 at the natural commutation points. The effect of inductance is shown in fig.2.34b. The natural commutation point is denoted by X, and D2 starts to conduct at this point. The

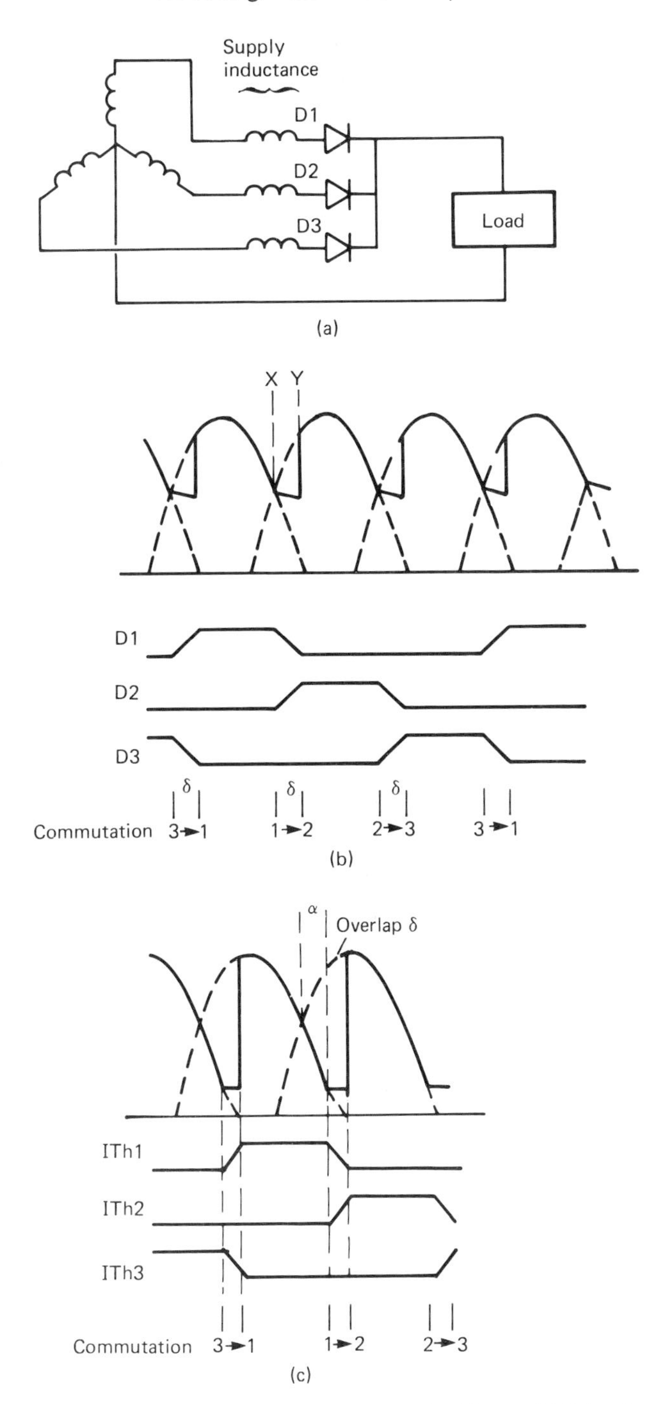

Fig.2.34 The effect of supply inductance. (a) Circuit. (b) Waveforms. (c) Controlled rectifier waveforms.

inductance, however, keeps current flowing in D1 until point Y. Between X and Y D1 and D2 both conduct, and the output voltage is the average of the two phase voltages giving a notch in the output. The time between X and Y is called the commutation angle, or overlap, and is typically a few degrees. If is often denoted by the Greek letter δ.

If the diodes are replaced by controlled thyristors, overlap gives the waveform of fig.2.34b. Again, a notch appears in the output waveform as two thyristors conduct together.

Overlap sets a maximum limit for α (and minimum limit for β) and prevents inversion to the theoretical value of $\alpha = 180°$. If the commutation between successive thyristors has not occurred before 180°, a commutation failure will occur and current transfer will not take place. In practice, a minimum value of β of about 10° must be set to ensure commutation.

2.9. Motor control

2.9.1. Simple single phase circuit

Section 2.4.6. showed that the speed of a DC motor can be controlled either by varying the armature voltage (or current) or the field strength. Figure 2.35a shows a single phase half wave controlled circuit, possibly the simplest motor control circuit possible. This operates with a fixed field, and speed is controlled by varying the armature current through Th1. The current is discontinuous at all times, which limits the effectiveness of the circuit.

The speed is measured by a tacho generator and compared with the set speed. The error is used to advance or retard the firing pulses to Th1 (usually with P and I control) to make the actual speed equal the set speed.

Waveforms are shown in fig.2.35b, and may be different from those expected. Applied volts are denoted by V and the back emf by E (it will be remembered that E is proportional to speed). Thyristor current is denoted by I. The thyristor cannot be fired until point A when the input volts exceeds the back emf. The firing circuit fires the thyristor at point B, and armature current continues until point C, when the input voltage falls below the back emf.

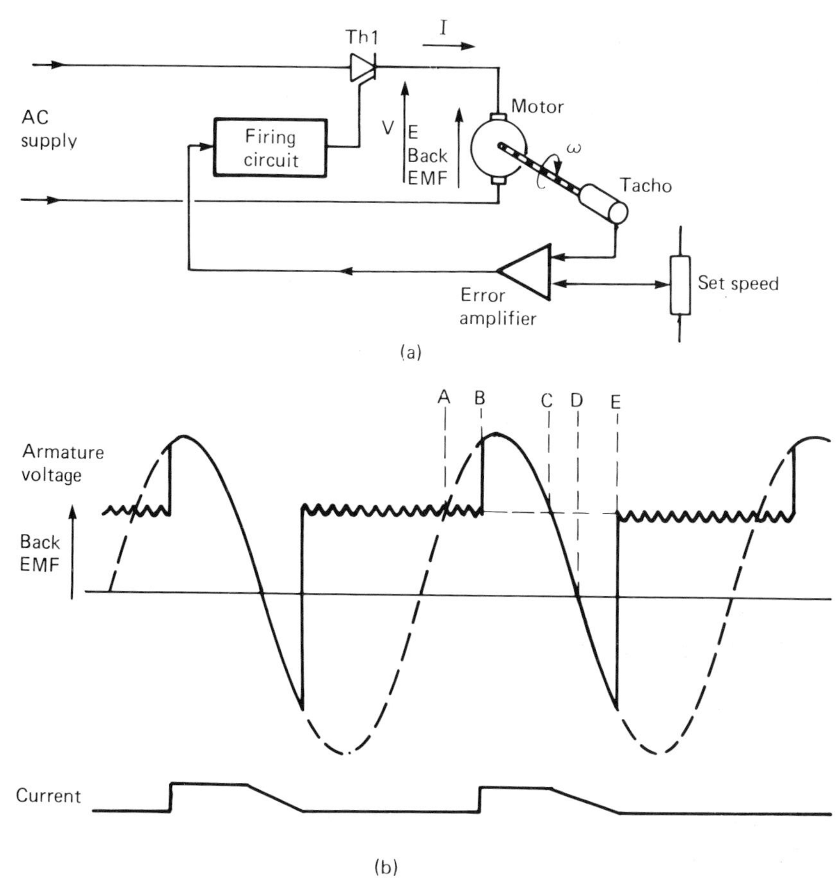

Fig.2.35 Simple motor speed control. (a) Circuit. (b) Waveforms.

Armature inductance, however, means that the current cannot fall to zero instantly, and Th1 continues to conduct, and the armature voltage follows the input. At point D the voltage across the armature reverses (and Th1 becomes an inverter). At point E the current falls to zero, and the armature voltage jumps to the back emf, where it remains to point A in the next cycle. The ripple between points E and A is caused by the commutator segments on the motor.

It is interesting to note that the circuit of fig.2.35 has a degree of self-regulation, even without the tacho feedback. If the load increases and the speed consequently falls, the back emf will also be lowered. This will move point C further back and increase the length of the armature current pulse, mitigating the effect of the increased load.

The circuit of fig.2.35 cannot provide a braking action in the same direction as it provides motoring action. Although the circuit inverts between points D and E in fig.2.35b, it is the inductive energy that is being returned, not the kinetic energy of the motor. If a reduction in speed is required, the gate pulses would be phased right back and the motor allowed to coast down under the action of the load and friction until the new speed was reached. At this time the pulses would be advanced sufficiently to maintain speed.

Figure 2.35 can, however, provide braking in the reverse direction as shown in fig.2.36. Reversed direction gives a

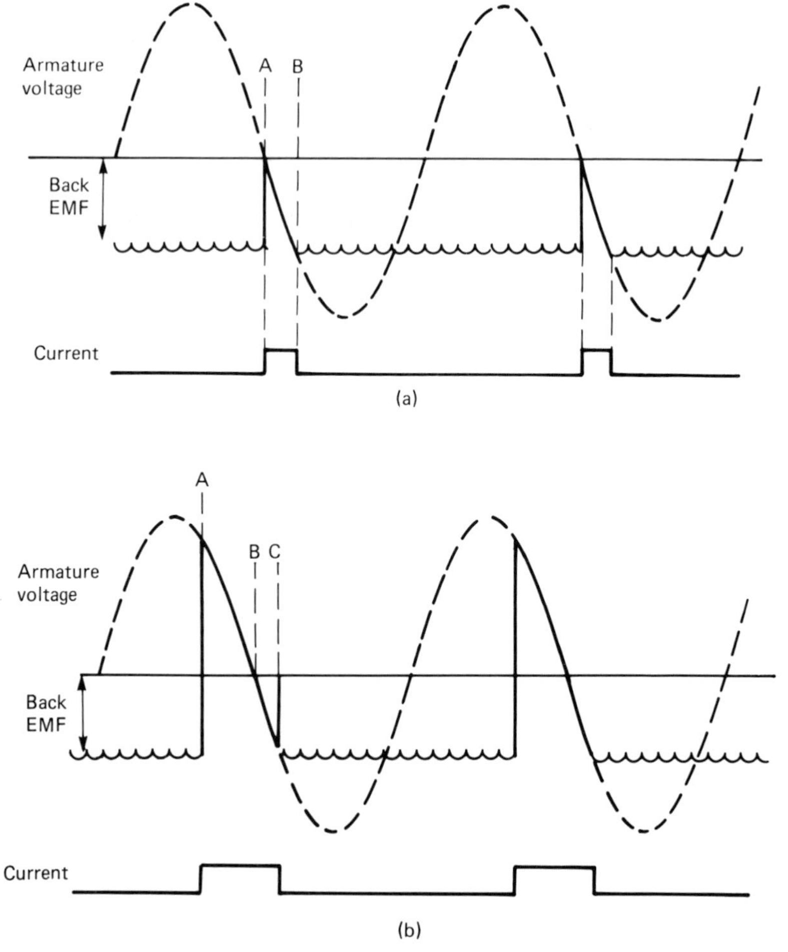

Fig.2.36 Dynamic braking of a motor. (a) Firing at zero crossing point. (b) Firing before zero crossing point.

reversed, i.e. negative, back emf as shown. If Th1 is fired at point A in fig.2.36a, current will be returned from the kinetic energy of the motor to the supply (V is negative, I is positive, so energy is flowing from the motor to the supply) until point B, where the supply volts goes more negative than the back emf (neglecting the effects of armature inductance). The circuit is acting as an inverter, albeit in a limited manner because of the discontinuous current.

If the firing point is advanced further as in fig.2.36b, the braking becomes more efficient. The circuit, however, is now acting as a rectifier from points A to B forcing current through the armature. Inversion still takes place from B to C, but on average the power transfer is from supply to motor (and the motor kinetic energy is converted to heat, not returned to the supply).

2.9.2. Single, two and four quadrant operation

Figures 2.35 and 2.36 are said to be two quadrant circuits. Let us examine what this means and the implications for motor control. Figure 2.37a shows a block representation of a controlled rectifier connected to a DC load. Assuming the load is capable of sustaining continuous current (either via inductance or, say, a generator), there are four possible operating conditions, shown in fig.2.37a. These are called the four operating quadrants. A simple uncontrolled rectifier operates in quadrant 1 for example.

In quadrants 1 and 3, power is transferred from the AC supply to the load. In quadrants 2 and 4, power is transferred from the load back to the supply.

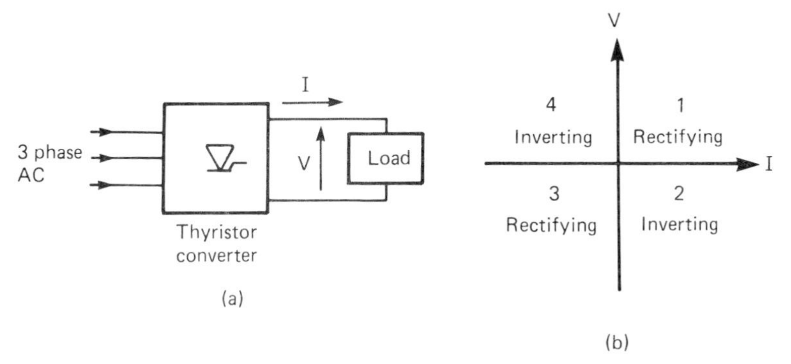

Fig.2.37 Four quadrant converters (a) Circuit diagram. (b) Four quadrants of operation.

No single controlled rectifier bridge can operate in more than two quadrants. Fully controlled rectifiers (with thyristors in each bridge position), such as figs.2.28, 2.30 and 2.31, can operate in quadrants 1 and 2 (or 3 and 4 by reversing the connections from the rectifier to the load).

Half controlled bridges (with mixed thyristors and diodes) such as fig.2.33 can only act as rectifiers and hence only operate quadrant 1 (or quadrant 3 by reversal of connections). Such circuits cannot invert.

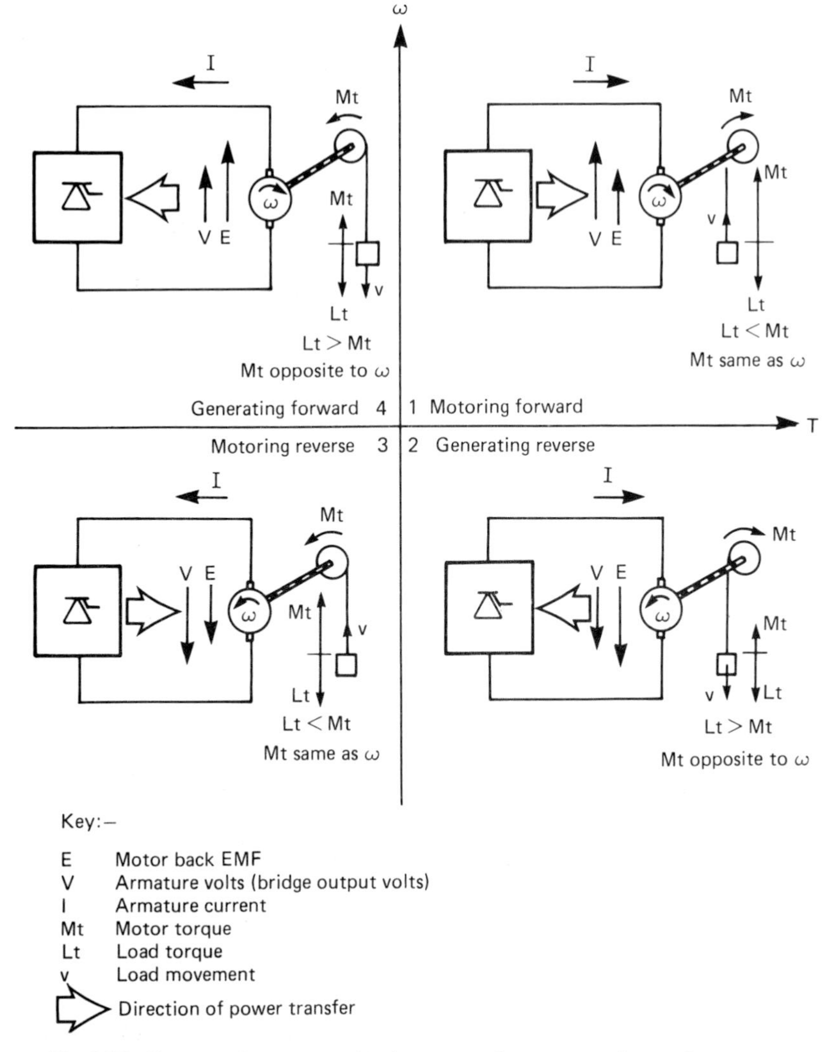

Fig.2.38 Four quadrant operation in terms of motor rotation and torque.

The four quadrants of fig.2.37 are redrawn as fig.2.38 in terms of motor rotation and motor torque. Note that for motoring quadrants, motor torque and rotation have the same sign, and Mt >Lt and V > E. Similarly for generating quadrants, torque and rotation are of opposite sign, and Mt <Lt and V <E.

It follows that a single quadrant converter can only motor, and a fully controlled (two quadrant) converter can only operate as a motor in one direction, and as a generator in the opposite direction. There are few applications for this type of control, the only common ones being shown in fig.2.39. The first is a bidirectional drive where material is wound back and forth through some process between coils (e.g. papermaking). The motors alternately drive and provide back tension (e.g. A

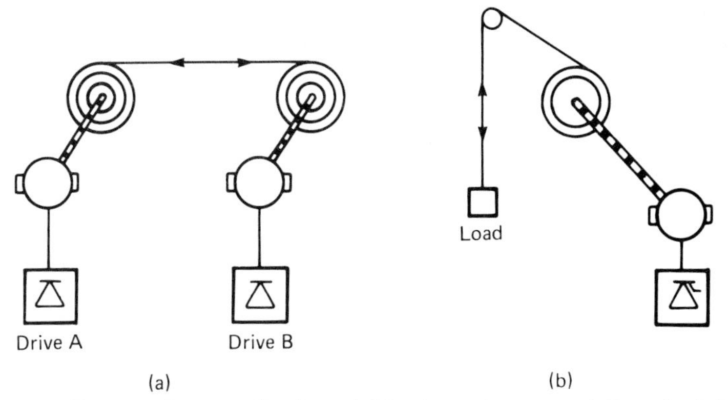

Fig.2.39 Two quadrant applications. (a) Back tension control. Transfer left to right; A inverts, B drives. Transfer from right to left; A drives, B inverts (b) Winch drive. Load rises with motor driving; falls with motor inverting.

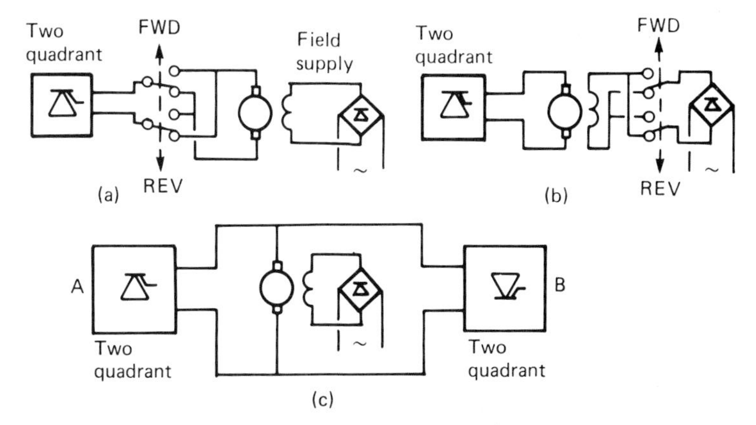

Fig.2.40 Various four quadrant arrangements. (a) Armature changeover. (b) Field changeover. (c) Dual converter system.

motoring and B generating). The second, and more common, is a winch, where the load falls with the motor generating and rises with the motor driving.

Most applications require operations in all four quadrants. This can be achieved by any of the schemes of fig.2.40. Circuit a uses a contactor to reverse the armature and circuit b the field. Both employ a two quadrant converter, and require external sequencing to control the contactor, and in particular ensure that switching only takes place at zero armature current.

Figure 2.40c uses two separate converters: A operating in quadrants 1 and 2, say, and converter B in quadrants 3 and 4. This configuration is discussed further in section 2.9.4.

2.9.3. Plugging

Quadrants 2 and 4 are often labelled 'braking'. Whilst it is true that generating does act as a brake, it is only efficient at high speeds, and the net torque reduces as the motor speed falls.

Quadrants 1 and 3 can also be used to perform very efficient braking, but the motor kinetic energy is dissipated as heat rather than being returned to the supply. Suppose, adopting the notation of fig.2.38, we have a motor spinning clockwise (positive ω) and we switch it into quadrant 1. V and E will be of opposite sense, and the current will rapidly bring the motor to a standstill (and even reverse its direction if the rectifier does not return α to 90° as the motor stops).

In practice, current control is essential to limit I. Normally the rectifier would start braking in quadrant 2, and shift into quadrant 1 as the motor speed falls. This action is sometimes called plugging. A typical situation occurs in the winch drive of fig.2.39b where the load is being lowered at a slow controlled speed. The converter will be rectifying to provide the necessary torque (and the motor will significantly heat up).

2.9.4. Dual converter systems

The principle of a dual converter system is shown in fig.2.41. Each of the converters can be considered to be a variable voltage source in series with a diode. Both converters can achieve a positive or negative output voltage, but the series diodes limit the

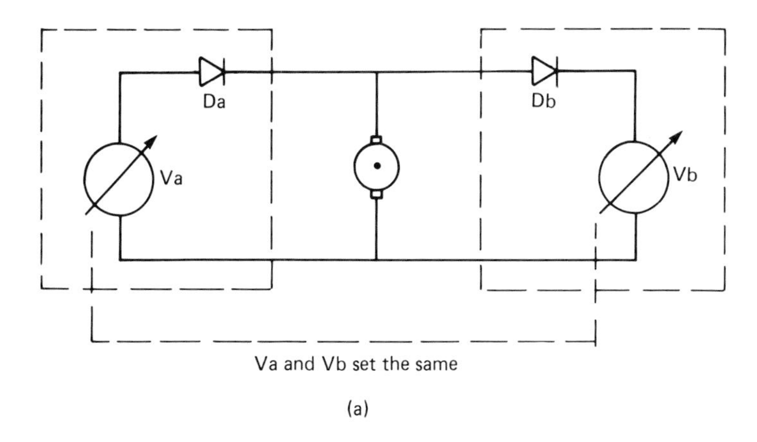

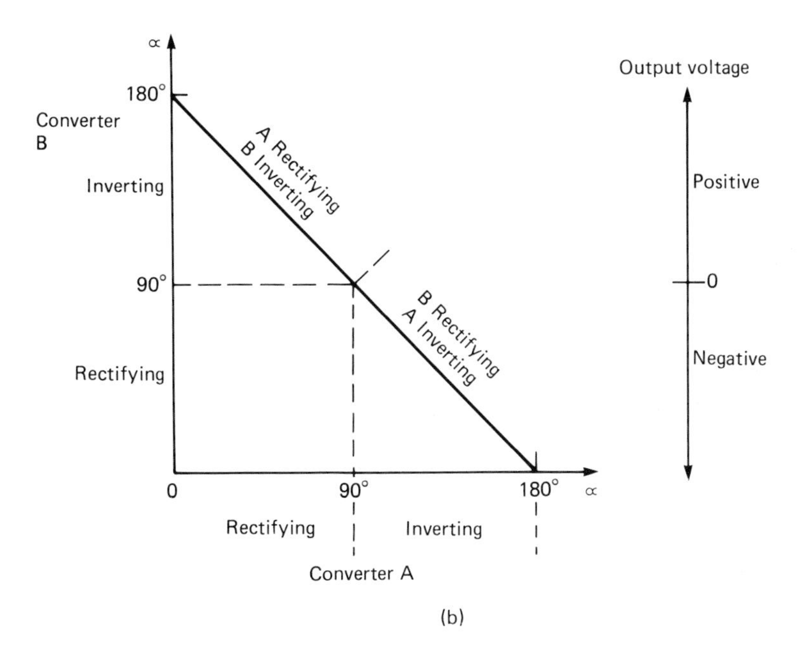

Fig.2.41 Operation of dual converter system. (a) Representation of dual converter system. (b) Relationship between the two converters.

direction of current flow. Both converters are externally controlled so they always have the same voltage.

Let us assume we are motoring forward with an armature voltage of +50 volts, and this gives clockwise rotation. Both V_a and V_b are set at 50 V but bridge A is rectifying (via D_a) and bridge B is blocked by D_b. We now reduce speed by dropping V_a and V_b to +25 V. D_a blocks bridge A, and the motor generates into bridge B until its speed falls to the point where bridge A conducts and starts rectifying, and bridge B is blocked.

If we wish to reverse direction, V_a and V_b will be reduced to zero, then increased negative, and bridge B will at first invert, then plug, then rectify until the motor achieves the required speed. For a reduced reverse speed, the motor inverts into bridge A until the speed falls to the required level.

Effectively the two converters are operating up and down the curve of fig.2.41c, and for any given armature voltage one bridge is rectifying and one inverting. Note that $\alpha A+\alpha B=180°$, and for zero armature volts both inverters have $\alpha=90°$.

It is not possible to make the voltages from the two bridges identical. The imbalance can be handled in two ways. A circulating current bridge allows current to flow between bridges, usually with inductors in the DC legs of each bridge.

A more common arrangement, though, is to use external logic to ensure only one bridge fires at once. This is known as an anti-parallel supressed half (APSH) bridge. The bridge selection logic requires current monitoring to ensure bridge selection changes only take place at zero current. There is inevitably a small delay during which time the motor coasts.

Suppose we are motoring forward on bridge A, and a speed reversal is called. A typical sequence could be:

(a) A phases right back (large α) and inverts until armature current falls to zero (energy stored in armature inductance). Motor is still travelling forward.
(b) Logic sees zero current; blocks bridge A, releases gate pulses on bridge B.
(c) B phases forward to invert from motor back emf. Note that current direction is now reversed. Motor deceleration is controlled by bridge B. As motor speed falls, B phases further forward with $\alpha < 90°$ near zero speed.
(d) Motor stops, and accelerates in reverse up to required speed on bridge B.

It is also instructive to follow the sequence where a speed reduction is called without a reversal of direction. Again let us start motoring forward on bridge A, and call for a speed reduction:

(a) A phases back and inverts ($\alpha>90°$) to reduce the armature current to zero.

(b) Logic sees zero current, blocks bridge A and releases bridge B. Current flow through motor is reversed.
(c) Motor back emf inverts into bridge B; motor speed falls under control of bridge B.
(d) When correct speed is reached bridge B decreases α to give zero current. When zero current is detected bridge B is blocked and A released.
(e) Bridge A now drives motor at new speed.

2.9.5. Armature voltage control and IR compensation

Equations 2.14 and 2.15 show that the back emf of a motor is proportional to speed, and the applied armature volts approximate to the back emf if the IR term is small. From this a simple speed control can be derived as shown in fig.2.42. The armature

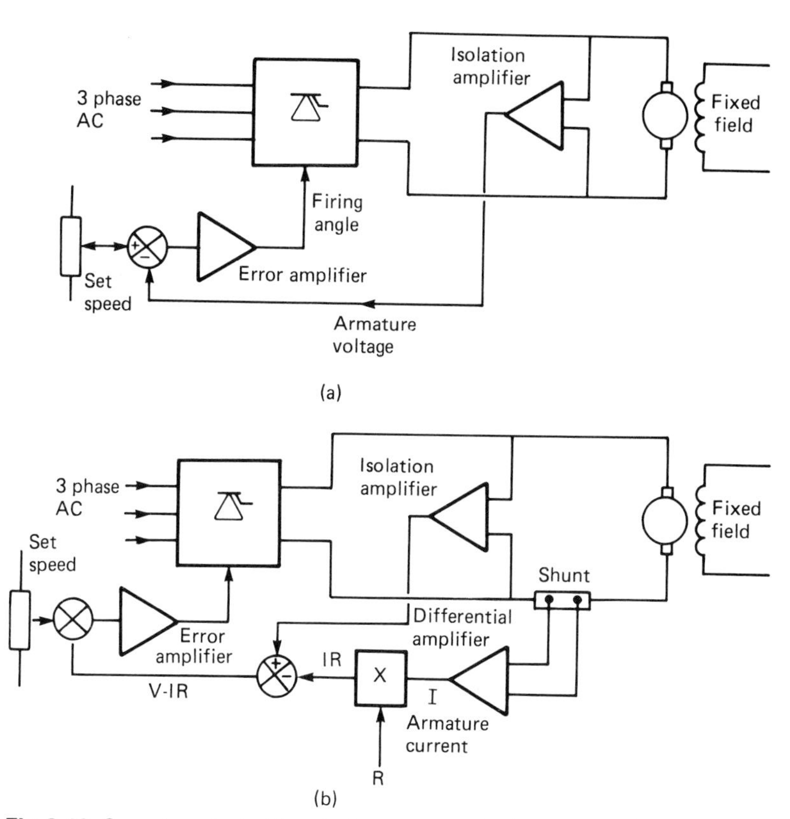

Fig.2.42 Speed control without a tachometer. (a) Armature voltage control. (b) IR compensation.

voltage is compared with the set speed. The resultant error is used to advance, or retard, the firing angle of the thyristor bridge and hence control the speed. Usually a P + I controller is used.

Figure 2.42a does not, however, compensate for load changes. Figure 2.42b measures both armature volts and current (V and I in equation 2.14) and computes E from the relationship $E = V - IR$. This can be done with a few DC amplifiers. E (and hence ω) are compared with the set speed and used to control the bridge firing angle. This, in theory, gives true speed control with load changes and is called IR compensation. In practice, if the IR term is too large the control becomes unstable, so it is normal to set the drive for about 5% speed 'droop' from no load to full load. Motor windings also experience severe temperature changes, which cause the value of R to change significantly with load and time. This further reduces the effectiveness of IR compensation. The technique is, however, simple and cheap and is widely used with small motors and reasonably constant loads.

2.9.6. Tacho feedback and current control

Figure 2.43 shows the commonest arrangement for a reversing thyristor drive, and represents an APSH circuit with tacho speed feedback. The set speed is represented by a voltage in, say, the range −10 V to +10 V. This is passed through a ramp circuit to limit the drive acceleration (a possible circuit is fig.1.26 in chapter 1).

The actual speed comes from a tachogenerator (typically 1 V per 100 rpm) and the resultant speed error presented to a P + I amplifier. The output from this amplifier is a current set point (limited to safe values by a limit circuit). The actual current is measured either by a shunt in the motor armature circuit or, as shown, by rectifying the output from current transformers on the AC supply. The latter is cheaper as it avoids the need for isolation amplifiers, but adds a lag into the current measurement via the filters on the rectifier.

Bridge selection logic enables the A/B firing circuits as described above, and controls the polarity of the current feedback signal (the output of the current feedback rectifier being unipolar). Gating is sequenced by the output of zero current detection circuit.

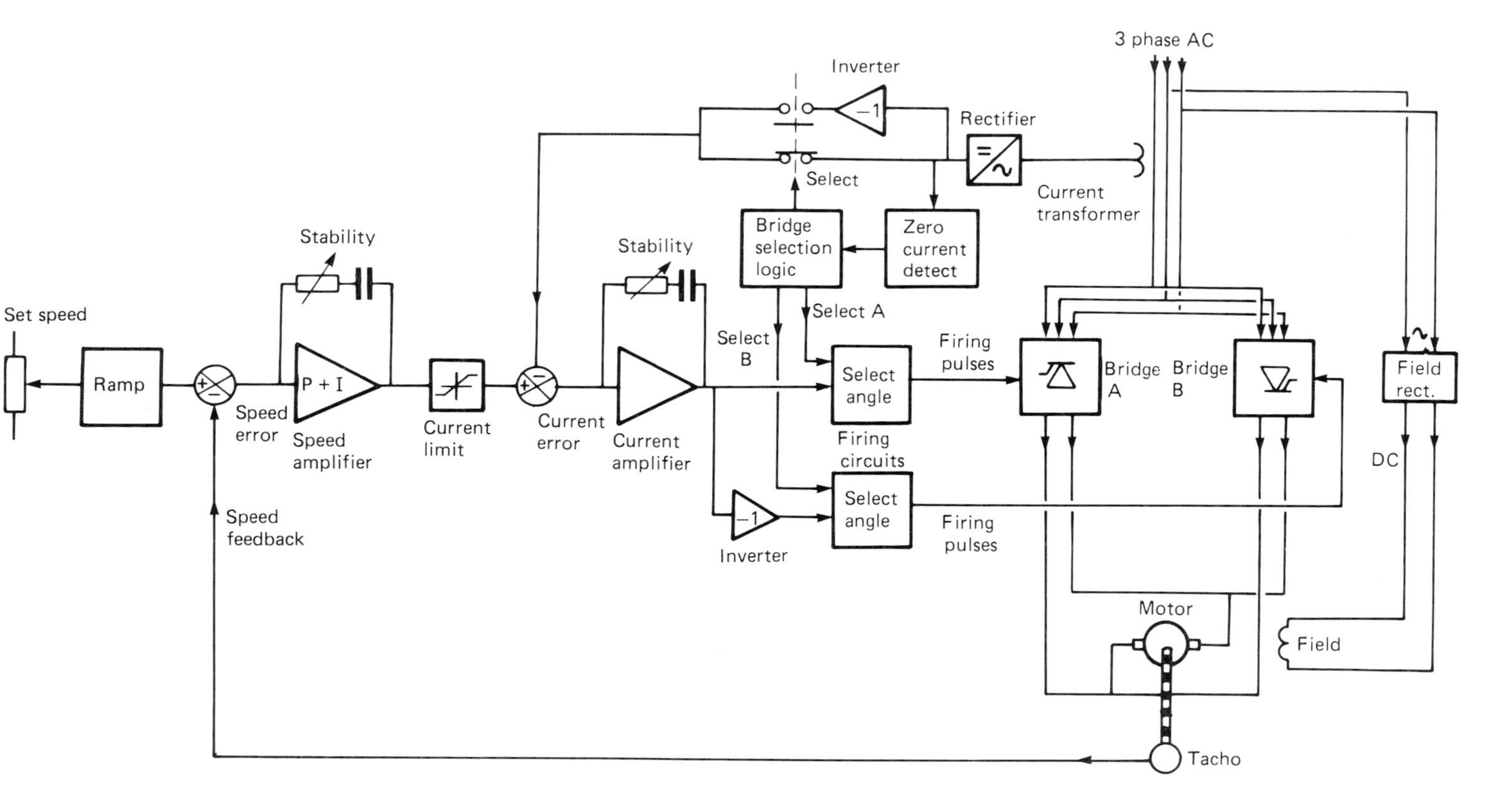

Fig.2.43 Block diagram of APSH drive.

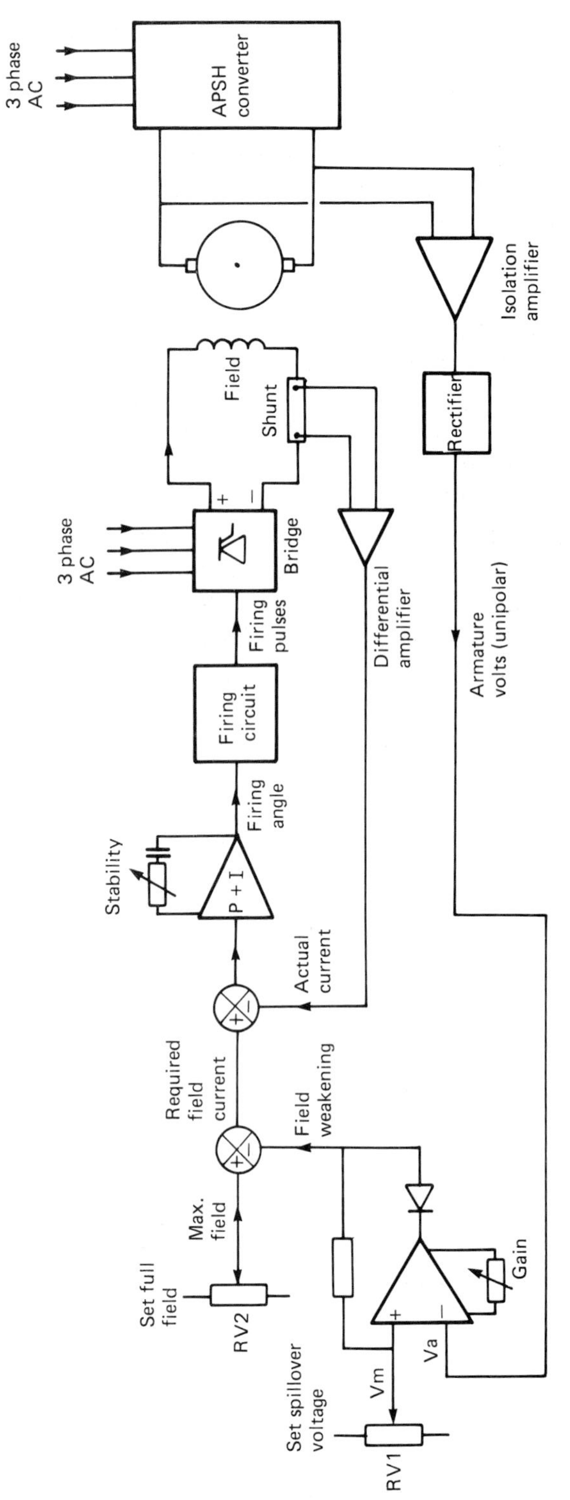

Fig.2.44 Field converter with field weakening.

2.9.7. Field weakening

Equation 2.15 shows that a motor speed can also be controlled by the field strength; for a given armature voltage a reduced field would cause an increase in speed. Any motor has a maximum armature voltage determined by the insulation and similar factors. This armature voltage will correspond to a no-load speed called the base speed at the specified field current.

Armature voltage/current can only be used to control the motor up to base speed, but the speed can be increased further by reducing the field strength once maximum armature voltage is reached. The circuit to achieve this is shown in fig.2.44.

The armature voltage is reduced to a reasonable level by an isolation amplifier, and rectified to give a unipolar DC signal. This is compared with the maximum armature voltage set on RV1 (often called the spillover voltage). As long as V_a is less than V_m, the output from the spillover amplifier will saturate positive, and the reference for the field controlled rectifier will be purely set by RV2. If V_a starts to rise above V_m, however, the spillover amplifier will reduce the current reference, causing the field current to reduce.

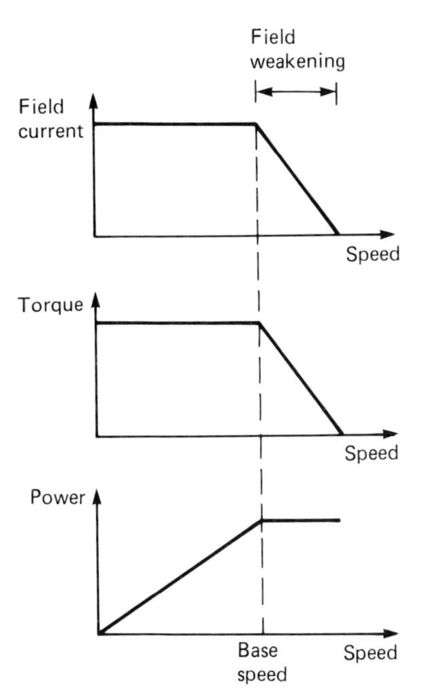

Fig.2.45 Effect of field weakening.

The reduced field will cause an increase in speed, and the main speed loop will reduce the armature voltage. The net result is that V_a stays nearly equal to V_m, and speed is controlled by field current above base speed.

Below base speed, the motor is current limited and the maximum available torque is fixed (as in fig.2.45). A motor accelerating in current limit, for example, is accelerating with constant torque. Above base speed, the maximum available power is fixed, and the maximum torque falls. Ultimate speed is limited either by centrifugal forces on the armature or commutator segments, or by commutation failures (leading to excessive sparking).

Commutation can be improved by having a current limit which reduces as speed rises above base speed.

2.9.8. Gate firing circuits

The firing circuits in a thyristor drive are required to convert a control signal, say 0 to 10 volts, into firing pulses over a 0° to 180° range for each bridge thyristor. Figure 2.46 shows the firing pattern for a typical three phase bridge, and it can be seen that each thyristor requires two pulses 60° apart to ensure operation with discontinuous current. Although, in theory, the pulses can be moved over a 180° range, in practice α is usually limited to the range 7° to 150° to avoid commutation failures.

One common circuit is shown in fig.2.47a. Zero crossing pulses are derived from one phase, and these are set by an RC differentiator to be 7° wide. These trigger a ramp, and the pulses are added to the ramp to set the front 7° limit. The 30° back limit is derived by detecting crossover points between phases, the resultant square wave also being added to the ramp.

The composite ramp is now compared with the control voltage to give a firing pulse at the point where the ramp and control voltage are equal. Pulses from two identical circuits on the other phases are combined in a diode matrix to give the six double pulses for the six thyristors. Pulse amplifiers fire the thyristors via pulse transformers. Waveforms are shown in fig.2.47b.

The circuit of fig.2.47 has a non-linear response between control voltage and bridge output (as shown by equations 2.23 and 2.24). A linear response can be obtained by replacing the

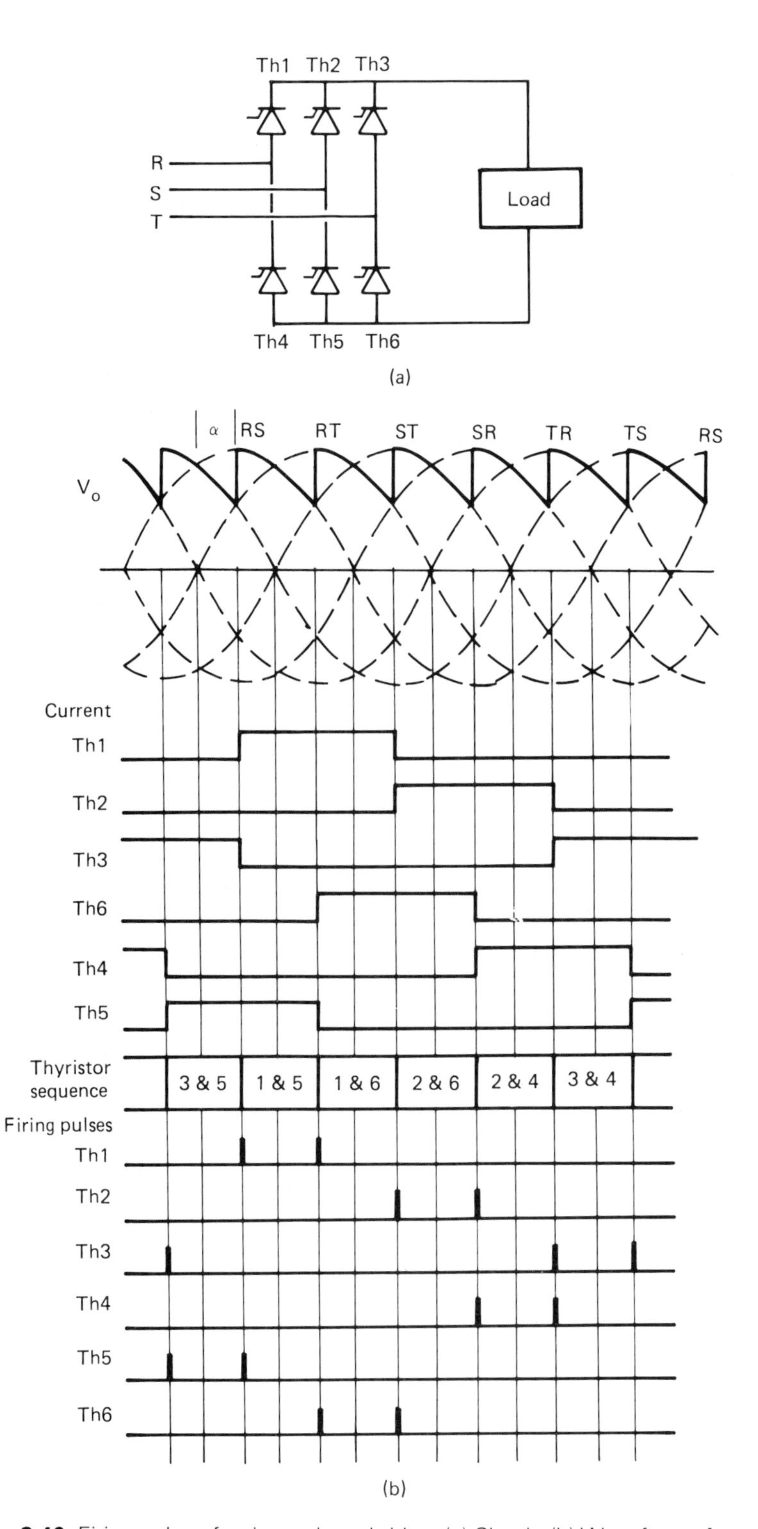

Fig.2.46 Firing pulses for three phase bridge. (a) Circuit. (b) Waveforms for $\alpha = 30°$.

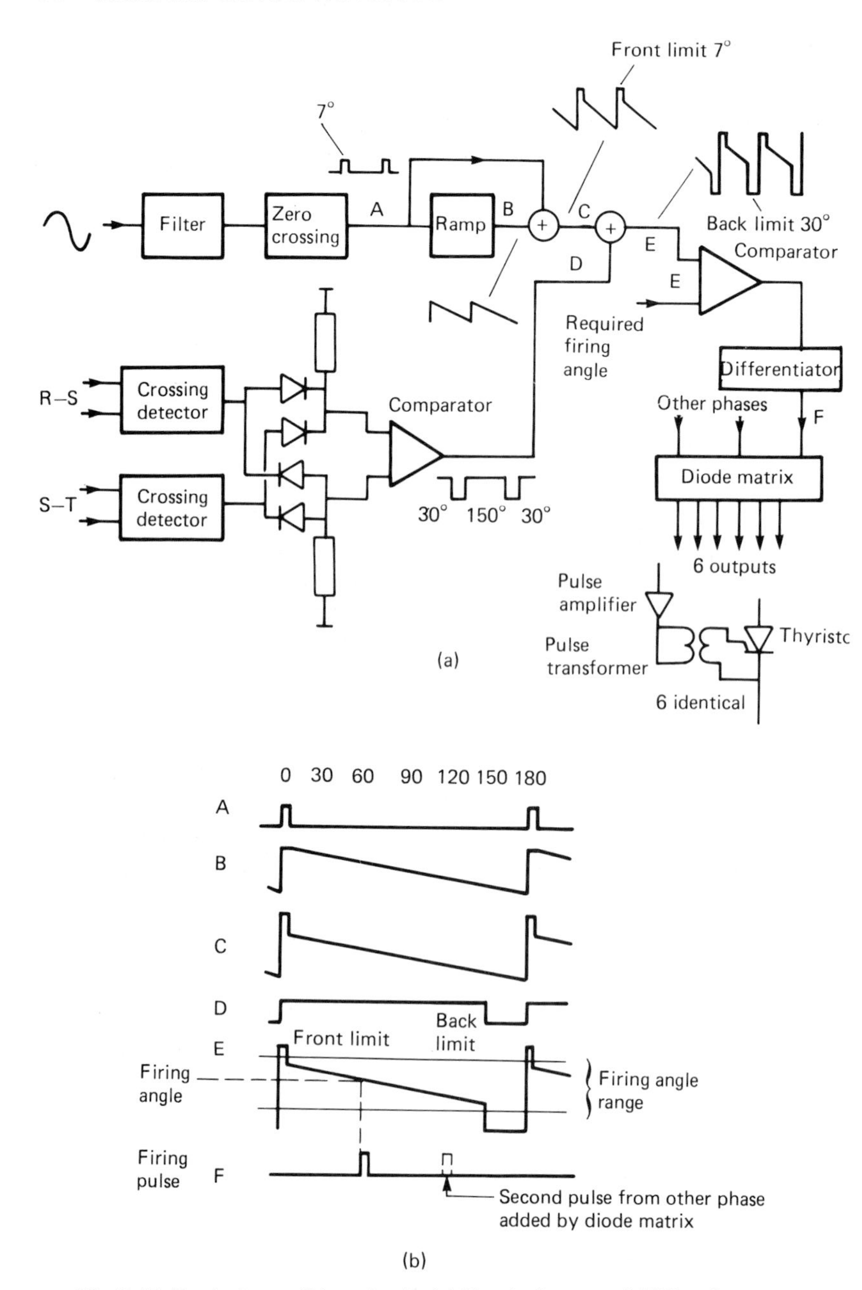

Fig.2.47 Typical gate firing circuit. (a) Circuit diagram. (b) Waveforms.

ramp of fig.2.43 with a cosine curve. This is simply obtained by phase shifting one supply phase with an RC network. Front and back stop pulses are added in a similar way to fig.2.47.

2.9.9. Ward Leonard control

The classical motor control scheme is the Ward Leonard system shown in fig.2.48, although this is only found nowadays on large motors with sudden load changes (such as mine winders or steel mill drives).

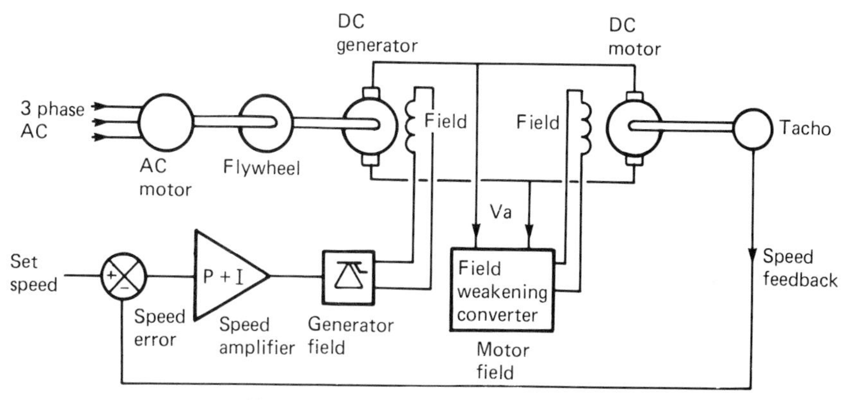

Fig.2.48 Ward Leonard control.

A DC generator is run at fixed speed by an AC motor, with a flywheel to absorb load changes. The output of the generator is attached directly to the motor armature, and controlled by the generator field. This is set by the low power signal from the speed error amplifier. The motor field is controlled by a spillover field weakening circuit similar to fig.2.44.

The only electronics in fig.2.48 are the low power field circuits, so the arrangement is very robust. The flywheel prevents impact loads being reflected into the supply. The disadvantages are the extra expense of the motor generator set, and a relatively slow response caused by the highly inductive field windings.

2.10. AC machines

2.10.1. Rotating magnetic fields

DC motors have excellent controllability, but are expensive and need regular inspection and maintenance; particularly the brushgear and commutator. In many applications, motors are required to run at a fixed speed (or have adjustments over a limited range), and accurate speed holding is not required. In these circumstances there is an overwhelming cost advantage in using an AC motor.

Almost all industrial AC machines are fed from three phase AC at around 415 V, and work by producing a rotating magnetic field. Consider first, for simplicity, the two phase circuit of fig.2.49. The two coils, denoted A and B, are fed with voltages displaced 90°, as shown in fig.2.49a. The resulting magnetic fields are shown in fig.2.49 b, c and d.

At time T1, there is only current in the A coil, producing a magnetic flux with direction parallel to the B coil. At time T2, 45° later, the currents in the two coils are equal, causing the flux direction to move as shown. Similarly, at time T3 the current in A

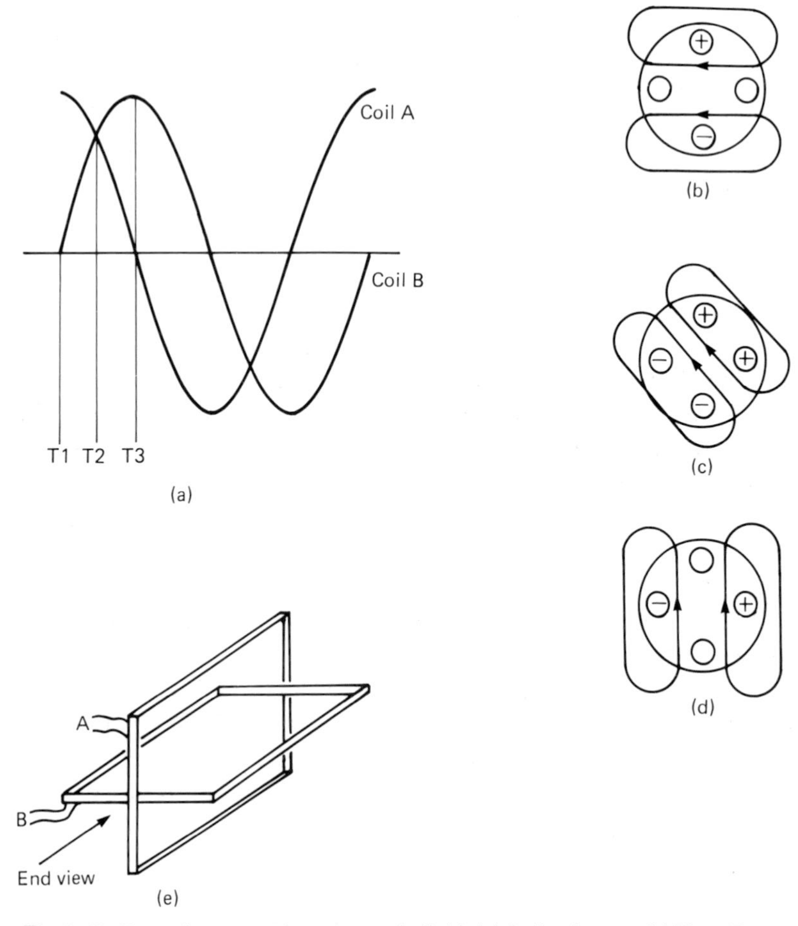

Fig.2.49 Two phase rotating magnetic field. (a) Coil voltages. (b) Time T1. (c) Time T2. (d) Time T3. (e) Coil arrangement.

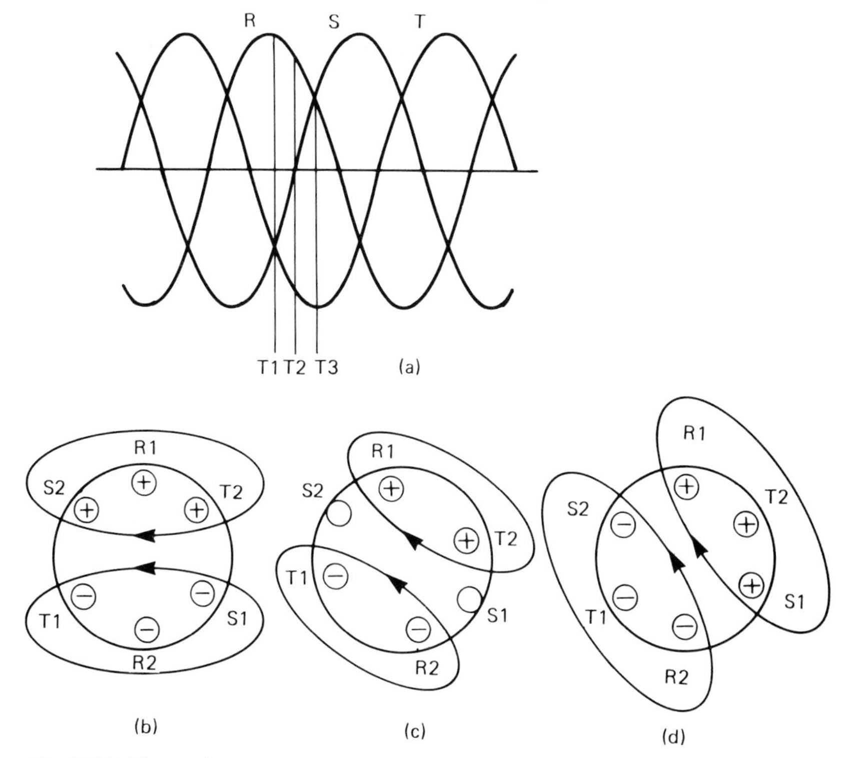

Fig.2.50 Three phase rotating field. (a) Coil voltages. (b) Time T1. (c) Time T2. (d) Time T3.

coil has fallen to zero and the flux direction rotates further. Examination of further points in time would show that the flux would rotate through 360° (one revolution) for one complete cycle of the AC supply.

Similar considerations apply to the three-phase circuit of fig.2.50. Here we have three coils spaced 120° apart fed with three-phase AC as in fig.2.50a. The arrangement again produces a rotating magnetic field, examples of which are shown at 30° intervals in fig.2.50b, c and d.

It can be seen that, for fig.2.50, the angle of rotation of the magnetic field is the same as the phase in fig.2.50a. At T3, for example, the three phases have advanced by 60°, and the field rotated by 60°. In one full mains cycle, the flux rotates through one revolution, i.e.:

$$n = f \tag{2.25}$$

where n is the rotational speed of the flux, and f the supply frequency.

The stator in fig.2.50 is wound for one pole pair. If p is the number of pole pairs, equation 2.25 can, by similar reasoning, be ammended to:

$$n = \frac{f}{p} \tag{2.26}$$

Values of n, in rpm, for 50/60 Hz supplies and different number of pole pairs are:

Pole pairs	50 Hz	60 Hz
1	3000	3600
2	1500	1800
3	1000	1200
4	750	900

It can be seen that the flux rotational speed decreases with the number of poles.

2.10.2. Synchronous machines

In fig.2.51 a permanently magnetised rotor is under the influence of a rotating magnetic field produced by a three phase stator, as

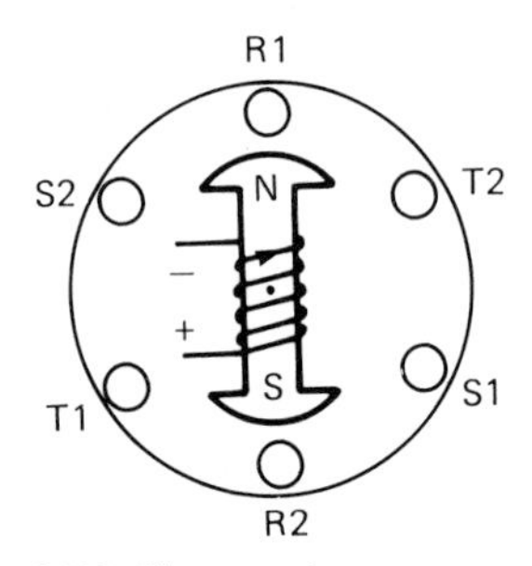

Fig.2.51 The synchronous motor.

described above. The rotor magnetisation can be achieved either by a permanent magnet, as shown, or from DC windings connected by slip rings.

The interaction between the stator and rotor fields produces a rotational torque which causes the rotor to rotate at the same speed as the field, but lagging by an angle which is a function of the torque required to rotate the rotor and load.

Figure 2.51 is called a synchronous motor because it rotates in exact synchronism with the stator field (and hence some multiple of the supply frequency according to the number of pole pairs). A four pole (two pole pairs) synchronous motor, for example, will run at 1500 rpm, regardless of load up to the point where the motor cannot supply the required torque, at which point it will stall.

Synchronous machines, however, are only widely used for clocks and timer applications where a permanently magnetised rotor is used. Large synchronous machines with DC rotors (upwards of 300 to 10,000 kW) are also used for some pumps and ID fans. The need for slip rings and heavy rotor windings, however, makes the induction motor, described below, more attractive for most industrial applications.

2.10.3. Induction motors

In fig.2.52, a stationary conductor is under the influence of a moving magnetic field. From Fleming's right-hand rule (generators), a current will be induced in the conductor. This current,

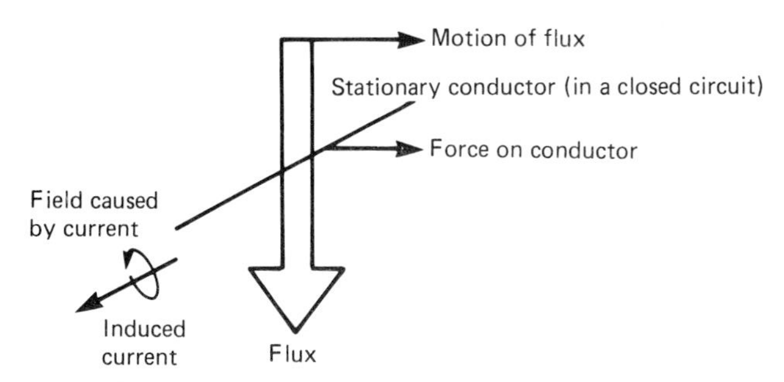

Fig.2.52 The principle of the induction motor.

however, produces a magnetic field which interacts with the external field to produce a force on the conductor (Fleming's left-hand rule, motors).

If the conductor is free, it will accelerate in the direction of the field movement. As its speed increases, however, its velocity relative to the field reduces and the induced current falls, reducing the accelerating force. The conductor will settle at a constant speed, less than the velocity of the field, where the force on the conductor balances the force required to propel the conductor and any external load.

In the simplest induction machine, called the squirrel cage motor, the rotor consists of copper, or aluminium, conductors shorted by end rings as in fig.2.53. A laminated iron rotor core increases the magnetic efficiency. The rotor is free to rotate inside a stator similar to fig.2.50 which produces a rotating magnetic field.

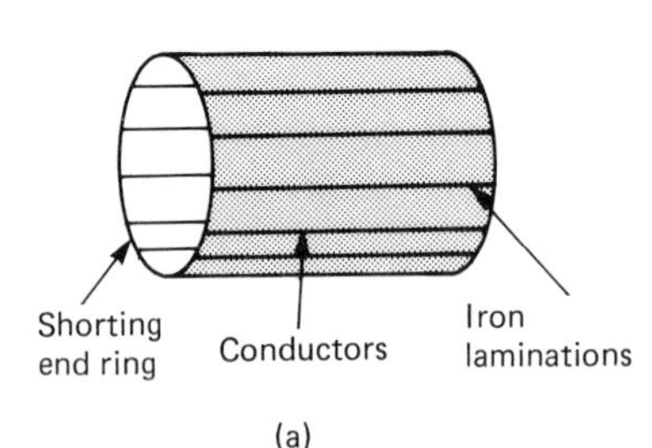

(a)

(b)

Fig.2.53 The squirrel cage induction motor. (a) Rotor construction. (b) Construction of a typical induction motor. (Photo courtesy of GEC small machines.)

Induced currents from the field cause the rotor to rotate as described above, with the rotor rotating at a lower than synchronous speed to produce torque. If the rotor speed is ωr and the supply frequency ω, the fractional slip is defined as:

$$s = \frac{\omega - \omega r}{\omega} \tag{2.27}$$

Slip increases with load torque, and will be typically 5% for a normal machine under design operating conditions.

At start-up, the rotor is stationary and the induced currents are large. As the motor speeds up, the induced currents (and frequency) fall. The induction motor has a large starting torque, and a correspondingly (sometimes embarrassingly) high starting current.

At a slip s, the induced rotor current has a frequency, sf, where f is the supply frequency. The rotor circuit impedance is therefore

$$Z = \sqrt{R^2 + (sX)^2} \tag{2.28}$$

where X is the inductive impedance at standstill (where the rotor current frequency and supply frequency match) and R the rotor resistance.

The rotor input power is ωT. Neglecting windage losses, this is split between useful output power, ωrT, and rotor iron losses, $3I^2R$ (for a three phase machine), i.e.:

$$\omega T = \omega rT + 3I^2R \tag{2.29}$$

$$T(\omega - \omega r) = 3I^2R \tag{2.30}$$

$$\frac{(\omega - \omega r)}{\omega} . \omega T = 3I^2R \tag{2.31}$$

$$\text{or } s\omega T = 3I^2R \tag{2.32}$$

i.e. slip × (input power) = rotor I^2R loss.

The electrical power dissipated in the rotor is $3I^2R$

$$= \frac{3s^2E^2}{R^2 + (sX)^2} \tag{2.33}$$

where E is the induced rotor emf at standstill. Substituting into equation 2.30 gives:

$$T(\omega - \omega r) = \frac{3\, s^2\, E^2\, R}{R^2 + (sX)^2} \tag{2.34}$$

$$\frac{T(\omega - \omega r)}{\omega} = \frac{3s^2E^2R}{\omega(R^2 + (sX)^2)} \tag{2.35}$$

$$sT = \frac{3E^2}{\omega} \cdot \frac{s^2R}{R^2 + (sX)^2} \tag{2.36}$$

$$\text{or } T = K\frac{sR}{R^2 + (sX)^2} \tag{2.37}$$

where K is a constant for the machine ($3E^2/\omega$).

For small values of slip, as is usually the case, the term $(sX)^2$ can be neglected, and:

$$T = \frac{Ks}{R} \tag{2.38}$$

i.e. for small slip, the torque is proportional to the slip.

The torque/slip relationship, equation 2.37, is shown for different relationships between R and X in fig.2.54a. Note that slip = 1 corresponds to a stationary rotor, and slip = 0 corresponds to synchronous speed. For practical machines, X>R, giving a speed/torque relationship similar to fig.2.54b. The

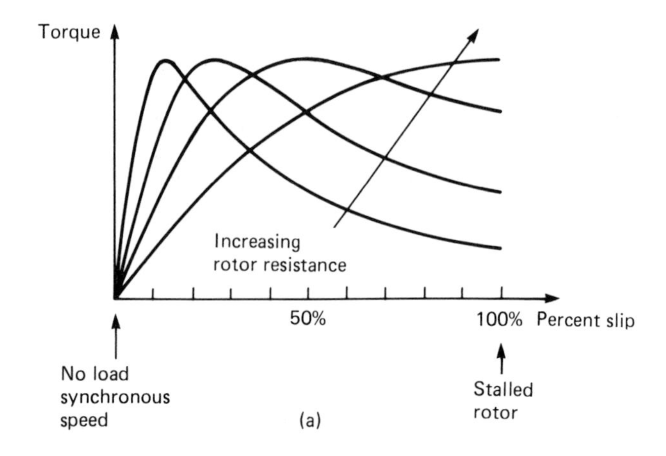

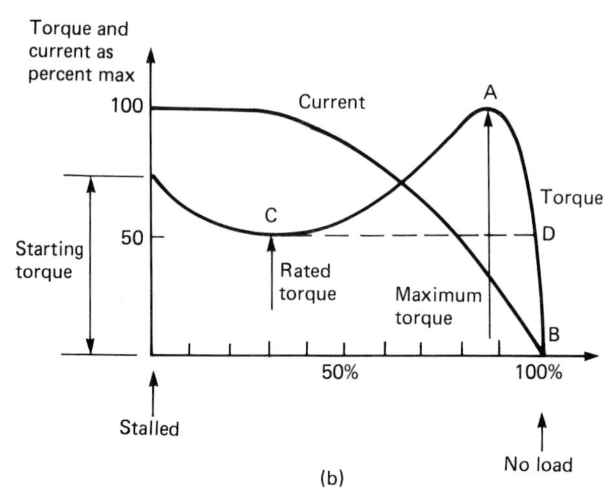

Fig.2.54 The performance characteristic of the induction machine. (a) Torque/slip relationship. (b) Torque-current/speed relationship.

maximum torque occurs for a slip of about 10%, and is roughly twice the minimum torque. The machine is normally operated in the region AB. The motor's nominal torque is the minimum torque, corresponding to point C. This also occurs at point D where the slip is about half that at point A. Note that the starting torque is roughly 150% of the minimum torque. Operation at speeds below point A is not recommended (other than for starting and stopping).

The induction motor has a very high initial current, typically two to three times the running current. Motors up about 5 kW can be started direct on line (DOL). Above this rating, inrush

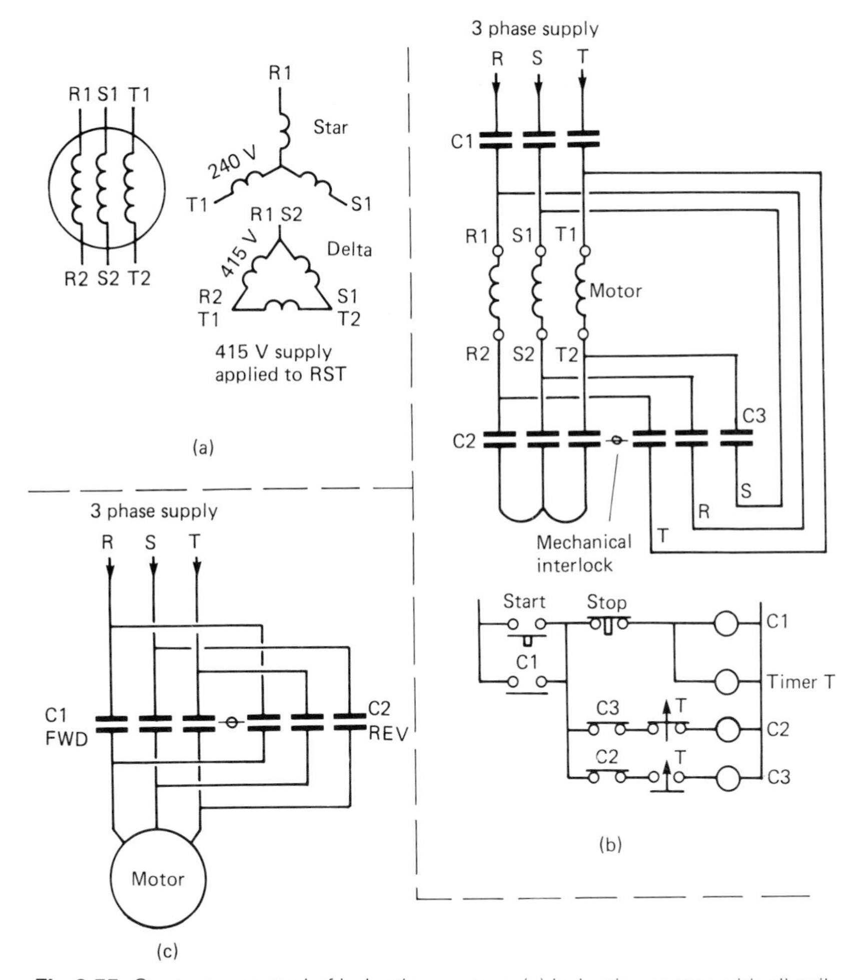

Fig.2.55 Contactor control of induction motors. (a) Induction motor with all coil ends brought out. (b) Star delta starter. (c) Reversing starter.

currents become embarrassingly large and star-delta starters are used. In fig.2.55, a motor is wound with all six stator coil ends brought out. These are switched initially in star (C1 and C2 energised) to start the motor at reduced coil voltage. When the motor speed has stabilised, C1 and C3 are energised to apply full coil volts, bring the motor to full operational speed and torque.

The direction of rotation of an induction machine is determined by the direction of rotation of the magnetic flux. This can be reversed by interchanging any two rotor connections. Figure 2.55c shows a typical contactor scheme where C1 and C2 interchange the S and T phases.

2.10.4. DC link inverters

The speed of rotation of an induction motor can be altered by:

(a) Altering the number of poles.
(b) Altering the supply frequency.
(c) Altering the rotor resistance (and hence the X/R ratio).

Pole switching is generally limited to two speed motors (e.g. 3000/1500 rpm) and the speed selection is coarse in the extreme. This section, and section 2.10.5, are concerned with method b.

The principle of the DC link inverter is shown in fig.2.56a. The incoming three phase supply is rectified by a controlled rectifier to give a DC voltage which can be adjusted by the control unit. The DC voltage is switched by SCRs or transistors to give a pseudo three phase output voltage at a frequency determined by the control unit. Reversing is obviously easily achieved electronically without the need for external contactors. A 'soft start' controlled acceleration can also be provided to reduce mechanical stresses.

Three switching patterns are commonly employed: the quasi square waves of fig.2.56b, c, and the pulse width modulation (PWM) waveform of fig.2.56d. In general, but by no means universally, quasi square waves operate best when the speed of a motor is to be increased above its nominal speed, and PWM operates best when the speed is to be decreased. Although PWM is more complex, recent advances in LSI integrated circuits and the ready availability of gate turn off (GTO) thyristors have led to a shift to PWM in modern drives.

The firing pattern for PWM inverters is obtained by comparing a reference sine wave from an oscillator with a triangular wave as shown in fig.2.56e.

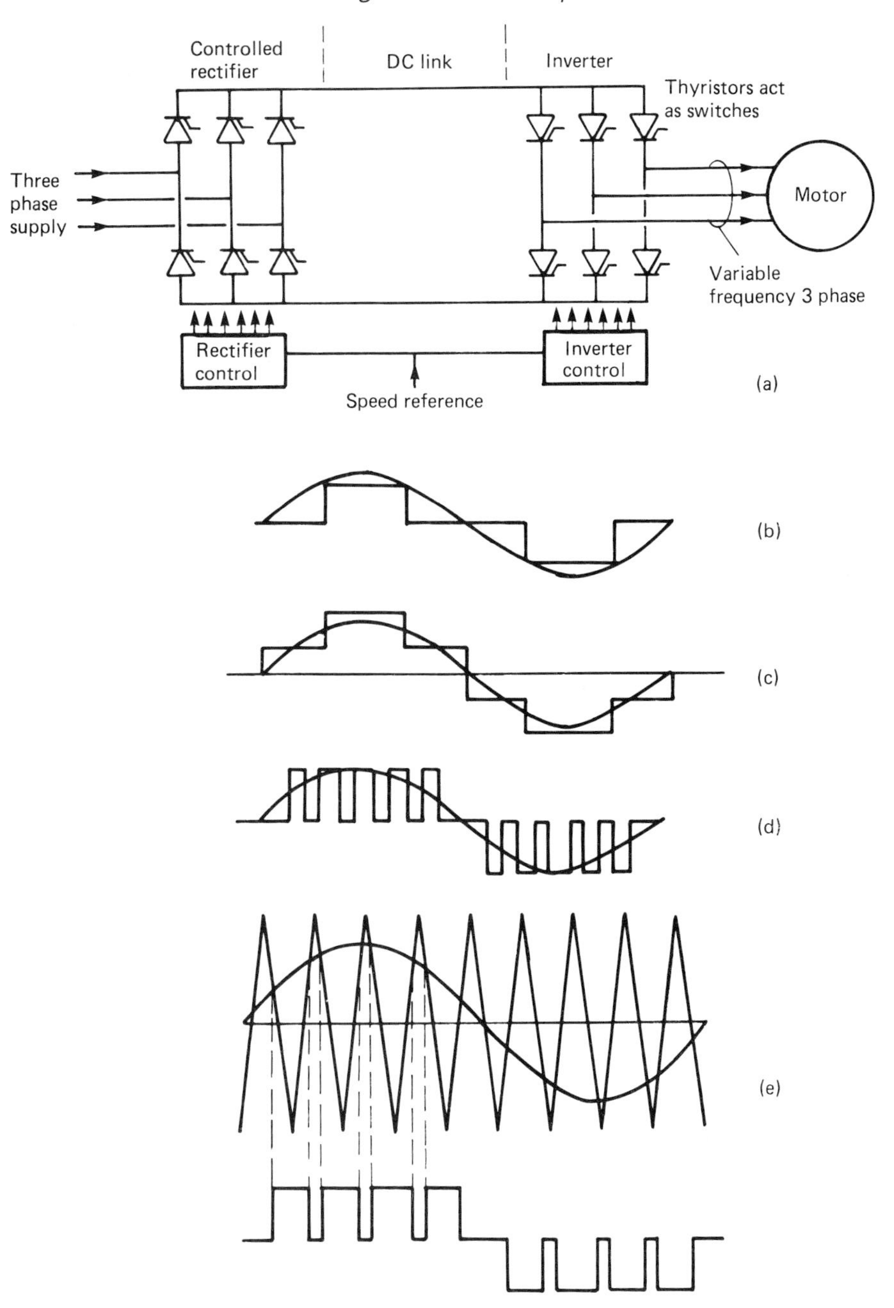

Fig.2.56 DC link inverters. (a) Circuit diagram. (b) Quasi square wave output wrt 0 V. (c) Quasi square wave output phase to phase. (d) Pulse width modulated (PWM) output (only needs uncontrolled rectifier). (e) Production of PWM signal by comparison of reference sine wave from oscillator with triangular wave. In practice, triangular wave has much higher frequency.

If the frequency alone is varied, the motor will overheat at low speeds. Consequently, the voltage must be reduced as shown in fig.2.57. Point B represents the nominal motor operating speed and voltage. In the region AB, the volts/frequency relationship is linear, and the motor operates at constant torque. Above B, the voltage cannot be increased, and the motor operates at constant power. The upper speed limit is set by centrifugal forces in the motor (typically 150% rated speed). At low speeds, cooling by fans driven direct off the motor shaft obviously cannot be used.

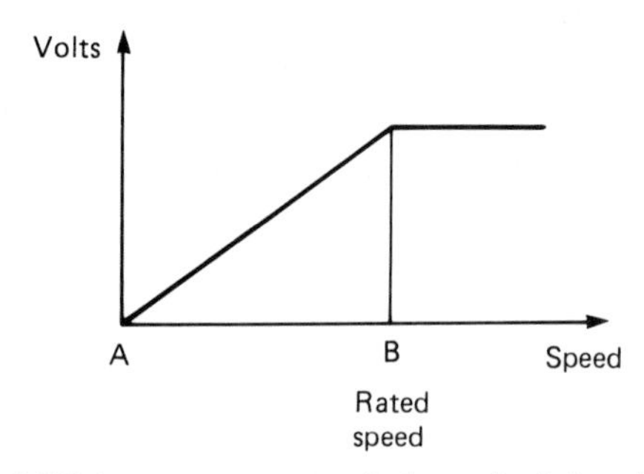

Fig.2.57 Inverter output volts/speed relationship.

The AC output voltage can be controlled by controlling the firing angle of the rectifier or (in PWM circuits) by varying the pulse width. The ability to operate with a fixed DC link voltage is one of the advantages of the PWM arrangement.

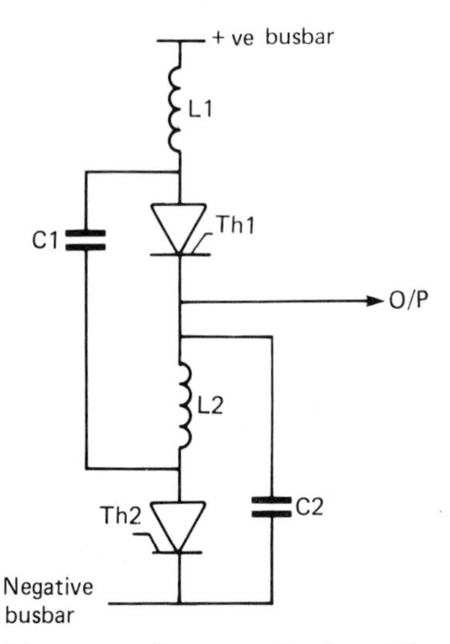

Fig.2.58 Typical switching stage. Commutation L and C reverse bias opposing thyristor when a thyristor is fired.

Figure 2.58 shows a typical thyristor switch stage. Once a thyristor is switched into conduction it will continue to conduct until its anode cathode voltage is reversed or the anode current falls to zero. Controlled rectifiers, described earlier, commutate naturally from the AC waveform. The inverter thyristors in fig.2.58 will not commutate naturally as they are switching DC. The components C1, L1 and C2, L2 force commutate the thyristors by reverse biasing a thyristor as its pair is fired. Alternatively GTO thyristors can be used, which have the advantage that commuting components are not required.

Transistors can also be used as switches. These do not need forced commutation, and as such are far simpler to use. At the time of writing, however, high current, high voltage transistors tend to be more expensive than thyristors of similar rating.

2.10.5. The cycloconverter

In fig.2.59a, a full APSH rectifier is connected to a load. The speed reference is derived from a variable low frequency sine wave oscillator, and the feedback is simply the DC output voltage

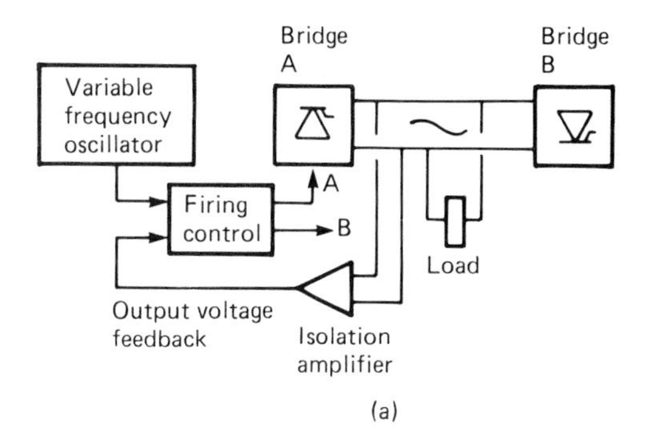

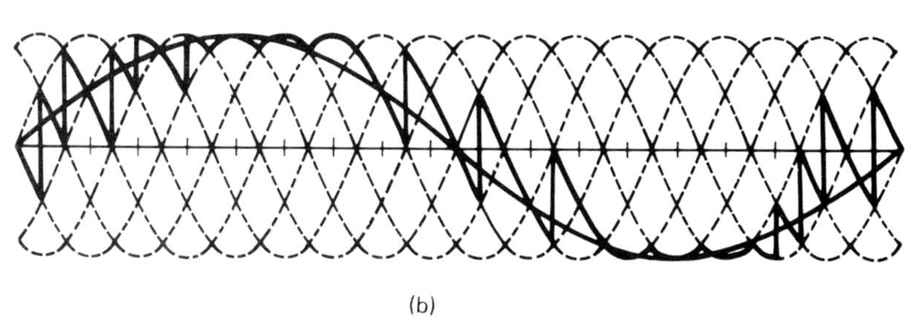

Fig.2.59 The cyclo converter. (a) Principle of operation. (b) Typical waveform.

(rather than from a tacho). The DC output will follow the input sine wave as shown in fig.2.59b. The circuit can give a variable frequency sine wave output up to about 60% of the supply frequency.

Figure 2.59 is called a cycloconverter, and as drawn provides a variable frequency single phase output. Three phase circuits are naturally more complex. Figure 2.60 shows a simple three phase arrangement.

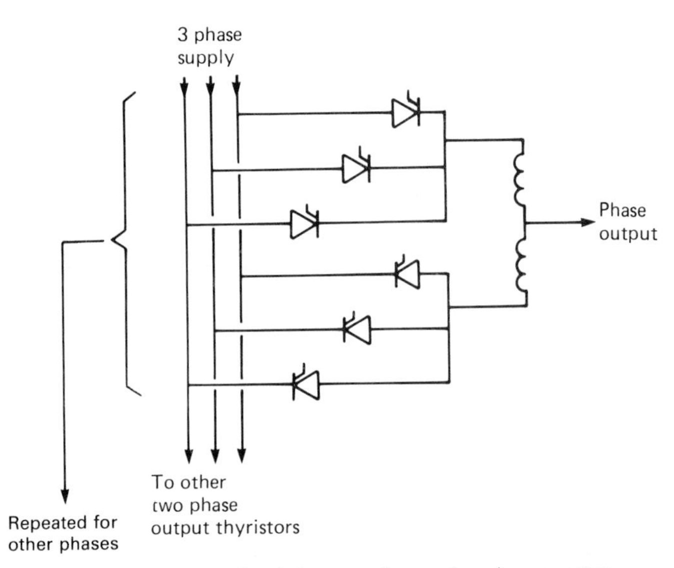

Fig.2.60 Thyristor bank for one phase of cycloconverter.

Early variable frequency drives tended to use the principle of the cycloconverter as it allowed manufacturers to economise by using assemblies from standard APSH drives. The advent of LSI devices, however, has made the construction of DC link inverters straightforward, and the large number of thyristors in a cycloconverter has reversed the cost advantage. The main advantage of the cycloconverter is its natural commutation and the consequential saving on forced commutation components.

2.10.6. Slip power recovery

If power is extracted from the rotor of an induction motor, the speed will reduce. Figure 2.61a shows a fairly crude way of achieving this. A wound rotor is used, whose connections are

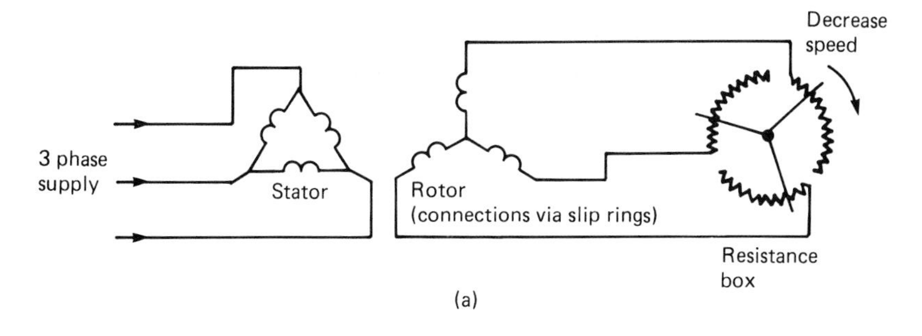

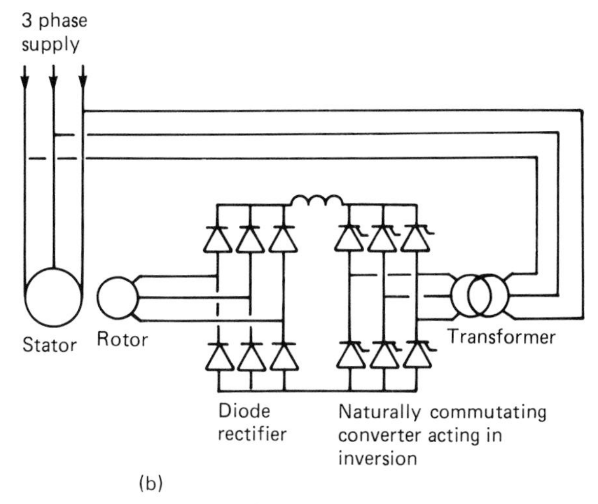

Fig.2.61 Slip power recovery speed control. (a) Speed control via resistance box. (b) Speed control via rectifier/converter.

brought out via slip rings. If these are shorted, the motor acts as a normal induction motor.

In fig.2.61a, the rotor connections are linked via three variable resistances. The higher the resistance, the higher the I^2R loss and the lower the speed. The speed will, however, be also load dependent, and as such is best suited for simple applications such as crane hoist control. The rotor resistors also dissipate a fair amount of heat.

Figure 2.61b is a more sophisticated version of fig.2.61a, which can be adapted to give closed loop control over a limited speed range. The output from the rotor (which is low frequency AC) is rectified and fed back to the supply via a naturally commutated controlled converter. This arrangement is called slip power recovery. It can, of course, only be used to reduce the speed in an induction motor.

2.10.7. Variable speed couplings

Altering the speed of a motor is not the only way to alter the speed of a load. In many applications, notably fans and pumps, it is often cheaper to run an AC motor at fixed speed and use a variable speed coupling.

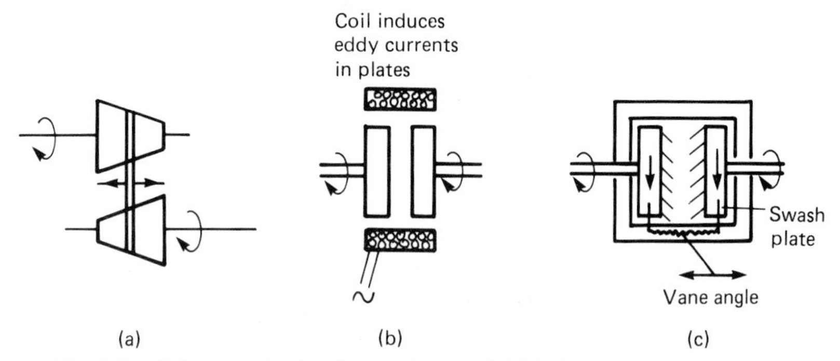

Fig.2.62 Other methods of speed control. (a) Infinitely variable coupling. (b) Eddy current coupling. (c) Hydraulic coupling.

Gearboxes are, of course, one solution, particularly with infinitely variable ratios as in fig.2.62a. Other solutions are the eddy current couplings shown in fig.2.62b where a low power field current controls the coupling between input and output and the hydraulic coupling shown in fig.2.62c where a vane angle controls the coupling.

2.11. The stepper motor

If an AC or DC motor is used for position control, some form of separate position transducer and control system must be used. The stepper motor combines the functions of both an output angular actuator and an angular position transducer. It is therefore possible to drive a stepper motor and know where the load is at all times without a separate position transducer. The torque available from stepping motors is, however, small, and they can only be used in applications such as small robots.

One form of stepper motor, called the variable reluctance motor, is shown in fig.2.63a. The rotor has a number of soft iron teeth which is unequal to the number of stator teeth. The stator has a series of coils which are driven as three (or more) separate

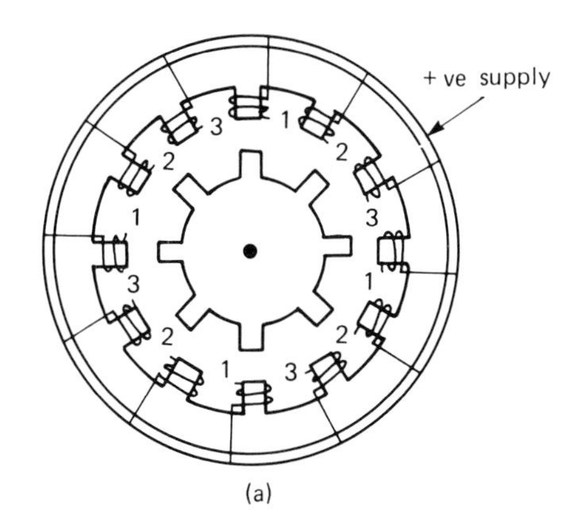

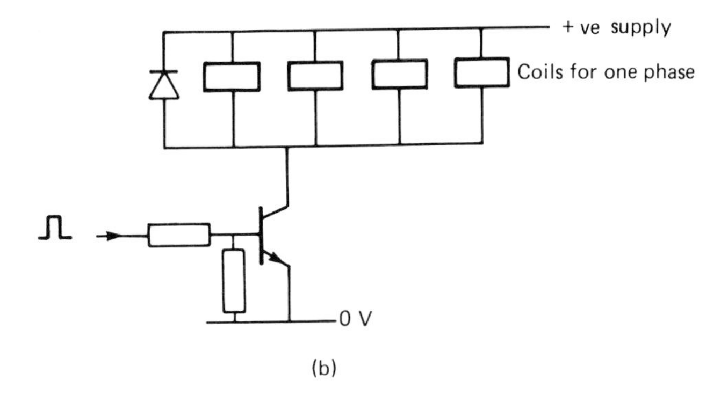

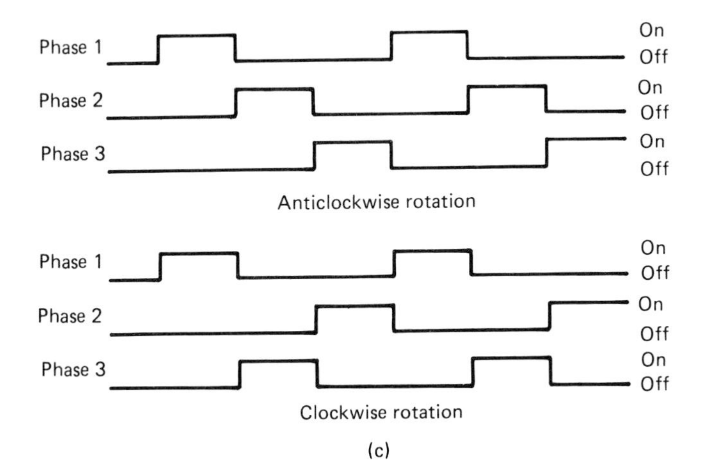

Fig.2.63 The stepper motor. (a) Variable reluctance stepper motor. (b) Coil driving. (c) Energisation sequence.

groups from the controlling logic. This is sometimes (incorrectly) called a three phase supply.

When a group is energised (phase 1 in the diagram), the rotor will align itself with the energised coils to give least magnetic reluctance. To move the rotor, the current group is de-energised, and the new group energised. The rotor rotates to a new position of minimum reluctance. If phase 2 were energised after phase 1, the rotor would rotate one step to the left, for example. Phase 3 after phase 1 would cause it to rotate one step to the right.

The number of steps per revolution is given by:

$$N = \frac{S.R}{S - R} \tag{2.39}$$

where S is the number of stator slots and R the number of rotor slots. Common step sizes are 7.5° and 1.8°.

The stepper motor of fig.2.63 needs to have one phase energised to hold the rotor in position. A variation, called a hybrid stepper motor, has a magnetically energised rotor with offset slots. This allows the rotor to hold its position without any power being applied, although the holding torque is again small.

Stepper motors are driven by quite simple circuits such as in fig.2.63b. Pulsing the outputs as in fig.2.63c will cause the motor to rotate a known angular distance.

2.12. Microelectronics and rotating machines

The microprocessor is appearing in an increasing number of drive control systems, along with a shift from analog to digital control systems. Figure 2.43 showed a typical APSH drive. Conventionally this would be implemented with an analog tacho, and a collection of DC amplifiers for the various operations.

In a digital drive, all the functions would be performed by a dedicated microcomputer, which would interface to the outside world via the usual analog and digital conversion circuits. Increasingly, analog tachos are being replaced by pulse tachos, giving speed holding characteristics of better than 0.01%. Speed can be determined by counting pulses for a fixed time, timing between pulses, or determining speed error directly from a phase locked loop (PLL) IC.

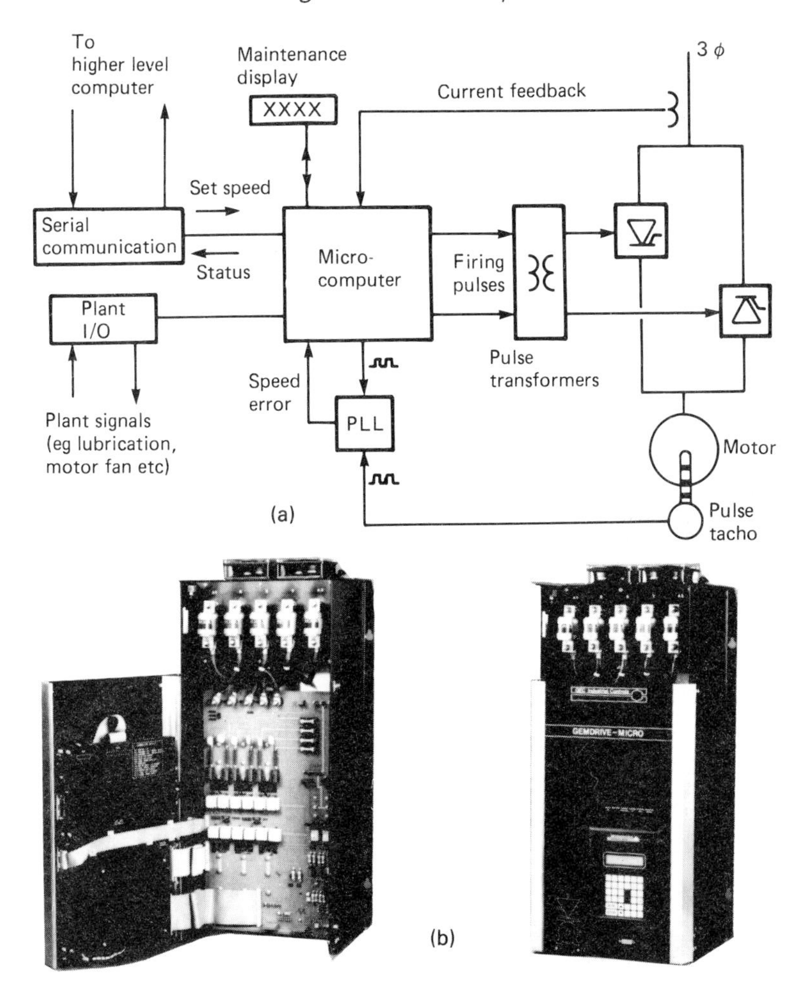

Fig.2.64 Microelectronics and thyristor drives. (a) Block diagram of digital drive. (b) Modern digital drive; the GEC Gemdrive Micro. Digital techniques are used for speed/current control, thyristor firing and interfacing to higher level computers/controllers. Drive set up and diagnostics are achieved via an operator's keypad. The interior view shows how little equipment is required for digital drives. (Photo courtesy of GEC Electrical Projects, Kidsgrove.)

Figure 2.64 shows a block diagram for a typical device. Note that this is part of a distributed computer system, receiving its speed reference and start commands via a serial link, and returning its status in a similar way. Figure 2.64b is the complete electronics for a digital drive. The only other equipment in the drive are the thyristor pulse transformers, power components (such as isolators) and the thyristor stacks themselves. All the functions of fig.2.43 are performed in software.

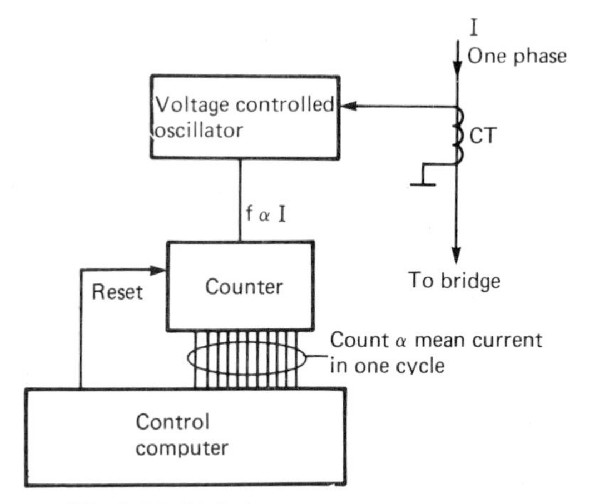

Fig.2.65 Digital current measurement.

Fig.2.66 Modern thyristor drives. Note how insulated encapsulated thyristors allow the heatsink to be used as an integral part of the chassis. (Photo courtesy of GEC Electrical Projects, Kidsgrove.)

Digital drives can also provide powerful diagnostic aids for maintenance and fault finding, such as a 'replay' of events leading to a drive trip, along with auto tuning of the speed and current control functions.

Current feedback in large drives is usually obtained from current transformers on the AC side of the thyristor stack. The signals from these must be rectified and smoothed to give a current signal that can be used by the current amplifier. This operation introduces a significant lag into the current loop. Figure 2.65 shows a digital current sensing method which introduces minimal lag, and can measure the current in one bridge leg. The outputs from the CTs are taken to voltage controlled oscillators. The oscillator outputs feed counters which are read and reset by the control microcomputer in synchronism with the supply frequency. The count is proportional to the mean current during the previous mains cycle.

Large scale integration (LSI) ICs are available for many drive functions. Increasingly, DC and VF drives are constructed from a small number of specialist ICs. This in turn leads to smaller, cheaper and more reliable drives, and the possibility of treating drives of up to about 200 amp rating as replaceable items. This tendency has been helped by the arrival of compact thyristor stacks with insulated mounting blocks, allowing the drive electronics to be neatly and compactly mounted directly on to the heatsink as in fig.2.66.

Chapter 3
Digital circuits

3.1. Introduction

3.1.1. Analog and digital systems

Signals in process control are conventionally transmitted as a pneumatic pressure or electrically as a voltage or current. A pneumatic signal of 3 to 15 psi, for example, could represent a liquid flow from 0 to 600 litres per minute, or an electric current of 4 to 20 mA could represent a temperature from −100°C to +400°C.

These signals are said to be continuously variable in that they can take any value between the two extreme limits. In the above temperature measurement, for example, a reading of 250°C would be represented by a current of 15.2 mA (given by 350 × 16/500 + 4). Similarly a flow of 540 litres per minute would be represented by a pressure of 13.8 psi in the flow measurement above.

In each of these examples an electrical or pneumatic signal is used as an analog of the process variable and the signal follows the process variable within the accuracy limits of the system. Such systems are called analog systems.

Digital systems are concerned with signals that can only take certain values. Most digital systems deal with electrical signals that can only have two values; 5 V or 0 V, for example. Many systems are inherently of this type: a light can be on or off, a valve open or shut, a motor running or stopped.

Figure 3.1. shows two possible approaches to liquid level measurement. In fig.3.1a a weighted float is mechanically linked to a potentiometer which gives an output voltage proportional to level. Within the resolution limits of the potentiometer, the

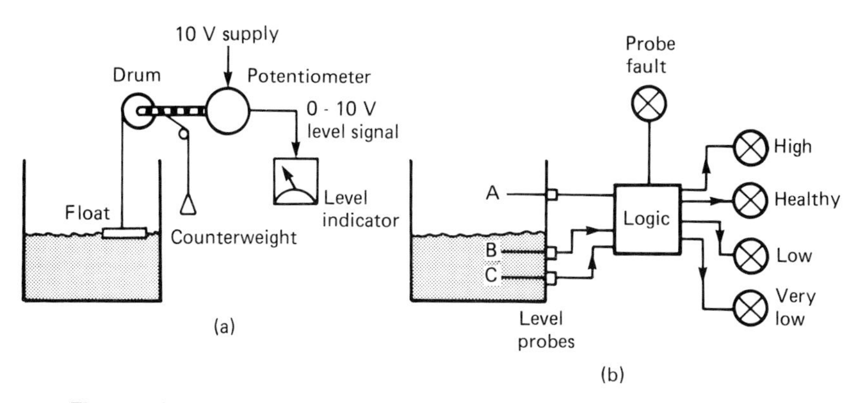

Fig.3.1 Comparison of analog and digital systems. (a) Analog system. (b) Digital system.

voltage can take any value between 0 and 10 V, so this is an example of an analog system.

Figure 3.1b uses three level switches A, B, C. These can be ON or OFF (ON being defined as the switch being submerged). There are four possible conditions:

State	A	B	C
Very low level	OFF	OFF	OFF
Low level	OFF	OFF	ON
Healthy	OFF	ON	ON
High level	ON	ON	ON

There are actually four other possible combinations of A, B, C; these all indicate a switch failure (e.g. A on, B off, C on). Figure 3.1b uses on/off signals and is an example of a digital system.

3.1.2. Types of digital circuit

Digital applications can, in general, be classified into three types. The simplest of these are called combinational logic (or static logic), and can be represented by fig.3.2a. Such systems have several digital inputs and one or more digital outputs. The output states are uniquely defined for every combination of input states, and the same input combination always gives the same output states. Figure 3.2a is a combinational logic system.

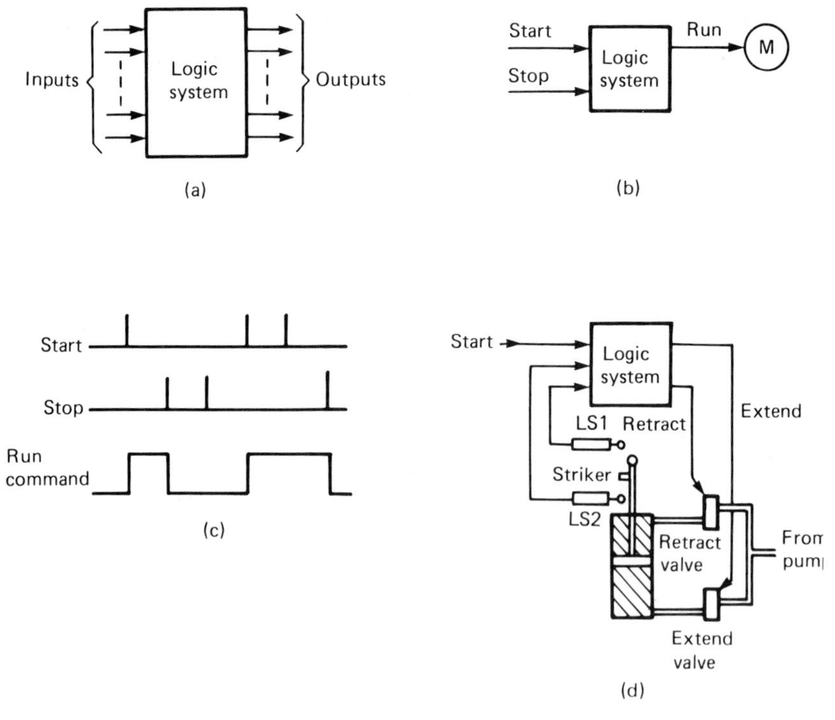

Fig.3.2 Types of digital system; combinational and sequencing logic. (a) Representation of combinational logic system: output states solely defined by input states. (b) A simple sequencing system, a motor starter. (c) Operation of motor starter. (d) Sequencing system.

A sequencing logic system is superficially similar to fig.3.2a, but the output states depend not only on the inputs but also on what the system was doing last (its previous state). Sequencing systems therefore have memory and storage elements. A very simple example is the motor starter of fig.3.2b. The start input causes the motor to start running and keep running even when the start signal is removed. The stop input stops the motor. Note that with neither signal present the motor could be running or stopped dependent upon which signal occurred last; the output state is not defined solely by the present input states.

Another sequencing example is shown in fig.3.2d. The digital circuit has three inputs – start and two limit switches – and two outputs – extend and retract. On a start signal the hydraulic ram extends until LSI is made, then retracts until LS2 is made, at which point the ram remains until another start signal is received. With no inputs present, the ram can be travelling out or in.

The final group of digital systems uses digital signals to represent, and manipulate, numbers. Such systems cover the range from simple counters and digital displays to complex arithmetic and computing circuits.

3.1.3. Logic gates

The simplest digital device is the electromagnetic relay, and it is useful to describe some of the fundamental ideas in terms of relay contacts. In fig.3.3a, the coil Z will energise when contact A AND contact B AND contact C are made. The series connection of contacts performs an AND function.

Similarly, in fig.3.3b the coil Z will energise when contact A OR contact B OR contact C are made. The parallel connection of contacts performs an OR function.

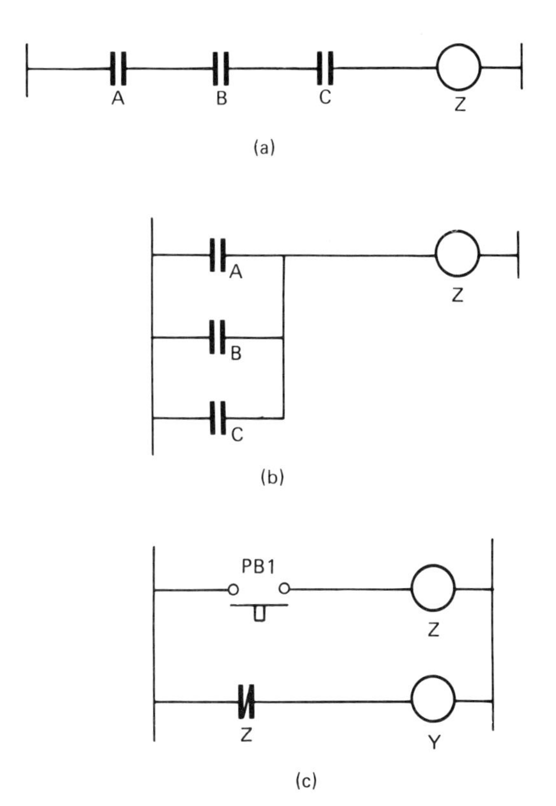

Fig.3.3 Simple relay logic. (a) AND combination. (b) OR combination. (c) Inversion.

In fig.3.3c, coil Z is energised when the push button is pressed. A normally closed contact of Z controls coil Y. When Z is energised Y is de-energised, and vice versa. The normally closed contact can be said to invert the state of its coil.

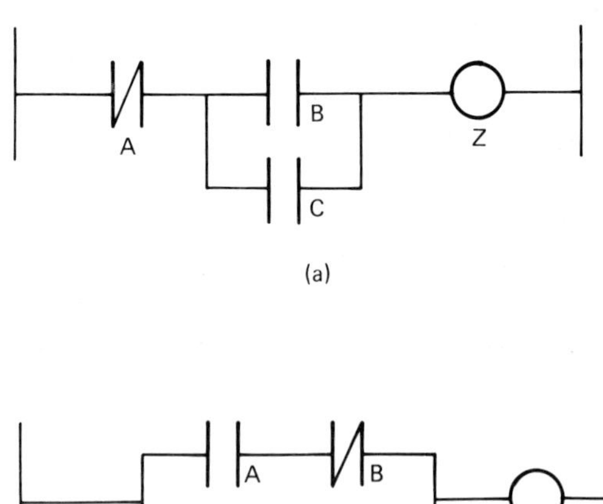

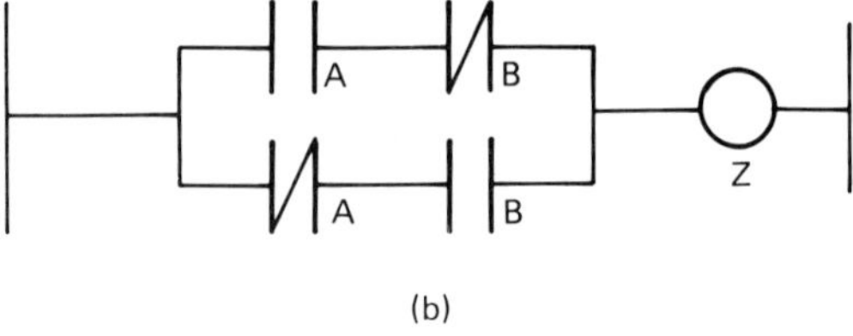

(b)

Fig.3.4 Further relay logic. (a) Z is energised when (A is not energised) AND (B energised or C energised). (b) Stairwell lighting circuit.

Combinational logic circuits are built round combinations of AND, OR and INVERT circuits. In fig.3.4a, for example, Z will be energised for:

(A not energised) AND (B energised OR C energised)

Such verbal descriptions are impossibly verbose for more complex combinations. Circuit operations are more conveniently expressed as an equation. Normally closed contacts are represented by a bar over the top of the contact name (e.g. $\bar{A}$, verbalised as A bar). The circuit of fig.3.4a can then be represented as:

$$Z = (B \text{ OR } C) \text{ AND } \bar{A}$$

Similarly the circuit of fig.3.4b (commonly used for stairwell lighting) can be represented by:

$$Z = (A \text{ AND } \bar{B}) \text{ OR } (\bar{A} \text{ AND } B)$$

These are known as Boolean equations, a topic discussed further in section 3.3.3.

Relays can perform all logic functions but are slow (typically 20 operations per second), bulky and power hungry. Electronic circuits performing similar functions are called logic gates. These work with signals that can only have two states. A signal in CMOS logic, for example, can be at 12 V or 0 V and could represent a limit switch made or open

The two logic states can be called high/low, on/off, true/false and so on. The usual convention, however, is to call the higher voltage 1 and the lower voltage 0. For a CMOS gate, therefore, 12 V is 1 and 0 V is 0.

Figure 3.5a shows the circuit of a simple AND gate. Neglecting diode drops, the output Z will be equal to the lower of the two input voltages. In other words, it will be a 1 if, and only if, both inputs are 1. This can be represented by fig.3.5b (which is called a truth table).

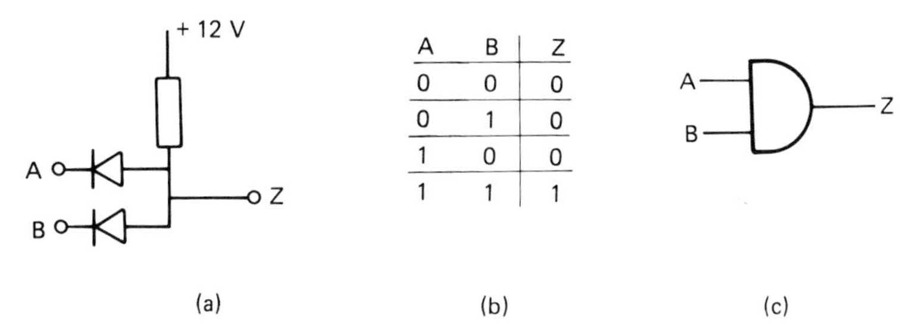

A	B	Z
0	0	0
0	1	0
1	0	0
1	1	1

Fig.3.5 The AND gate. (a) Circuit. (b) Truth table. (c) Logic symbol.

On circuit diagrams it is clearer to use logic symbols rather than the actual circuit diagram. The symbol for an AND gate is shown in fig.3.5c; the output Z is 1 when A AND B are both 1.

In fig.3.6a the output Z will be equal to the higher of the two inputs (again neglecting diode drops). Z will therefore be 1 if either input is 1, giving the truth table of fig.3.6b. The logic symbol for an OR gate is shown in fig.3.6c.

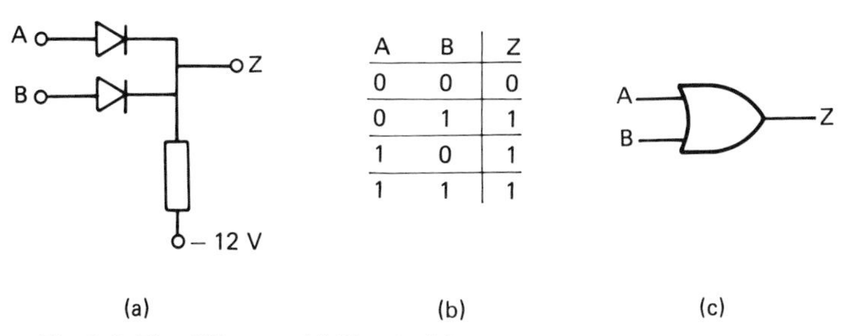

A	B	Z
0	0	0
0	1	1
1	0	1
1	1	1

Fig.3.6 The OR gate. (a) Circuit. (b) Truth table. (c) Logic symbol.

The invert function is given by the simple saturating transistor of fig.3.7a. When A is 0, the transistor is turned off and the output Z is pulled to a 1 state by the collector load resistor. When A is 1, the transistor is saturated on taking Z to 0 V; a 0. The circuit behaves as the truth table of fig.3.7b and has the logic symbol of fig.3.7c.

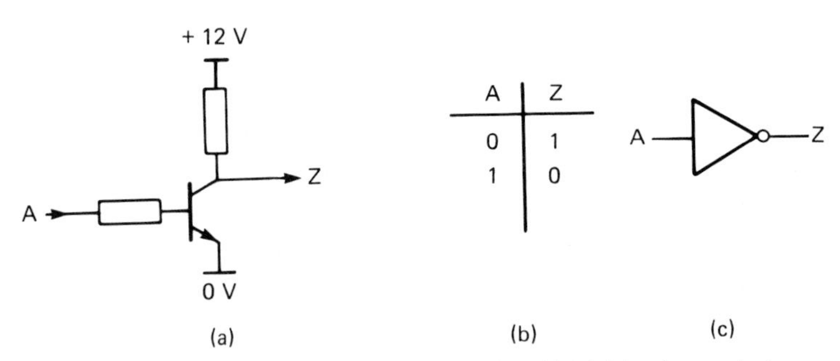

A	Z
0	1
1	0

Fig.3.7 The inverter. (a) Circuit. (b) Truth table. (c) Logic symbol.

Combinational logic circuits can be drawn purely in terms of AND gates, OR gates and inverters (although other gate types are more commonly used for reasons given later in section 3). The stairwell lighting circuit of fig.3.4b is drawn with logic symbols in fig.3.8a. This behaves as the truth table of fig.3.8b which shows that Z is 1 if only one input is 1. This circuit is known as an exclusive OR and is sufficiently common to merit its own logic symbol, shown in fig.3.8c.

If an inverter is used after an AND gate as in fig.3.9a, the truth table of fig.3.9a is produced. This arrangement is called a NAND

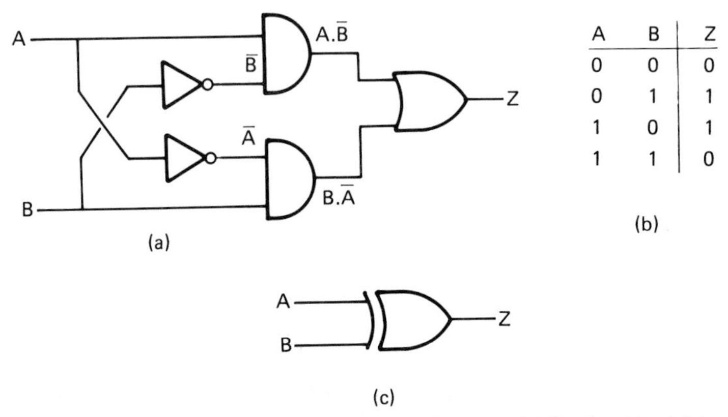

A	B	Z
0	0	0
0	1	1
1	0	1
1	1	0

Fig.3.8 The exclusive OR gate. (a) Logic diagram. (b) Truth table. (c) Logic symbol.

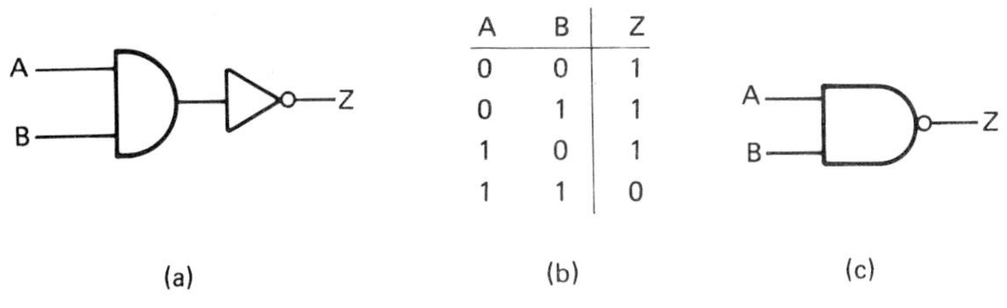

A	B	Z
0	0	1
0	1	1
1	0	1
1	1	0

Fig.3.9 The NAND gate. (a) Logic diagram. (b) Truth table. (c) Logic symbol.

gate (for NOT-AND) and has the logic symbol of fig.3.9c. The NAND gate is probably the commonest logic gate.

Adding an inverter to an OR gate as in fig.3.10a gives the truth table of fig.3.10b. This is known as a NOR gate (for NOT-OR) and is given the logic symbol of fig.3.10c.

Note that the logic symbols for NAND/NOR gates are similar to those for AND/OR gates with the addition of a small circle on the output. The circle denotes an inversion operation.

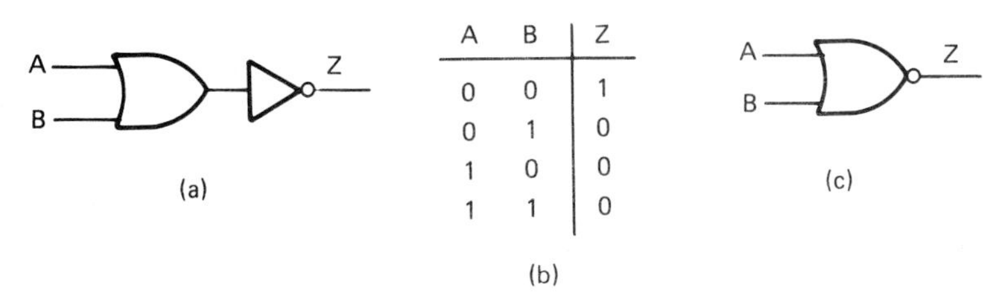

A	B	Z
0	0	1
0	1	0
1	0	0
1	1	0

Fig.3.10 The NOR gate. (a) Logic diagram. (b) Truth table. (c) Logic symbol.

The illustrations for AND/OR/NAND/NOR gates show two inputs. In reality these gates can have any required number of inputs (up to eight being readily available in commercial logic families). The exclusive OR gate inherently has only two inputs and the inverter, of course, has only one.

3.2. Logic families

3.2.1. Introduction

The circuits of figs.3.5 to 3.7, whilst illustrating the principles of logic gates, have many shortcomings. The voltage drops across the diodes would lead to severe degradation of the voltage levels after several gates, and the operating speed would be limited by the time constant formed by the load resistors and stray capacitance.

Most logic circuits are constructed from integrated circuits, and have high operating speed and well defined levels. Two logic families (TTL and CMOS) are widely used in industrial applications and a third family (ECL) may be encountered where very high speed is required. Before these are described, we must first examine how the various factors of a logic gate's performance are specified.

3.2.2. Speed

A logic gate such as the inverter of fig.3.11a does not respond instantly to a change at its input. For infinitely fast input signals the output will be delayed and the edges slowed, as shown in fig.3.11b.

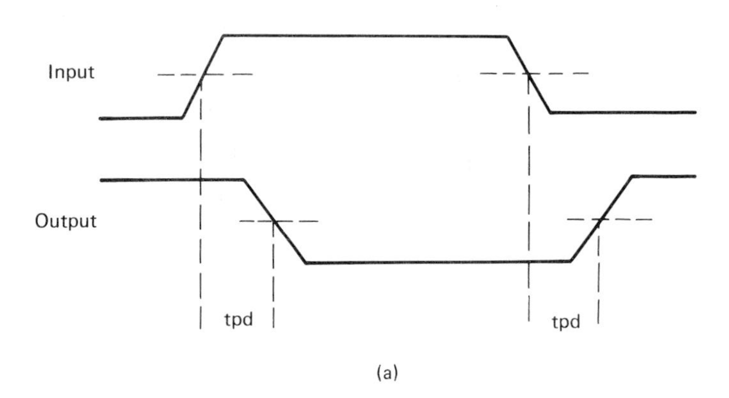

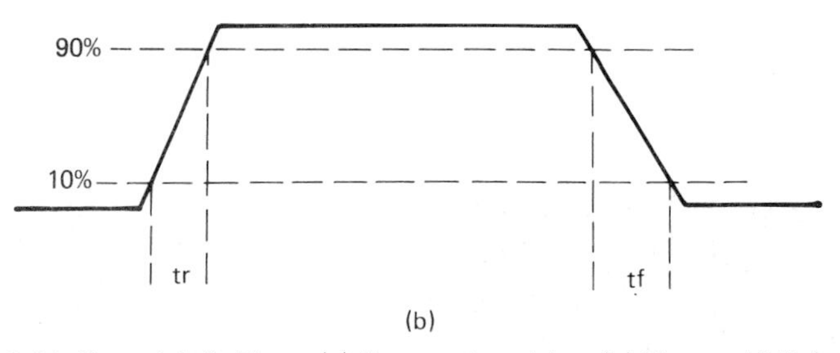

Fig.3.11 Speed definitions. (a) Propagation delay. (b) Rise and fall times.

The delay is called the propagation delay and is defined from the mid point of the input signal to the mid point of the output signal. Typical values are around 5 nS for TTL.

The edge speeds are defined by the rise time (for the 0 to 1 edge) and the fall time (for the 1 to 0 edge). These are measured between the 10% and 90% points of the output signal. Typical values are 2 nS for TTL.

Propagation delays and rise/fall times determine the maximum speed at which a logic family can operate. TTL can operate in excess of 10 MHz, CMOS around 5 MHz and ECL at over 500 MHz (although considerable care needs to be taken with board layout at speeds over 10 MHz).

Power consumption is related to speed, as increased speed is obtained by reducing RC time constants formed by stray capacitance, and by using non-saturating transistors. CMOS, for example, has a power consumption of about 0.01 mW per gate compared with ECL's figure of 60 mW/gate.

3.2.3. Fan in/fan out

The output of a logic gate can only drive a certain load and remain within specification for speed and voltage levels. There is therefore a maximum number of gate inputs a given gate output can drive.

A simple gate input is called a standard load, and is said to have a fan in of one. A gate output's drive capability is called its fan out, and is defined in unit loads. A TTL gate output, for example, can drive ten standard gate inputs and correspondingly has a fan out of ten.

Some inputs appear as a greater load than a standard gate. These are defined as a fan in of an equivalent number of standard gate inputs. An input with a fan in of three, for example, looks like three gate inputs.

Obviously the sum of all the fan in loads connected to a gate output must not exceed the gate's fan out.

3.2.4. Noise immunity

Electrical interference may cause 1 signals to appear as 0 signals, and vice versa. The ability of a gate to reject noise is called its noise immunity.

Defining noise immunity is more complex than it might at first appear, but the method usually adopted is that shown in fig.3.12a. The voltages given are those for a TTL gate which has a nominal 1 voltage of 3.5 V and a nominal 0 voltage of 0 V.

Next we define how for an output 1 can fall to (2.4 V) and a 0 rise to (0.4 V). These are respectively termed V_{OH} and V_{OL}. Finally we define how low a gate's input 1 can fall and an input 0 rise without allowing its output to go between V_{OH} and V_{OL}. These volts are called V_{IH} (2.0 V) and V_{IL} (0.8 V). The noise immunity is then the smaller of $V_{OH}-V_{IH}$ or $V_{IL}-V_{OL}$. For TTL the figure is 0.4 V. This is a worse case value, a more typical noise immunity being about 1.2 V.

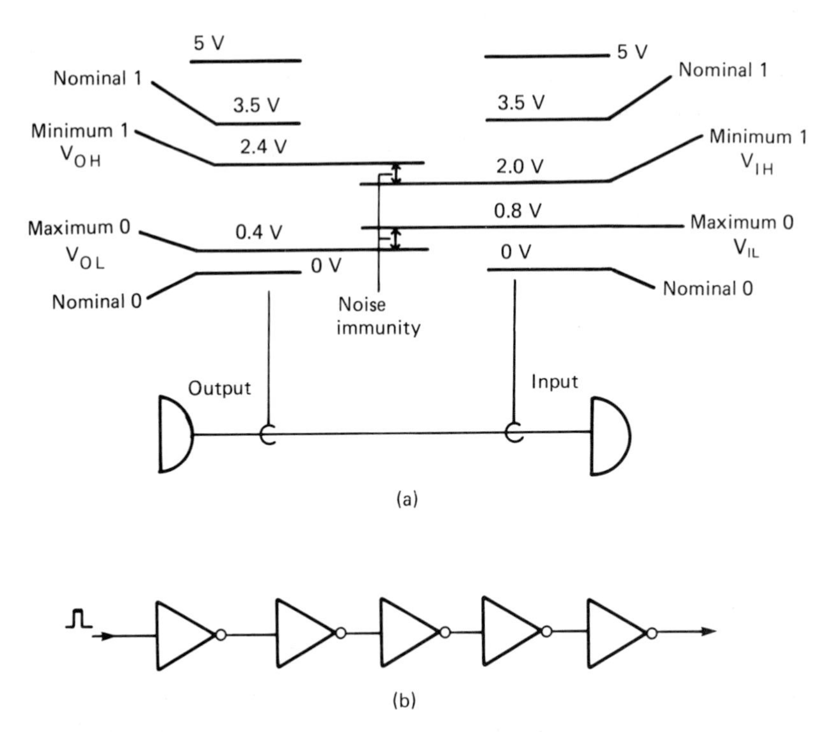

Fig.3.12 Definitions of noise immunity. (a) DC noise margin. (b) Testing for AC noise immunity.

A figure sometimes quoted is the AC noise margin. This is defined as the largest pulse that will not propagate down a chain of gates similar to fig.3.12b. This gives a more favourable figure than fig.3.12a, but is a more realistic test.

3.2.5. Transistor transistor logic (TTL)

TTL is probably the most successful logic family. TTL is NAND based logic, the circuit of a two input NAND gate being shown in fig.3.13. The rather odd-looking dual emitter transistor can be considered as two transistors in parallel or three diodes, as shown.

If both inputs are high, Q2 is turned on by current from R1 supplying base current to Q3. The output is therefore nominally 0 V. With either input low, Q1 is turned on, Q2 turned off and Q4 pulls the output high to a nominal 3.5 V.

The output transistors Q3, Q4 are called a totem pole output and play a significant part in increasing the operating speed. When the output is a 0, Q3 acts as a saturated transistor. When the output is a 1, Q4 acts as an emitter follower. Both states have low output impedances which reduce RC time constants with stray capacitance.

There are at least six versions of TTL with differing speeds and power consumption. Schottky versions use Schottky diodes within the gate to reduce hole storage delays. The six common types are:

Name	Suffix	Prop. delay (nS)	Power/ gate (mW)
Normal	None	10	10
Low power	L	33	1
High speed	H	6	22
Schottky	S	3	22
Low power Schottky	LS	9.5	2
Advanced low power Schottky	ALS	5	2

All TTL is part of the so-called 74 series (originally conceived by Texas Instruments) having the same pin arrangements on all the ICs. They can also be intermixed although care must be taken because of the different input loadings and output capabilities (an LS gate input, for example, looks like 0.5 of a normal gate input). All run on a 5 V supply and use logic levels of 3.5 V and 0 V. The suffix in the above table appears as part of the device identification; a 74LS06, for example, is a low power Schottky gate.

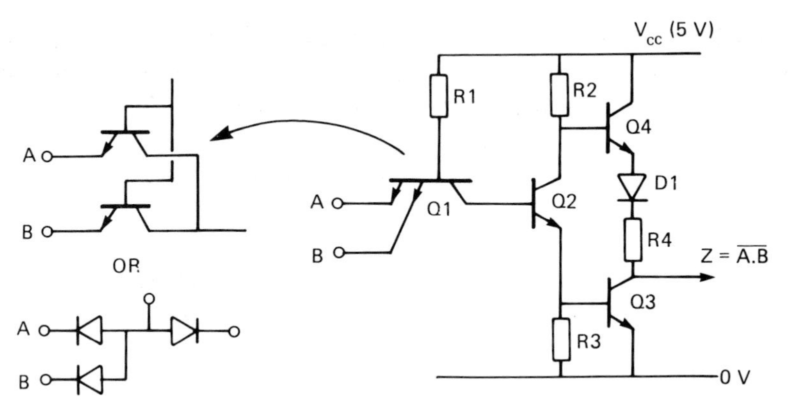

Fig.3.13 Transistor transistor logic (TTL).

3.2.6. Complementary metal oxide semiconductor (CMOS) logic

CMOS is virtually the ideal logic family. It can operate on a wide range of power supplies (from 3 to 15 V), uses little power (approximately 0.01 mW at low speeds), has high noise immunity (about 4 V on a 12 V supply) and very large fan out (typically in excess of 50). It is not as fast as TTL or ECL but its maximum operating speed of 5 MHz is adequate for most industrial purposes (too high a maximum speed can actually be a disadvantage as it makes a system more noise prone).

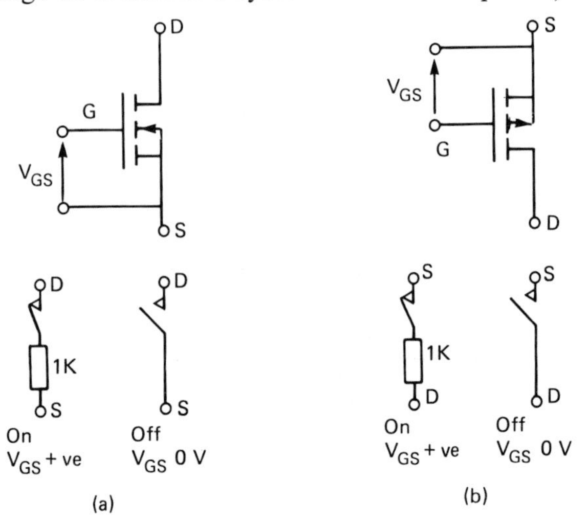

Fig.3.14 Metal oxide semiconductor (MOS) transistors. (a) n channel. (b) p channel.

CMOS is built around the two types of field effect transistors shown in fig.3.14. From a logic point of view these can be considered as a voltage operated switch. These switches can be used to manufacture logic gates.

Figure 3.15a shows how an inverter can be implemented. With A low, Q1 is turned on and Q2 off. With A high, Q2 is turned on and Z is low.

Similarly a NAND gate can be constructed as in fig.3.15b. If A or B is low, Z will be high because one of the parallel pair Q1, Q2 will be on, and one of the series pair Q3, Q4 will be off. The

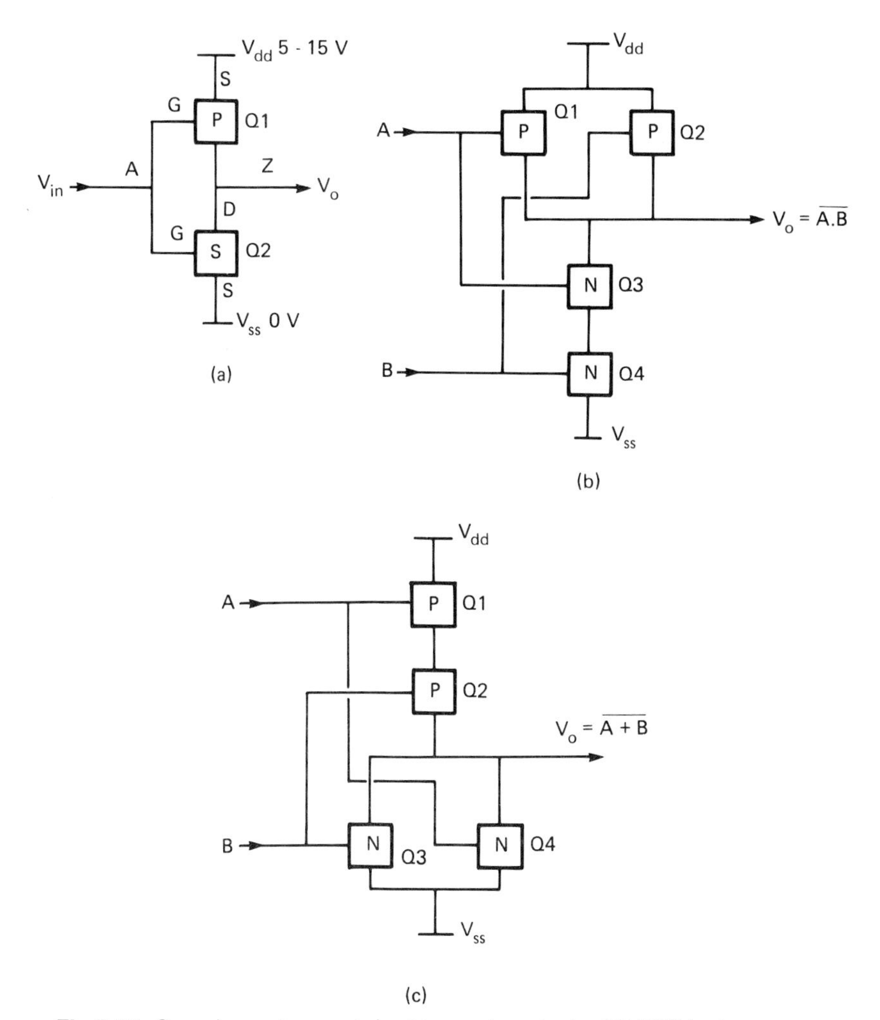

Fig.3.15 Complementary metal oxide semiconductor (CMOS) logic gates. (a) CMOS inverter. (b) CMOS NAND gate. (c) CMOS NOR gate.

output Z will be low only when both A and B are high when Q1, Q2 are both off and Q3, Q4 are both on.

Figure 3.15c shows a CMOS NOR gate. If A or B is high, one of Q3 or Q4 will be on, taking the output Z low (with one of Q1, Q2 off). When both A and B are low, Q1, Q2 will both be on and Q3, Q4 off, taking the output high.

The high input impedance of FETs can present handling problems, and early devices could be irreparably damaged by static electricity from, say, nylon clothing or leakage currents from unearthed soldering irons. Modern CMOS now includes protection diodes and can be treated like any other component.

Another effect of the high input impedance is the tendency for unused inputs to charge to an unpredictable voltage. All CMOS inputs must go somewhere; even unused inputs on unused gates on multigate packages must go to a supply rail (thereby forcing a 1 or 0 state).

CMOS is generally sold in the so-called 4000 series which is a rationalisation of the original RCA COSMOS and Motorola McMOS ranges. A B suffix denotes buffered signals and improved protection; needless to say, the B devices are better suited for industrial systems. An interesting, and useful, variation is the 74C series of CMOS, which is CMOS logic that is pin compatible (but not electrically compatible) with 74 series TTL.

3.2.7. Emitter coupled logic (ECL)

ECL is the fastest commercially available logic family, and with care it can operate at 500 MHz. At such speeds, however, extreme care needs to be taken with the circuit board layout to avoid crosstalk and power supply induced noise.

ECL acquires its speed from the use of non-saturating transistors and high power levels (around 60 mW per gate compared with the CMOS figure of 0.01 mW). Figure 3.15 shows an ECL NOR gate, which is based superficially on a long tailed pair.

The logic levels in ECL are −0.8 V and −1.6 V (giving a rather poor noise immunity of 0.25 V). Q1 sets a bias voltage of −1.2 V at the base of Q2.

If A or B is at −0.8 V, Q3 or Q4 will pass current (but not saturate) and the voltage at Q5 base will fall, taking the output

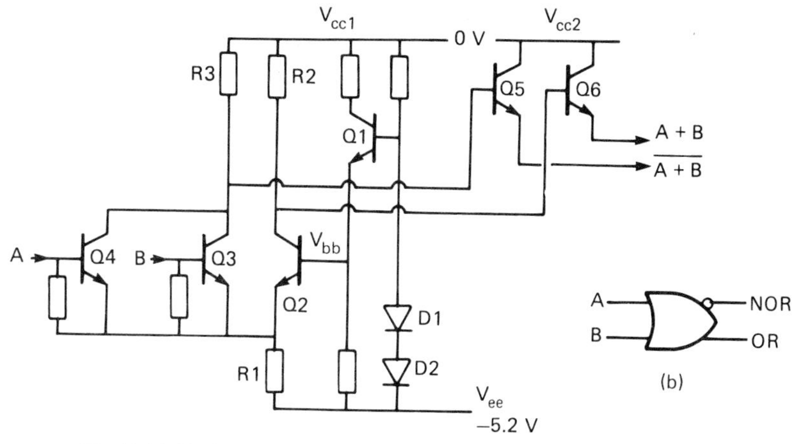

Fig.3.16 Emitter coupled logic. (a) Circuit diagram. (b) Symbol.

low to −1.6 V. Q5 is an emitter follower, giving a low impedance output. When both A and B are at −1.6 V, Q2 will pass current and Q3, Q4 will be off. R3 takes Q5 base high giving an output voltage of −0.8 V.

ECL is very fast, but its odd voltage levels, strict wiring and power supply requirements, and poor noise immunity preclude its use in industrial applications except where very high speed is needed.

3.2.8. Industrial logic families

Many sequencing applications only require operating speeds of at most 100 Hz. Industrial logic families, such as Mullard's NORBIT or the German Sigmatronic, are designed to replace relay panels. These devices are not based on integrated circuits, but are usually constructed from discrete components encapsulated in epoxy resin.

Although physically larger than IC logic such as TTL or CMOS, industrial logic families are virtually indestructible (usually able to withstand 240 V AC on inputs and power supplies), and the slow speed gives very high noise immunity. They are well suited to applications where the technical expertise of maintenance staff is low because devices can usually be changed solely with a screwdriver or by unplugging snap connectors.

3.3. Combinational logic

3.3.1. Introduction

Combinational logic is based around the block diagram of fig.3.17a. Such systems have several inputs and one or more outputs. The output states are uniquely defined for each and every combination of inputs and the 'block' does not contain any device such as storage, timers or counters. We therefore have n inputs I1 to In and Z outputs O1 to Oz. In systems with multiple outputs it is usually easier to consider each separately as in fig.3.17b, allowing us to consider the circuit as Z blocks, each different but represented by fig.3.17c.

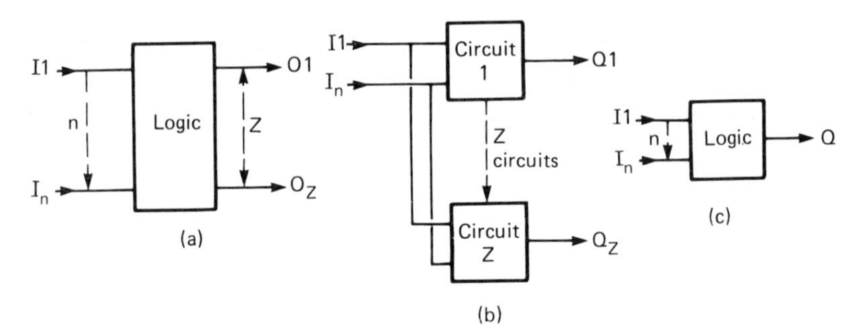

Fig.3.17 Combinational logic block diagrams. (a) Generalised problem. (b) Separate circuits. (c) One circuit.

The number of possible input states depends on the number of inputs:

for two inputs there are four input combinations
for three inputs there are eight input combinations
for four inputs there are sixteen input combinations

and so on. Not all of these may be needed. There are frequently only a certain number of input combinations that may occur because of physical restrictions elsewhere in the system.

The design of combinational logic systems first involves examining all the input states that can occur and defining the output states that must occur for each and every input state. A logic design to achieve this is then constructed from the gates described in section 3.1.3 (and given again in fig.3.18). In many systems the design can be done in an intuitive manner, but the rest of this section describes more formal design procedures.

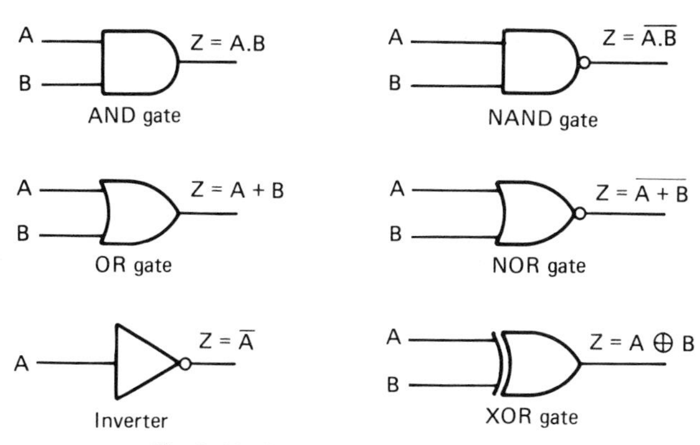

Fig.3.18 Summary of basic gates.

Few real-life systems need pure combinational logic; most need storage and similar dynamic functions. Such systems can be analysed and designed by considering them as smaller subsystems linked together. The design of dynamic systems is discussed in section 3.8.

3.3.2. Truth tables

A truth table is a useful way of representing a combinational logic circuit, and can be used to design the circuit needed to achieve a desired function.

Suppose we have three contacts monitoring some event (overpressure in a chemical reactor, for example) and we wish to construct a majority vote circuit. If the three switches are called A,B,C and the majority vote Z, this would have the truth table:

A	B	C	Z	
0	0	0	0	
0	0	1	0	
0	1	0	0	
0	1	1	1	←
1	0	0	0	
1	0	1	1	←
1	1	0	1	←
1	1	1	1	←

It can be seen that Z is 1 for:

$\bar{A}$ and B and C
or A and $\bar{B}$ and C
or A and B and $\bar{C}$
or A and B and C

The desired logic function can then be constructed directly from the truth table as in fig.3.19. In general, the circuit derived from a truth table will consist of a set of AND gates whose outputs are OR'd together.

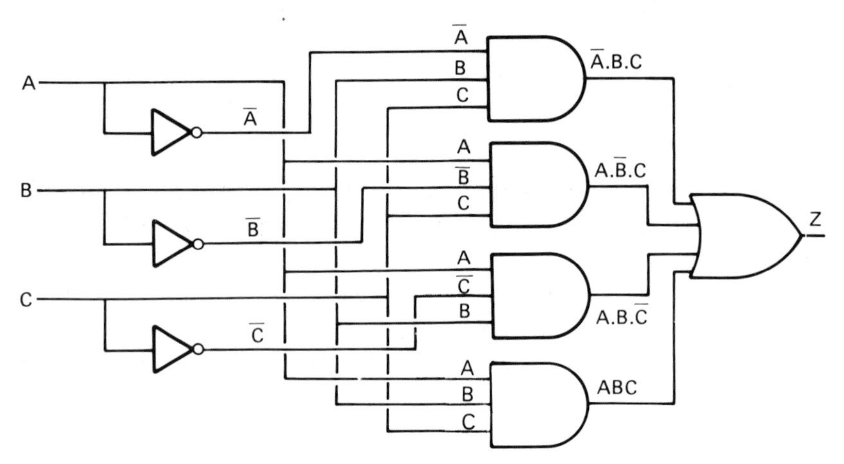

Fig.3.19 Implementation of majority vote circuit direct from truth table.

A truth table always gives a design which works and is logically correct, but does not always give a circuit which uses the minimum combination of gates. To do this we need one of the other techniques described below.

The form of a logic design derived from a truth table is always a series of AND gates whose outputs are OR'd together (fig.3.19 is a typical example). This form of circuit is known as a sum of products (see section 3.3.3 below), and one of the reasons for the popularity of NAND gates is that a sum of products expression can be formed purely with NAND gates.

Consider the expression:

Z = (A and B) OR (C and D)

This has the simple circuit of fig.3.20a, which obviously fulfils the logic function. Consider, however, the totally NAND based circuit of fig.3.20b. Straightforward, if laborious, testing of all

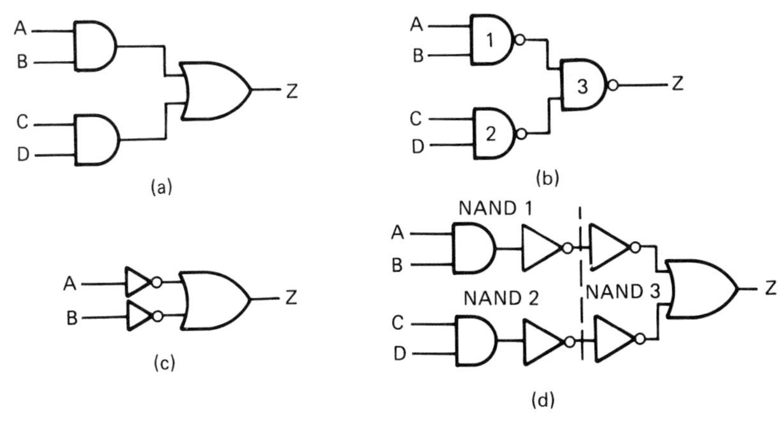

Fig.3.20 Logic circuits based solely on NAND gates. (a) Required logic. (b) NAND gate equivalent. (c) Representation of a NAND gate. (d) Operation of a circuit (b).

sixteen possible input states will show that it behaves identically to fig.3.20b. In some mysterious way, the right-hand NAND gate is behaving as an OR gate.

This rather surprising fact is a result of De Morgan's theorem, described in the next section. Intuitively, however, we can see the reason by drawing up the truth table for the OR gate preceded by inverters as in fig.3.20c:

A	B	Z
0	0	1
0	1	1
1	0	1
1	1	0

This is the same as a NAND gate, so a NAND gate can, with legitimacy, be drawn as fig.3.20c.

The circuit of fig.3.20b could now be drawn as fig.3.20d with the ingoing NANDs drawn as ANDs followed by an inverter, and the outgoing NAND by the arrangement of fig.3.20c. Obviously the intermediate inverters cancel, leaving the equivalent circuit of fig.3.20a.

3.3.3. Boolean algebra

In the nineteenth century a Cambridge mathematician and clergyman, George Boole, devised an algebra to express and manipulate logical expressions. His algebra can be used to represent, design and minimise combinational logic circuits.

The AND function is represented by a dot (.), so:

$$Z = A.B$$

means Z is 1 when A is 1 AND B is 1. Often the dot is omitted, e.g. Z = AB.

The OR function is represented by an addition sign (+), so:

$$Z = A+B$$

means Z is 1 when A is 1 OR B is 1.

The invert function is represented by a bar $\overline{\ }$, so

$$Z = \overline{A}$$

means Z takes the opposite state to A.

Boolean algebra allows complex expressions to be written in a concise manner. Figure 3.4b, for example, is:

$$Z = (A.\overline{B}) + (\overline{A}.B)$$

and fig.3.19 is:

$$Z = (\overline{A}.B.C) + (A.\overline{B}.C) + (A.B.\overline{C}) + (A.B.C)$$

Boolean algebra can also be used to simplify expressions. To achieve this, a series of rules are used. The first eleven of these are self-obvious (or can be visualised by considering the equivalent relay circuits):

(a) $A.A=A$
(b) $A+A=A$
(c) $A.1=A$
(d) $A.0=0$
(e) $A+1=1$
(f) $A+0=A$
(g) $\overline{\overline{A}}=A$
(h) $A.\overline{A}=0$
(i) $A+\overline{A}=1$
(j) $A+B=B+A$
(k) $A.B=B.A$

The next two laws allow us to group brackets around variables with the same operator:

(l) $(A+B)+C=A+(B+C)=A+B+C$
(m) $(A.B).C=A.(B.C)=A.B.C$

The next two laws are called the absorptive laws, and tell us what happens if the same variable appears with AND and OR operators:

(n) $A+A.B=A$
(o) $A.(A+B)=A$

The above two laws are not immediately obvious, and are shown in relay form in fig.3.21.

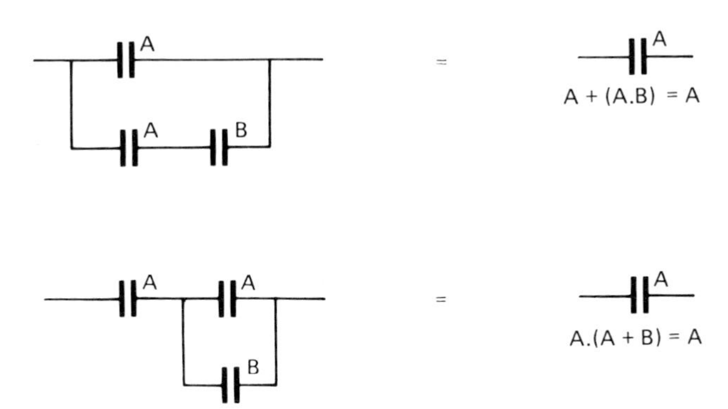

Fig.3.21 Relay demonstration of absorptive laws.

The next laws (called the distributive laws) tell us how to factorise Boolean equations:

(p) $A+B.C=(A+B).(A+C)$
(q) $A.(B+C)=A.B+A.C$

In general, Boolean expressions can be expressed in two forms. The first form, called product of sums (or P of S), brackets OR terms and ANDs the results, e.g.:

$$Z = (A+\overline{B}).(B+C+D).(\overline{A}+\overline{D})$$

The second form, called sum of products (S of P), groups AND terms and ORs the results, e.g.:

$$Z = AB\overline{D} + \overline{B}C + A\overline{D}$$

Truth tables, described above in section 3.3.2, inherently give an S of P result.

The complementary function of a Boolean expression yields the inverse of the expression (i.e. where the expression yields 1, the complement yields 0). The expressions $A+B$ and $\overline{A}.\overline{B}$, for example, can be shown to be complementary by simply constructing their truth tables.

The last two laws, known as De Morgan's theorem, show how to form the complement of a given expression (and give one way to interchange S of P and P of S forms):

(r) $\overline{A.B.C.....N} = \overline{A} + \overline{B} + \overline{C} + \ldots \overline{N}$
(s) $\overline{A+B+C+....+N} = \overline{A}.\overline{B}.\overline{C}....\overline{N}$

In its formal representation, De Morgan's theorem appears rather daunting. It can be more easily expressed:

'To form the complement of an expression:

(1) Replace each '+' in the original expression with '.' and vice versa.
(2) Complement each term in the original expression.'

For example, the expression $\overline{A}+B.C$ is complemented as below:

Step 1: replace '+' by '.' and '.' by '+' giving
$\overline{A}.(B+C)$
Step 2: complement each term
$A.(\overline{B}+\overline{C})$

which is the complement of $\overline{A}+B.C$ (as can be verified by trying all eight possible input states).

De Morgan's theorem explains the behaviour of fig.3.20b. The output Z is given by:

$$Z = \overline{(\overline{A.B}).(\overline{C.D})}$$

i.e. the complement of $(\overline{A.B}).(\overline{C.D})$.

Applying De Morgan's theorem to $(\overline{A.B}).(\overline{C.D})$ gives the complement form to be (A.B)+(C.D), hence:

$$Z = (A.B) + (C.D)$$

which is the required expression.

Boolean algebra can be used to minimise logical expressions, but the method is rarely obvious, and it is easy to make errors with double bars and swopping of '.' and '+'. Minimisation by Boolean algebra makes good examination questions, but is rarely used in practice. One example will suffice. Consider the expression:

$$Z = ABC + A\overline{B} + (\overline{\overline{A}\overline{C}})$$

Applying De Morgan's theorem to the right-hand term gives:

$$\begin{aligned} Z &= ABC + A\bar{B}(\bar{\bar{A}}+\bar{\bar{C}}) \\ \text{but } \bar{\bar{A}} &= A \text{ and } \bar{\bar{C}} = C \text{ giving} \\ Z &= ABC+A\bar{B}(A+C) \\ Z &= ABC+AA\bar{B}+A\bar{B}C \\ Z &= ABC+A\bar{B}+A\bar{B}C \\ \text{we observe } ABC+A\bar{B}C &= AC(B+\bar{B}) = AC \\ \text{hence } Z &= AC+A\bar{B} \end{aligned}$$

which is the minimal form

An easier way to achieve the same minimisation result is to use the graphical Karnaugh map, described below.

3.3.4. Karnaugh maps

A Karnaugh map is an alternative way of presenting a truth table. The map is drawn in two dimensions; two, three and four variable maps are shown in fig.3.22.

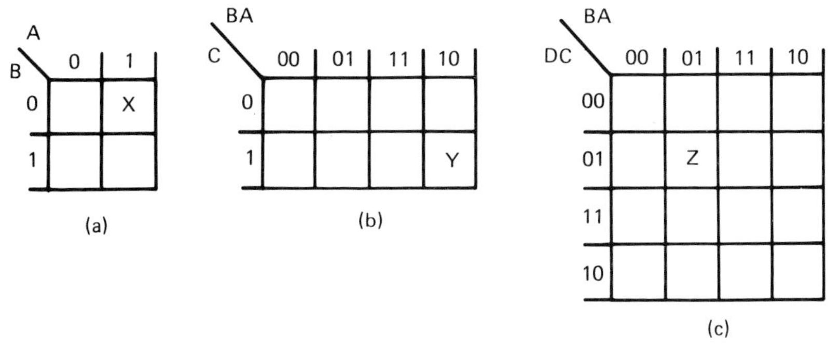

Fig.3.22 Karnaugh maps. (a) Two variable map. (b) Three variable map. (c) Four variable map.

Each square within the map represents one line on the truth table. For example:

square X represents A=1, B=0 which can be written $A\bar{B}$
square Y represents A=0,B=1,C=1 which can be written $\bar{A}BC$
square Z represents A=1,B=0,C=1,D=0 which can be written $A\bar{B}C\bar{D}$

The essential feature of a Karnaugh map is the way in which the axes are labelled. It will be seen that only one variable

DC \ BA	00	01	11	10
00	0	0	0	0
01	0	1	1	0
11	0	1	1	0
10	0	0	0	0

Fig.3.23 Representation of Z = AC on four variable map.

changes between horizontally adjacent squares in any row, and only one variable changes between vertically adjacent squares in any column.

The use of this feature is not immediately apparent, but consider fig.3.23. The truth table contains four terms giving a 1 output. These are:

$$A\bar{B}C\bar{D},\ ABC\bar{D},\ A\bar{B}CD,\ ABCD$$

so we could write (quite correctly):

$$Z = A\bar{B}C\bar{D} + ABC\bar{D} + A\bar{B}CD + ABCD$$

Examination of the map, however, shows that the D variable and B variable change without affecting the output. The circled squares, in fact, represent AC, so the above expression can be simplified to:

$$Z = AC$$

This result could, of course, also have been obtained (with great effort) by Boolean algebra.

C \ BA	00	01	11	10
0	0	0	1	0
1	0	1	1	1

Fig.3.24 Plot of $Z = A.B.C + A.B.\bar{C} + A\bar{B}C + \bar{A}BC$.

In a three variable map such as fig.3.24, each cell represents some combination of the three variables. On a four variable map such as fig.3.23, each cell represents some combination of the four variables.

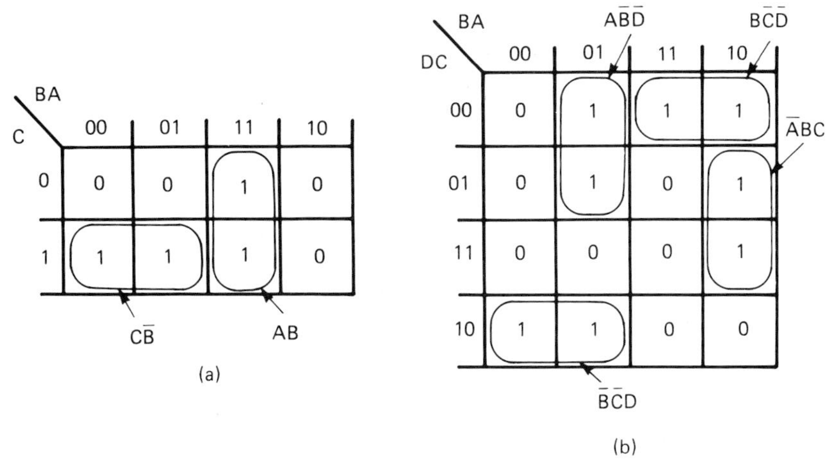

Fig.3.25 Grouping of 2 adjacent cells. (a) Three variable. (b) Four variable.

Groups of two adjacent cells on a three variable map represent some combination of *two* of the three variables. In fig.3.25a, groupings for AB and C$\bar{B}$ are shown. This map represents:

$$Z = AB + C\bar{B}$$

Two adjacent cells on a four variable map represent some combination of three of the four variables. In fig.3.25b, groupings for $\bar{A}BC$, $B\bar{C}\bar{D}$, $A\bar{B}\bar{D}$ and $\bar{B}\bar{C}D$ are shown. This map represents:

$$Z = \bar{A}BC + B\bar{C}\bar{D} + A\bar{B}\bar{D} + \bar{B}\bar{C}D$$

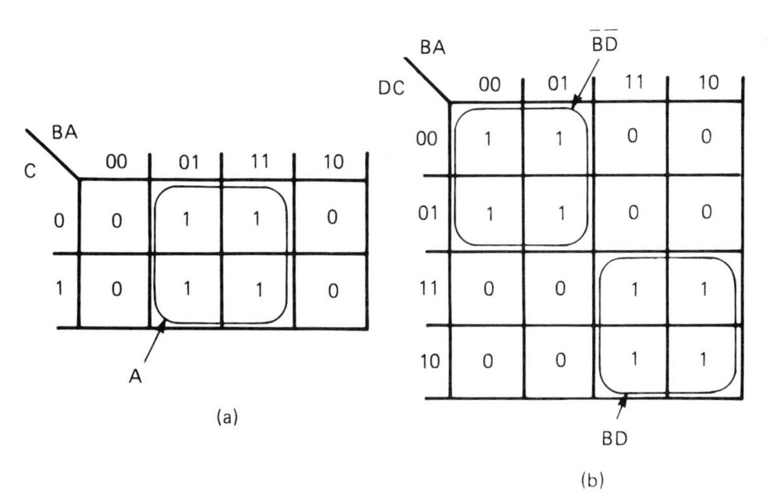

Fig.3.26 Grouping of 4 adjacent cells. (a) Three variable. (b) Four variable.

Groups of four adjacent cells on a three variable map represent a single variable. The group in fig.3.26a represents the variable A, hence:

$$Z = A$$

Groups of four adjacent cells on a four variable map represent some combination of two of the four variables. The groups in fig.3.26b represent $\bar{B}\bar{D}$ and BD. The map represents

$$Z = \bar{B}\bar{D} + BD$$

A group of eight adjacent cells on a four variable map represents a single variable. The groups in fig.3.27 represent C and $\bar{B}$, so:

$$Z = C + \bar{B}$$

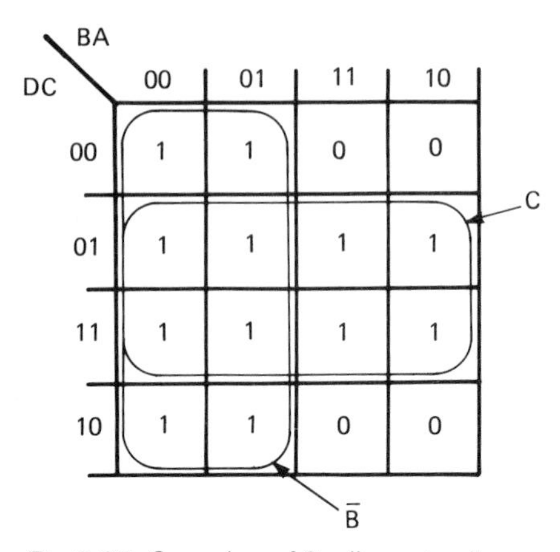

Fig.3.27 Grouping of 8 adjacent cells.

It is important to realise that top and bottom edges are considered adjacent, as are right and left sides. Grouping can therefore be made around the tops and sides, as in fig.3.28 which represents:

$$Z = \bar{A}C + A\bar{C}$$

Consider the expression:

$$Z = \bar{A}B\bar{C}\bar{D} + \bar{A}BC\bar{D} + ABC\bar{D} + A\bar{B}C\bar{D} + A\bar{B}CD + ABCD + \bar{A}BCD + \bar{A}B\bar{C}D$$

This has eight terms which are plotted in fig.3.29a (equally, this could be derived from eight lines of a truth table). Using the grouping ideas outlined above, these can be regrouped as

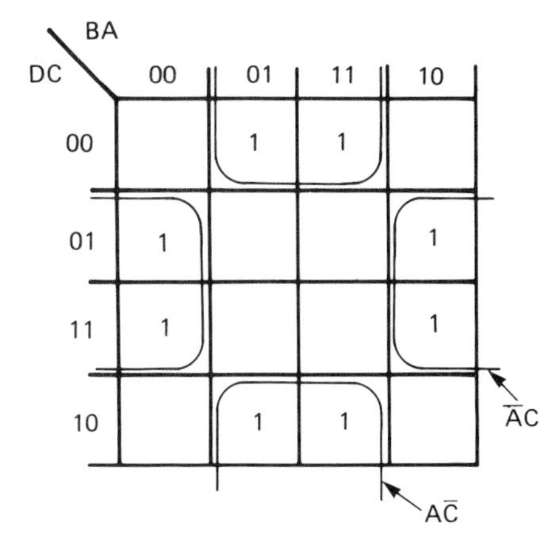

Fig.3.28 Adjacency of top/bottom and sides.

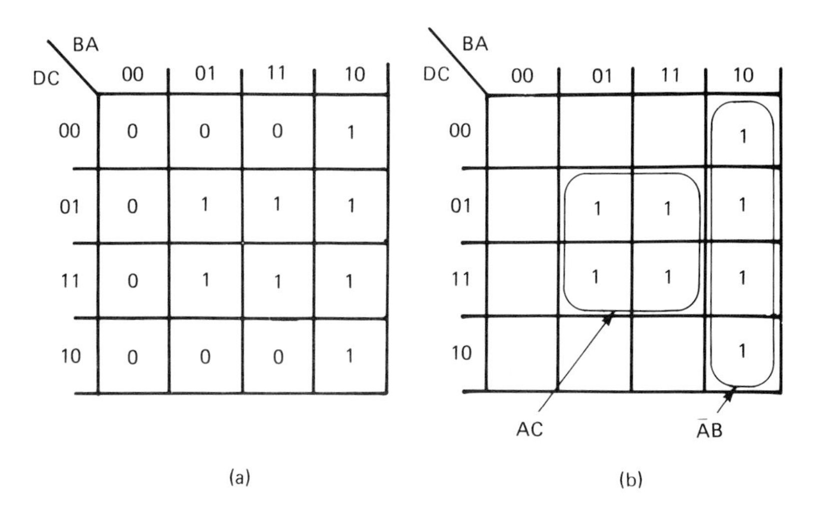

Fig.3.29 Grouping for minimal expression. (a) Four variable. (b) Minimal grouping.

fig.3.29b with just two terms AC and $\bar{A}B$. The expression for Z above becomes:

$$Z = AC + \bar{A}B$$

The same result could, of course, have been obtained by lengthy analysis by Boolean algebra.

The rules for minimisation using Karnaugh maps are simple and straightforward:

(1) Plot the Boolean expression or truth table on to the Karnaugh map.

(2) Form new groups of 1s on the map. Groups must be rectangular and contain one, two, four or eight cells. Groups should be as large as possible and there should be as few groups as possible. Do not forget overlaps and possible round-the-edge groupings.
(3) From the map, read off the expression for each group. The minimal expression is then obtained in S of P form, and can be directly implemented in AND/OR gates (as in fig.3.19) or NAND gates (as in fig.3.20b).

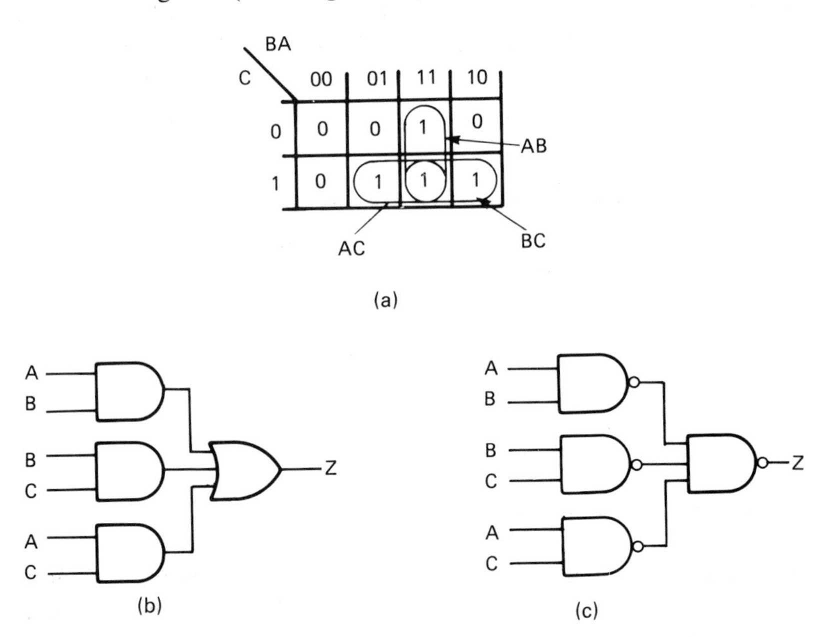

Fig.3.30 Design of majority vote circuit. (a) Karnaugh map of majority vote circuit. (b) AND/OR implementation. (c) NAND implementation.

In section 3.3.2 we designed a majority vote circuit (fig.3.19) direct from a truth table. This is plotted on to a Karnaugh map in fig.3.30a, and grouped on fig.3.30b. It will be seen that this has three terms giving the (simpler) circuit of fig.3.30c.

3.3.5. *Integrated circuits*

Many complex functions are available in IC form, and a circuit designer should aim to minimise cost and the number of IC packages rather than the number of gates. A minimisation exercise, whether by Boolean algebra or Karnaugh map, should always be preceded by a search of an IC catalogue for a suitable off-the-peg device.

3.4. Storage

3.4.1. Introduction

Most logic systems require some form of memory. A typical relay circuit is the motor starter circuit of fig.3.31 which 'remembers' which of the two operator push buttons was pressed last. The memory is achieved by the latching contact of relay A.

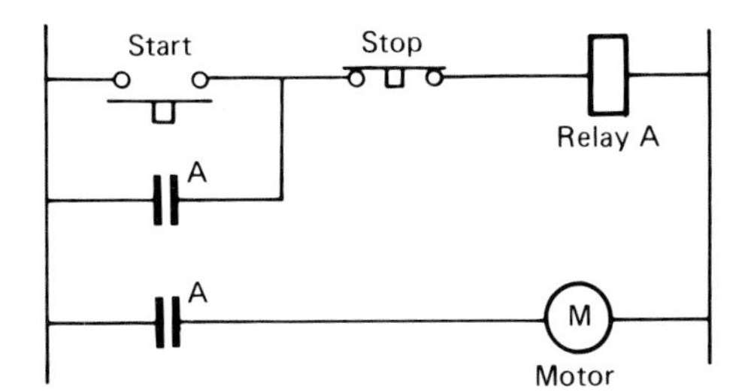

Fig.3.31 A relay storage circuit, a motor starter.

3.4.2. Cross coupled flip flops

The logical equivalent of fig.3.31 is the cross coupled NOR gate circuit of fig.3.32a. Assume both inputs are 0, and output Q is at a 1 state. The output of gate a will be 0, and the two 0 inputs to gate b will maintain Q in its 1 state. The circuit is therefore stable.

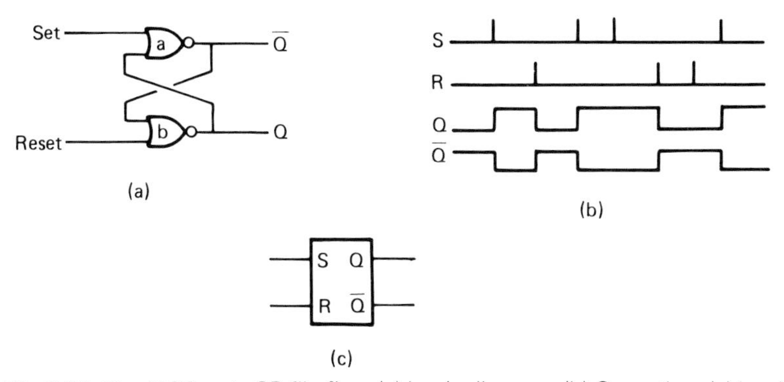

Fig.3.32 The NOR gate SR flip flop. (a) Logic diagram. (b) Operation. (c) Logic symbol.

If the reset input is now taken to a 1, Q will go to a 0 and Q to a 1. Similar analysis to that above will show that the circuit is stable in this state, even when the reset input goes back to a 0.

The set input can be used now to switch the Q output to 1 and the $\overline{Q}$ back to 0. The set and reset inputs cause the output to change state, with the outputs indicating which input was last at a 1 state, as summarised by fig.3.32b. If both inputs are 1 together both outputs go to a 0, but this condition is normally disallowed.

The cross coupled NOR gate circuit is called an RS flip flop, and is shown on logic diagrams by the symbol of fig.3.32c.

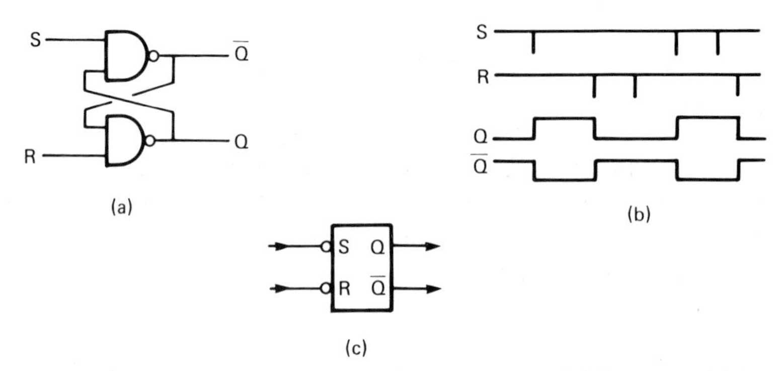

Fig.3.33 The NAND gate SR flip flop. (a) Logic diagram. (b) Operation. (c) Logic symbol.

It is also possible to construct a cross coupled flip flop from NAND gates, as in fig.3.33a. Analysis will show that this behaves in a similar way to fig.3.32, but the circuit remembers which input last went to a 0, as shown in fig.3.33b. The logic symbol for a NAND based RS flip flop is shown in fig.3.33c; the small circles on the input show that the flip flop responds to 0 inputs.

3.4.3. The transparent latch

The transparent latch (known also as a hold/follow latch) is used to freeze digital data. It is constructed as in fig.3.34 (and is usually obtained in IC form, such as the TTL 7475). With the enable input at 1, the output Q follows the input A. When the enable input goes to a 0, Q indicates the state of A at the instant the enable went from 1 to 0.

Transparent latches are typically used to transfer data from, say, a fast counter to a display.

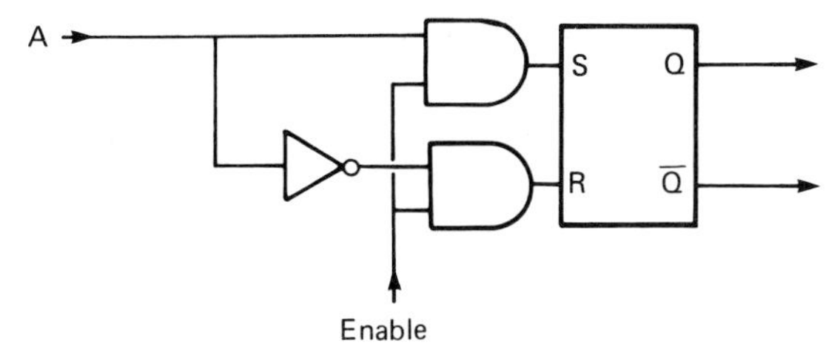

Fig.3.34 The transport latch.

3.4.4. The D type flip flop

The D type flip flop shown in fig.3.35a has a single data input (D), a clock input and the usual Q and $\overline{Q}$ outputs. Superficially this is similar to the latch memory above, but the clock operates in a more subtle way. The operation of a typical D type flip flop is shown in fig.3.35b. The clock samples the D input when the clock input goes from 0 to 1, but the output changes state when the clock goes from 1 to 0. The significance of this is explained below in section 3.4.6.

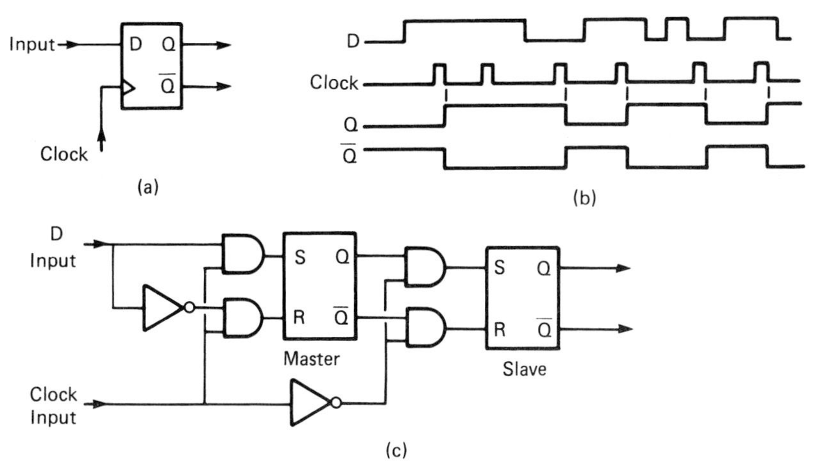

Fig.3.35 The D type flip flop. (a) Logic symbol. (b) Operation. (c) Logic diagram for master/slave circuit.

There are several ways in which a D type flip flop can be implemented. A common circuit uses the master/slave arrangement of fig.3.35c. When the clock input is 1, the D input sets, or resets, the master flip flop. When the clock input is 0, the state of the master flip flop is transferred to the slave flip flop (and the outputs take up the state of D when the clock input was 1). Note that the master flip flop is isolated from the D input whilst the clock is 0.

Although it would be feasible to construct a master/slave flip flop from discrete gates, IC D types (such as the TTL 7474 or the CMOS 4013) are readily available.

3.4.5. The JK flip flop

In section 3.4.2 the NOR based RS flip flop was described, and it was stated that the input state R=S=1 was normally disallowed. The JK flip flop is a clocked RS flip flop with additional logic to cover this previously disallowed state. The clock input acts as described above for the D type flip flop, i.e. sampling the inputs on one edge, and causing the outputs to change on the other.

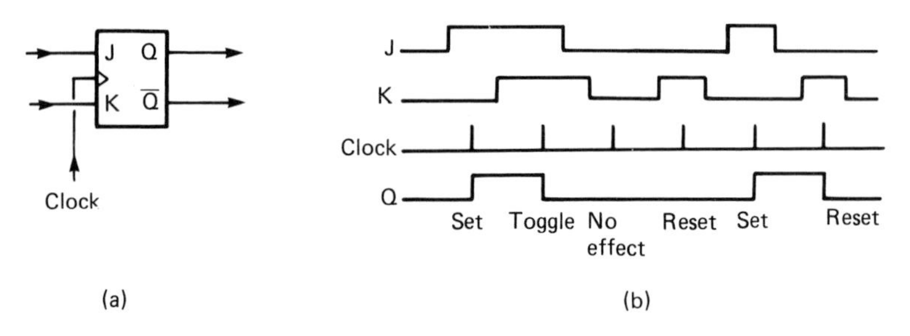

Fig.3.36 The JK flip flop. (a) Logic symbol. (b) Operation.

The outputs after a clock pulse for J=1, K=0; J=0, K=1; J=0, K=0 are as would be expected for a clocked RS flip flop. If J=K=1, the outputs toggle; that is, the states of the Q and $\bar{Q}$ interchange. This action is summarised in fig.3.36b and in the table below.

J	K	Output after clock pulse		Comment
		Q	$\bar{Q}$	
0	0	No change		Q = old Q, $\bar{Q}$ = old $\bar{Q}$
0	1	0	1	Reset
1	0	1	0	Set
1	1	Toggle		Q = old $\bar{Q}$, $\bar{Q}$ = old Q

In data sheets, the above would be represented:

J	K	CK	Q_{n+1}
0	0	Ω	Q_n
0	1	Ω	0
1	0	Ω	1
1	1	Ω	$\bar{Q}_n$

The toggle state is the basis for counters, described in section 3.7.

3.4.6. Clocked storage

The D type and JK flip flop described above are examples of clocked storage. The advantages and implications of this are probably not immediately obvious.

In all bar the simplest systems, data is often required to be moved around from one storage position to another. In fig.3.37,

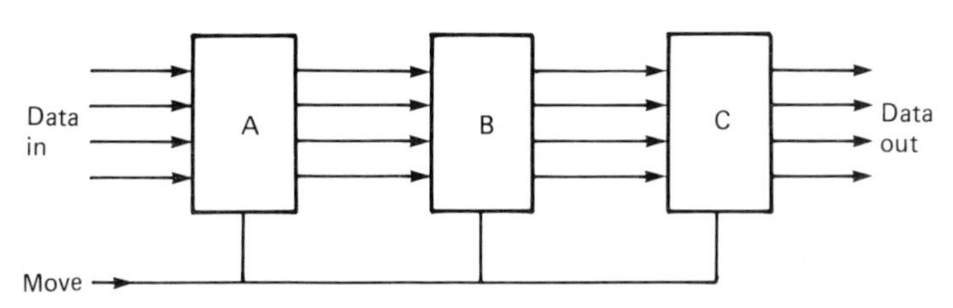

Fig.3.37 Clocked storage.

for example, data is to be moved through stores A, B, C in an orderly manner. If transparent latches were used along with a single enable as shown, the data would shoot straight through all the stages. If clocked storage is used, the data will sequence from A to B to C, moving one position for each clock pulse.

3.5. Timers and monostables

Control systems often need some form of timer; a gas igniter might operate under a pilot flame for 5 seconds, say, before a flame failure detector is enabled, or a nitrogen purge of a reactor vessel undertaken for 2 minutes before the reagents are admitted. Timing functions in logic circuits are provided by devices called monostables or delays.

There are many types of delay, although all can be considered, as fig.3.38a, to consist of an input, Q and $\bar{Q}$ outputs, and an RC network which determines the delay period.

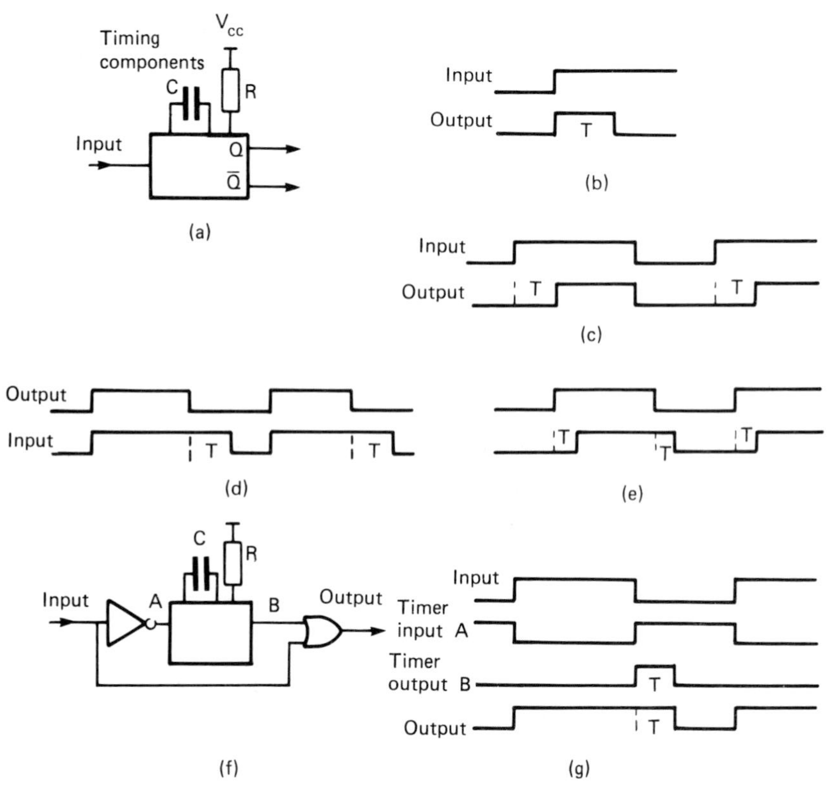

Fig.3.38 Timers and monostables. (a) General representation of a timer. (b) One shot timer. (c) Delay on. (d) Delay off. (e) Delay on and off. (f) Delay off circuit using a monostable. (g) Waveforms for circuit (f).

The commonest timer, often called the one shot or monostable, gives an output pulse, of known duration, for an input edge. The user can select which edge (0–1 or 1–0) triggers the circuit. In fig.3.38b a 0–1 edge is used. Monostables are the basis of all other delay circuits and are widely available (74121, 74122 in TTL, 4047, 4098 in CMOS).

Pure delays are shown in fig.3.38c, d and e, and these can be constructed by adding gates to monostable outputs. Figure 3.38f,g shows the circuit for a delay off.

A variation of the monostable is the retriggerable monostable. In most monostable circuits the timing logic ignores further input edges once started. In a retriggerable monostable each edge sets the timing circuit back to the start again. The actions of a retriggerable and normal monostable are compared in fig.3.39.

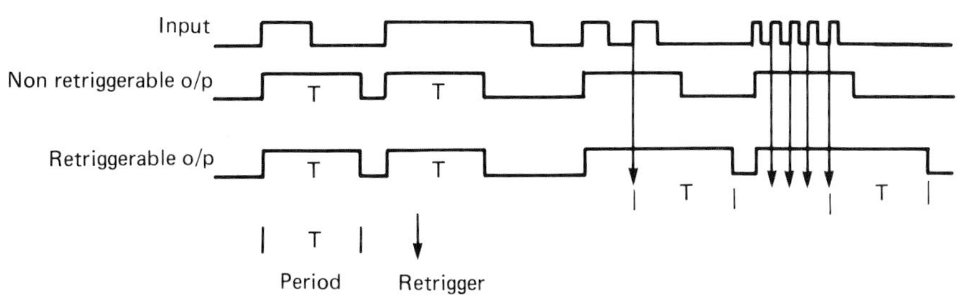

Fig.3.39 Retriggerable monostables.

Retriggerable monostables are commonly used as a low speed alarm with pulse tachos or as a watchdog protection for process control computers.

Monostables can also be used to construct digital oscillators (called clocks). Two monostables connected as in fig.3.40 will give a stable easy-to-adjust pulse train.

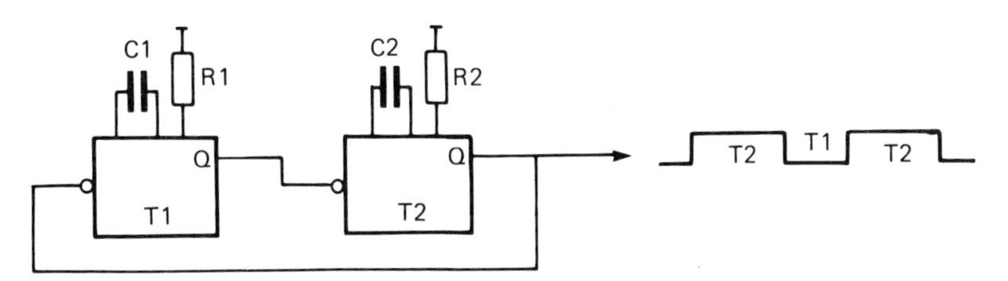

Fig.3.40 Oscillator built from two monostables.

3.6. Arithmetic circuits

3.6.1. Number systems, bases and binary

In previous sections, logic signals have been assumed to represent events such as printer ready, or low oil level. Digital signals can also be used to represent, and manipulate, numbers.

We are so used to the decimal number system that it is hard to envisage any other way of counting. Normal everyday arithmetic is based on multiples of ten. For example, the number 9156 means:

$$\begin{aligned}
&\text{9 thousands} = 9 \times 10 \times 10 \times 10\\
&\text{plus 1 hundred} = 1 \times 10 \times 10\\
&\text{plus 5 tens} = 5 \times 10\\
&\text{plus 6 units} = 6
\end{aligned}$$

Each position in a decimal number represents a power of ten. Our day-to-day calculations are done to a base of ten because we have ten fingers. Counting can be done to any base, but of special interest are bases 8 (called octal), 16 (called hex for hexadecimal) and two (called binary).

Octal uses only the digits 0 to 7; the octal number 317, for example, means

$$\begin{aligned}
&3 \times 8 \times 8 = \text{decimal } 192\\
&\text{plus } 1 \times 8 = \text{decimal } 8\\
&\text{plus } 7 = \text{decimal } 7\\
&\text{Total} = \text{decimal } 207
\end{aligned}$$

Hex uses the letters A–F to represent decimal ten to fifteen, so hex C52, for example, means

$$\begin{aligned}
&12 \times 16 \times 16 = \text{decimal } 3072\\
&\text{plus } 5 \times 16 = \text{decimal } 90\\
&\text{plus } 2 = \text{decimal } 2\\
&\text{Total} = \text{decimal } 3164
\end{aligned}$$

Binary needs only two symbols, 0 and 1. Each position in a binary number represents a power of two and is called a bit (for BInary digiT), the most significant to the left as usual, so 101101 is evaluated:

$$\begin{aligned}
&1 \times 2 \times 2 \times 2 \times 2 \times 2 = 32\\
&\text{plus } 0 \times 2 \times 2 \times 2 \times 2 = 0\\
&\text{plus } 1 \times 2 \times 2 \times 2 = 8\\
&\text{plus } 1 \times 2 \times 2 = 4\\
&\text{plus } 0 \times 2 = 0\\
&\text{plus } 1 = 1\\
&\text{Total} = \text{decimal } 45
\end{aligned}$$

Similarly 1101011 is evaluated:

$1 \times 64 = 64$
$1 \times 32 = 32$
$0 \times 16 = 0$
$1 \times 8 = 8$
$0 \times 4 = 0$
$1 \times 2 = 2$
$1 = 1$
Total = decimal 107

Conversion from decimal to binary is achieved by successive division by two, noting the remainders. Reading the remainders from the top (LSB–least significant bit) to bottom (MSB–most significant bit) gives the binary equivalent. For example, decimal 23:

23
11 r 1 (LSB)
5 r 1
2 r 1
1 r 0
0 r 1 (MSB)

Decimal 23 is binary 10111.

Similarly decimal 75:

75
37 r 1 (LSB)
18 r 1
9 r 0
4 r 1
2 r 0
1 r 0
0 r 1 (MSB)

Decimal 75 is binary 1001011.

Octal and hex give a simple way of representing binary numbers. To convert a binary number to octal, the binary number is written in groups of three (from the LSB) and the octal equivalent written underneath; for example, 11010110:

grouped in threes	11	010	110
octal	3	2	6

Hex conversion is similar, but groupings of four are used. Taking the same binary number 11010110:

grouped in fours	1101	0110
hex	D	6

The octal number 326 and the hex number D6 are both representations of the binary number 11010110.

3.6.2. Binary arithmetic

Consider the decimal sum:

$$\begin{array}{r} 345 \\ +\ 272 \\ \hline 617 \end{array}$$

This is evaluated in three stages:

5 + 2 = 7 no carry
4 + 7 = 11 one down (as result) plus carry
3 + 2 + carry = 6

At each stage we consider three 'inputs': two digits and a possible carry from the previous stage. Each stage has two outputs: a sum digit and a possible carry to the next, more significant stage. A single digit adder can therefore be considered, as in fig.3.41a. Several single digit adders can be cascaded, as in fig.3.41b, to give an adder of any required number of digits. Note the carry out of the most significant stage becomes the most significant digit.

Binary addition is similar, except that there are only two possible values for each digit. If fig.3.41a is a binary adder, there are eight possible input combinations:

Inputs			Outputs	
Digit 1	Digit 2	Carry	Sum	Carry
0	0	0	0	0
0	1	0	1	0
1	0	0	1	0
1	1	0	0	1
0	0	1	1	1
0	1	1	0	1
1	0	1	0	1
1	1	1	1	1

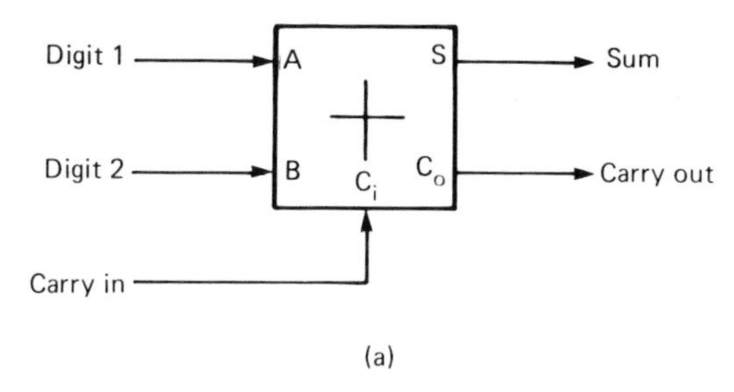

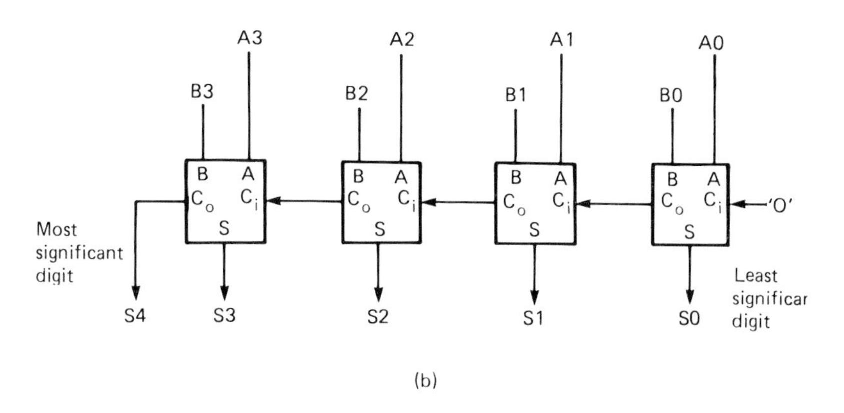

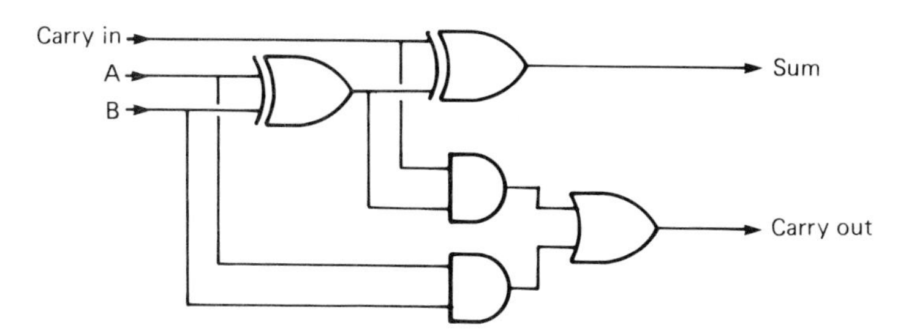

Fig.3.41 Digital adder circuits. (a) Representation of one digit adder. (b) Four digit adder constructed from four one digit adders. (c) One bit binary adder.

An example of binary arithmetic is:

```
  1 0 1 1 0 1 0
  0 1 0 1 0 1 1
  -------------
1 0 0 0 0 1 0 1    Sum (result)
 1 1 1 1 0 1 0     Carry
```

The implementation of the adder truth table is a simple problem of combinational logic; one possible solution is shown in fig.3.41c. In practice, of course, adders such as the TTL 7483 are readily available in IC form.

Negative numbers are generally represented in a form called twos complement. The most significant digit represents the sign, being 1 for negative numbers and 0 for positive numbers. The value part of the number is complemented and 1 added. For example, +12 in twos complement is 01100 (the MSB 0 indicating a positive number).

To get the twos complement for −12 we complement 1100 giving 0011, set the MSB to 1 giving 10011, then add 1 giving 10100 which is the twos complement representation of −12. Similarly:

+43	0101011
Complement	1010100
Add 1	1010101, which is −43

In each case, addition of the positive and negative number will give the result zero, e.g.:

```
  +43    0 1 0 1 0 1 1
  −43    1 0 1 0 1 0 1
         -------------
       1 0 0 0 0 0 0 0
```

The top carry is lost, giving the correct result of zero.

Twos complement representation allows subtraction to be done by adding a negative number, for example 12 − 3:

```
    0 1 1 0 0    +12
    1 1 1 0 1    −3
    ---------
  1 0 1 0 0 1
```

The top bit is lost, giving the correct result of +9.

3.6.3 Binary coded decimal (BCD)

A single decimal digit can take any value between 0 and 9. Four binary digits are therefore needed to represent one decimal digit. In BCD, each decimal digit is represented by 4 bits. For example:

9	4	0	7	6
1001	0100	0000	0111	0110

BCD is not as efficient as pure binary. In pure binary 12 bits can represent 0 to 4095, compared with 0 to 999 in BCD. BCD, however, has advantages where decimal numbers are to be read from decade switches or position measuring encoders.

3.6.4. Unit distance codes

Figure 3.42 shows a possible application of binary coding. The position of a shaft is to be measured to 1 part in 16 by means of an optical grating moving in front of four photocells. The photocell outputs give a binary representation of the shaft angular position.

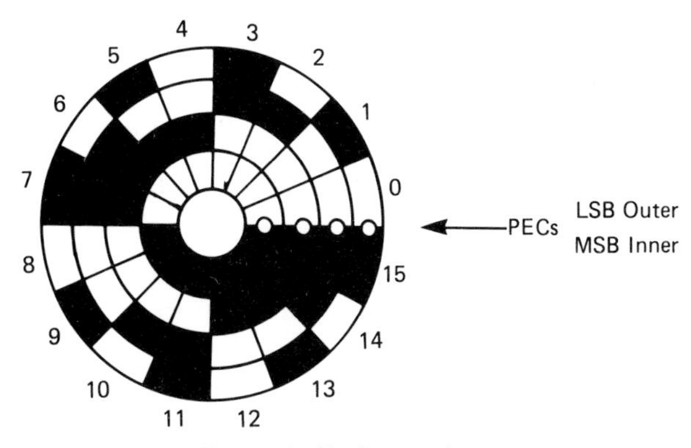

Fig.3.42 Shaft encoder.

Consider what may happen as the shaft goes from position 7 (0111) to position 8 (1000). It is unlikely that all the cells will switch together, so we could get:

0111→0000→1000 or
0111→1111→1000

or any other combination of 4 bits. These possible incorrect intermediate states can be avoided by using a code in which only 1 bit changes between adjacent positions. Such codes are called unit distance codes.

The commonest unit distance code is the Gray code, shown in 4 bit form below. It will be noted that the code is reflected about the centre. Sometimes the term 'reflected' code is used for unit distance codes.

Decimal	Gray		
0	0000		
1	0001		
2	0011		
3	0010		
4	0110		
5	0111		
6	0101		
7	0100	decimal	↑ symmetrical
8	1100	cyclic	↓
9	1101		
10	1111		
11	1110		
12	1010		
13	1011		
14	1001		
15	1000		

A unit distance code can be constructed to any even base by taking an equal number of combinations above and below the centre point of a Gray code. A decimal version (called the XS3

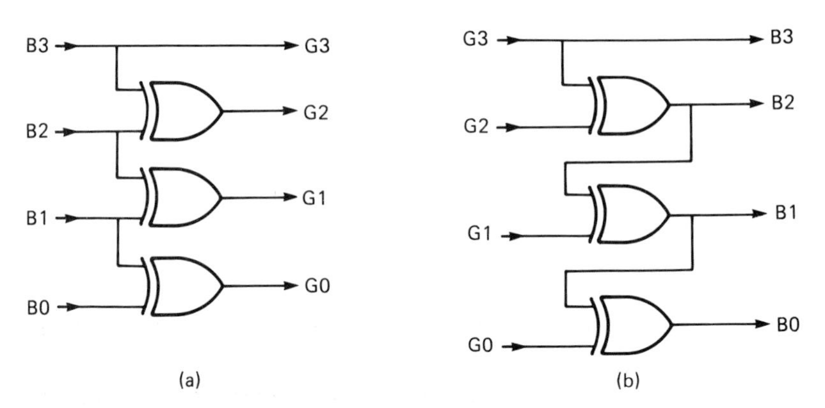

Fig.3.43 Binary/Gray conversion. (a) Binary to Gray. (b) Gray to binary.

cyclic BCD code) is shown above. In this code 0 is 0010, 1 is 0110, 2 is 0111 and so on to 9 which is 1010.

Conversion between binary and Gray code is straightforward, and is achieved with XOR gates as shown in fig.3.43a and b.

3.7. Counters and shift registers

3.7.1. Ripple counters

Counters are used for two basic purposes. The first, and obvious, use is the counting, or totalising, of external events. Batch counters, traffic recorders, frequency meters and such devices all use counters for totalisation. The second use of counters is the division of a frequency to give a new, lower frequency. A visual display unit (VDU), for example, is built around a timing chain which produces frequencies from several MHz down to 50 Hz from a single oscillator.

The 'building block' of all counters is the toggle flip flop which changes state each time its clock input is pulsed. Usually the toggling occurs on the negative edge, as shown in fig.3.44a. A toggle flip flop can be constructed from JK or D type flip flops, as shown in fig.3.44b,c.

If the Q output of a toggle flip flop is connected to the clock input of the next stage as shown in fig.3.45a, a simple binary counter can be constructed to any desired length. Figure 3.45 is a three bit counter, with A the LSB and C the MSB. This counts:

Pulse	C	B	A
0	0	0	0
1	0	0	1
2	0	1	0
3	0	1	1
4	1	0	0
5	1	0	1
6	1	1	0
7	1	1	1

Another pulse will take it to state 0 again. It can be seen that fig.3.45 is counting up.

To count down, the $\overline{Q}$ outputs are connected to the following stage as in fig.3.46a, and the signal outputs taken from the Q lines. Examination of fig.3.46b will show that this counter is counting down.

There are two limitations to the speed at which a counter chain similar to figs.3.45 and 3.46 can operate. The first is the maximum speed at which the first (fastest) stage can toggle. This

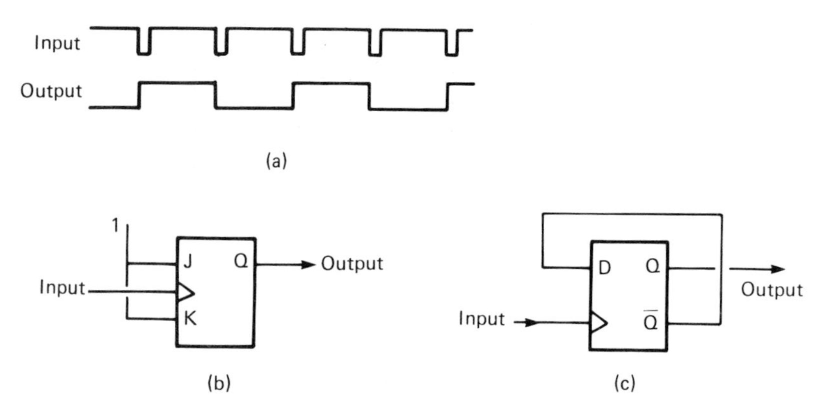

Fig.3.44 The toggle flip flop. (a) Counter operation. (b) JK based toggle flip flop. (c) D type based toggle flip flop.

is typically 20 MHz for a TTL device. The second restriction is not so obvious.

Consider the case of an 8 bit counter going from 01111111 to 10000000. The LSB toggling causes the next to toggle, and so on to the MSB. The change has to propagate through all 8 bits of the

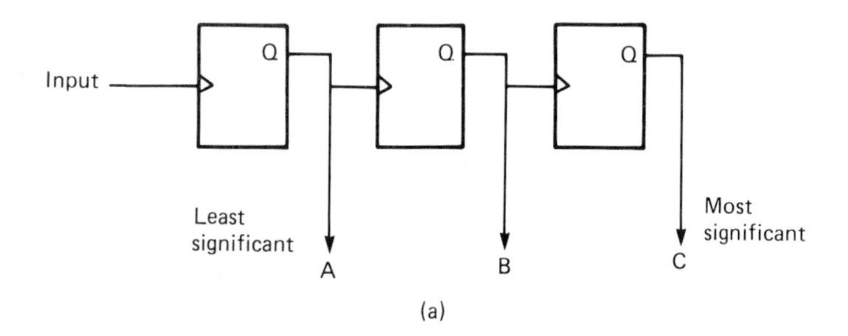

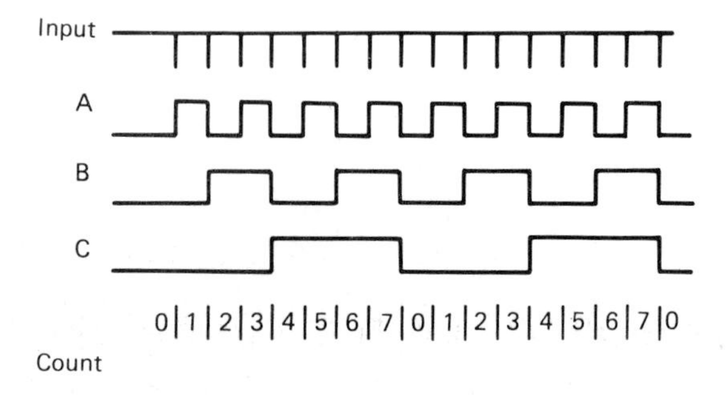

(b)

Fig.3.45 Simple 3 bit binary counter. (a) 3 bit binary counter constructed from toggle flip flops. (b) Counter operation.

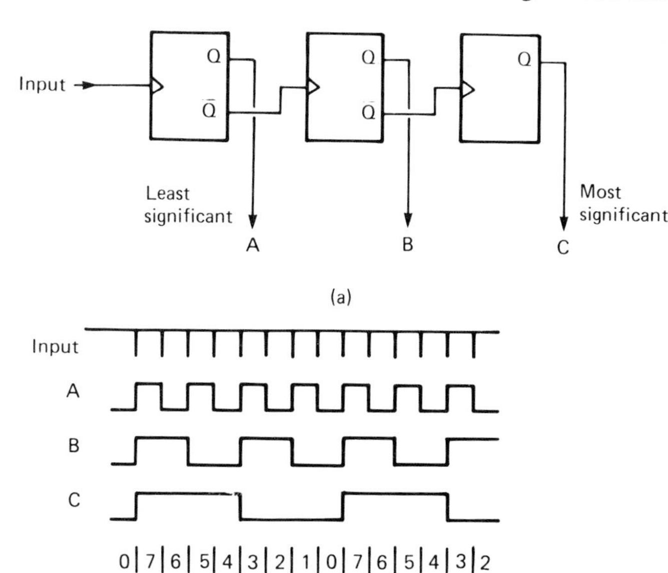

Fig.3.46 3 bit down counter. (a) Down counter logic diagram. (b) Counter operation.

counter, so circuits similar to figs.3.45 and 3.46 are called ripple counters.

During the 'ripple' the counter will assume invalid states and cannot be sensibly read. Obviously the propagation delay through all the stages should be considerably less than the input period. High speed applications use synchronous counters, described below.

In both figs.3.45 and 3.46 the frequency of output C is precisely one eighth of the input frequency. A simple ripple counter can act as a frequency divider. If we define:

$$N = f_{in}/f_{out}$$

then $N = 2^m$ for m binary stages.

It will also be seen that the output of any stage of a binary counter has equal mark space ratio regardless of the input mark/space, provided the input frequency is constant.

Although it is feasible to construct ripple counters with D type and JK flip flops, it is usually more cost effective to use MSI ICs such as the TTL 7493 4 bit counter or the CMOS 4024 7 bit counter. These incorporate features such as a reset line to take the counter to a zero state.

3.7.2. Synchronous counters

Ripple counters are limited in both speed and length by the cumulative ripple through propagation delay and also temporarily exhibit invalid outputs. Although these limitations are not important in slow speed applications, they can cause difficulties in high speed counting.

These restrictions can be overcome by the use of a synchronous counter where all required outputs change simultaneously. There is no ripple propagation delay through the counters and no transient false count stages. The only speed restriction is the toggling frequency of the first stage.

The building block of a synchronous counter is the JK flip flop/AND gate arrangement of fig.3.47a. If the T input is 1, the JK flip flop will toggle on the receipt of a clock pulse. If the T input is 0, the flip flop will not respond to a clock pulse. The carry output is 1 if T is 1 and Q is 1.

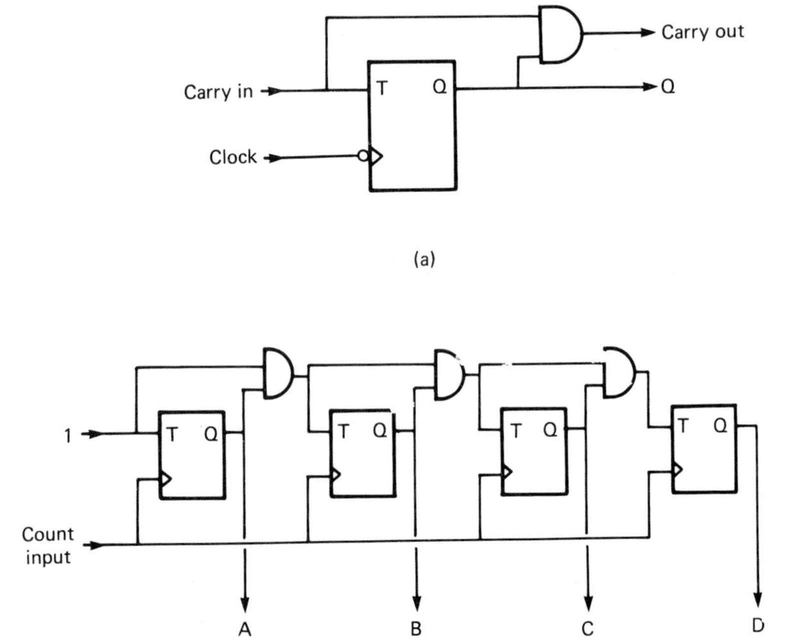

Fig.3.47 Synchronous counters. (a) Basic circuit for synchronous counter. (b) Series connected synchronous counter.

A synchronous up counter is constructed as in fig.3.47b, which is simply the circuit of 3.47a repeated. Note that the clock input is common to all stages, and the carry from one stage is the T input of the next.

It will be seen that the T inputs T_b, T_c, T_d will be 1 when all the preceding outputs are 1. T_c will be 1, for example, when A and B are both 1. This is the condition when a counter stage should toggle, taking DCBA from, say, 0011 to 0100.

It is also possible to construct a synchronous down counter by counting the AND gate input of 3.47a to the $\overline{Q}$ output rather than the Q, and observing the counter state on the Q output (superficially similar to fig.3.46.). A synchronous up/down counter with selectable direction can be constructed as in fig.3.48.

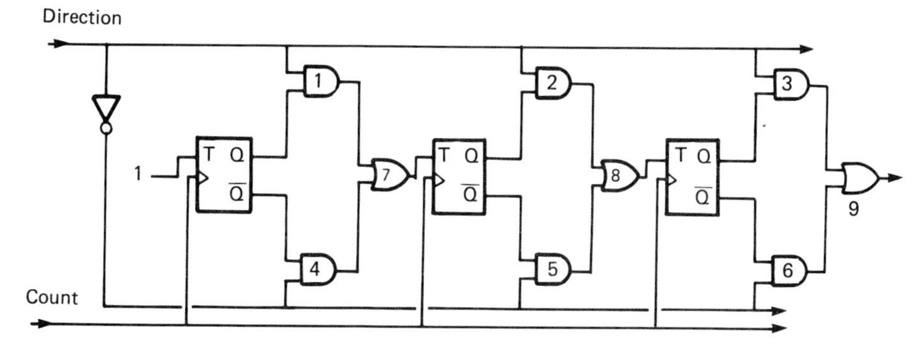

Fig.3.48 Up/down synchronous counter.

If the direction line is a 1, gates 1, 2, 3 are enabled, the Q outputs pass to the next stage and the counter counts up. If the direction line is a 0, gates 4, 5, 6 are enabled, the $\overline{Q}$ outputs pass to the next stage and the counter counts down.

3.7.3. Non-binary counters

Counting to non-binary bases is often required; a BCD count is probably the most common requirement. When the required count is a subset of a straight binary count (as BCD is) the circuit of fig.3.49a can be used. The counter output is decoded by external logic. When the counter reaches the desired maximum count the decoder output forces the counter to its zero state (which is 0000 for a BCD counter, but need not be for other counters).

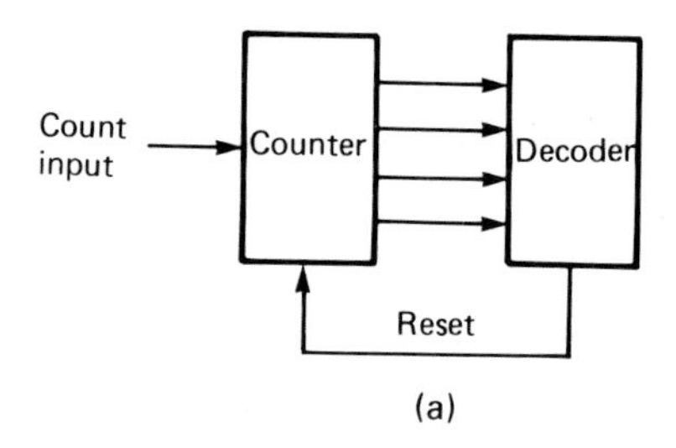

(a)

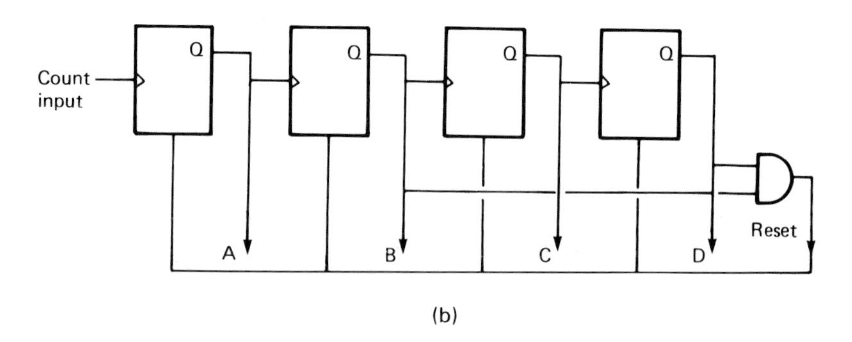

(b)

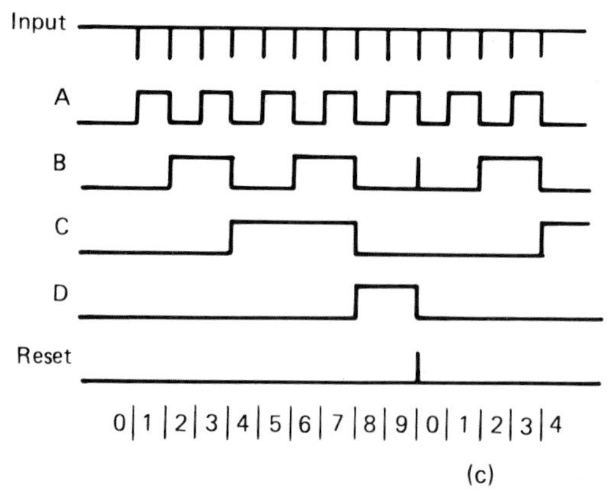

(c)

Fig.3.49 BCD counter. (a) Principle of operation. (b) Logic diagram. (c) Counter operation.

A single BCD stage constructed on these principles is shown in fig.3.49b. The circuit shown is a ripple counter, but could equally well be a synchronous counter. The gate detects a count of ten (binary 1010) and resets the counter to zero via direct reset inputs on the JK flip flops. Waveforms are shown in fig.3.49c.

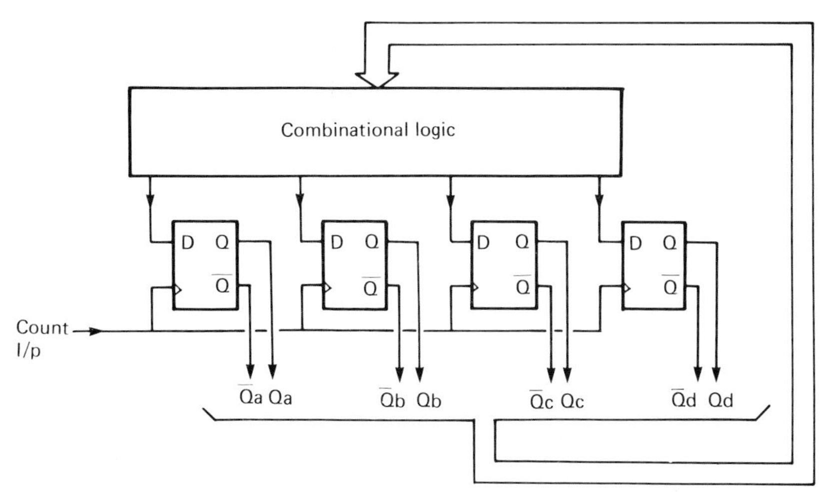

Fig.3.50 Generalised counter design with D type flip flops.

Where a non-binary count is needed (e.g. a Gray code count), it is best to use synchronous counters and an arrangement similar to fig.3.50. This is drawn for D type flip flops, but JK based design is similar.

A combinational logic network looks at the counter outputs and sets the D inputs for the next state. If the counter, say, was

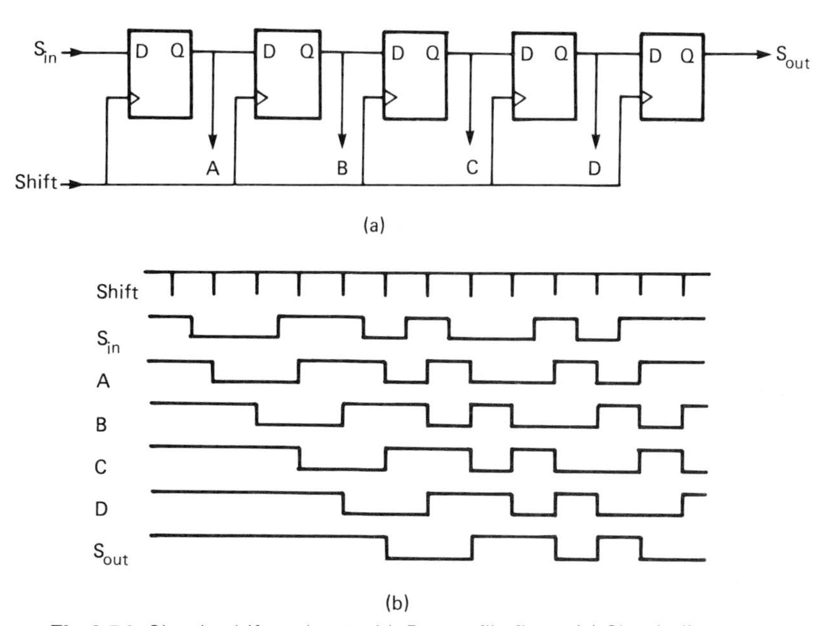

Fig.3.51 Simple shift register with D type flip flops. (a) Circuit diagrams. (b) Waveforms.

required to step from 1101 to 0011, the combinational logic output to the D inputs would be 0011 for an input of 1101. Effectively there are four combinational circuits in the network, one for each D input.

3.7.4. Shift registers

A simple shift register is shown in fig.3.51a. Data applied to the serial input, S_{in}, will move one place to the right on each clock pulse as shown on the timing diagram of fig.3.51b.

Shift registers are used for parallel/serial and serial/parallel conversions. They are also the basis of multiplication and division circuits as a shift of one place towards the MSB is equivalent to a multiplication by 2, and one place towards the LSB an integer division of 2.

3.8. Sequencing and event driven logic

Many logic systems are driven by randomly occuring external events, and follow a sequence of operations. In such systems, the

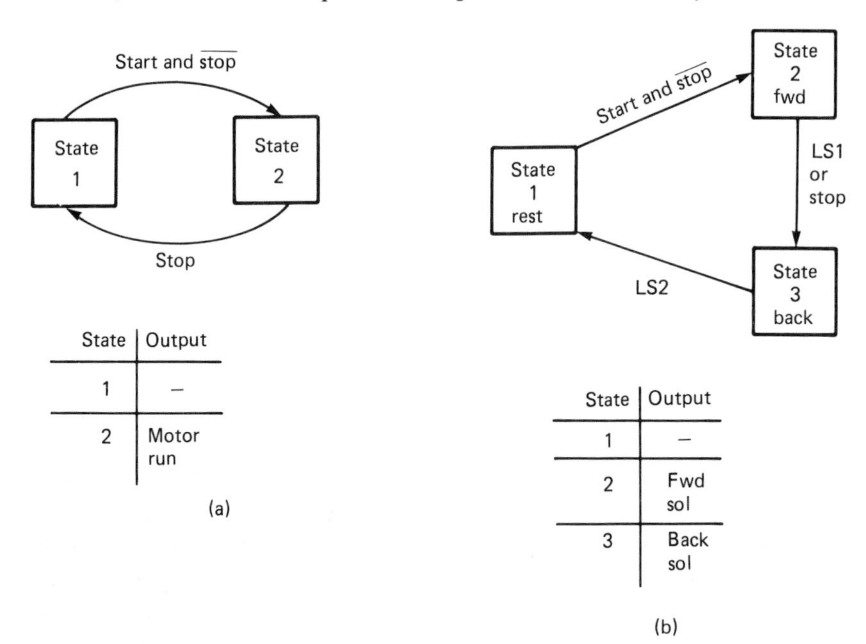

Fig.3.52 Simple state diagrams. (a) State diagram for motor starter. (b) State diagram for hydraulic ram of fig.3.2 (d).

output states do not depend solely on the input states, but also on what the system was doing last. These types of systems are said to be sequencing or event driven logic; simple examples were the motor starter and the hydraulic ram of fig.3.2b and d.

Sequencing logic is designed using a state diagram. This shows the possible conditions the system can be in, the signals that are required to move from one state to the next and the outputs in each state. Figure 3.52a shows the state diagram for the motor starter and fig.3.52b the state diagram for the hydraulic ram.

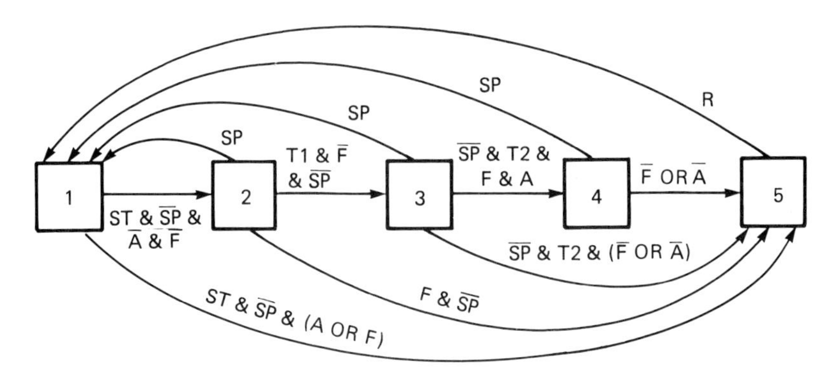

Inputs: Start PB (ST), Stop PB (SP), Flame Present (F), Reset PB (R), Timer 1 Complete (T1), Timer 2 Complete (T2), Air Flow SW (A)

Outputs:

State	Description	Air	Pilot valve	Ignition	Gas valve	Start Timer 1	Start Timer 2	Alarm Bell
1	Off	0	0	0	0	0	0	0
2	Air purge	1	0	0	0	1	0	0
3	Ignition	1	1	1	0	0	1	0
4	On	1	1	0	1	0	0	0
5	Alarm	1	0	0	0	0	0	1

Fig.3.53 State diagram for gas burner.

A more complex example is shown in fig.3.53, which is a state diagram for a gas burner control. When the start PB is pressed, a 15 second air purge is given (set by timer 1). The pilot valve is opened, and the igniter started for 4 seconds (timer 2). If, at the end of this time, the flame detector shows the flame to be lit, the main gas valve is opened. At any time the stop button terminates the sequence. A non-valid signal from the flame detector (i.e.

flame present in states 1 and 2 or no flame in state 4) puts the system to an alarm state, as does the incorrect signal from the air flow switch. Note that these are checked for being 'unfrigged' at the start of the sequence.

Event driven logic is built around flip flops, usually one for each state. The flip flop corresponding to state 4 is shown in fig.3.54a, and is set by the required conditions from state 3 and reset by the possible next states (1 and 5). Outputs are simply obtained by ORing the necessary states. The pilot output, shown in fig.3.54b, is simply state 3 OR state 4.

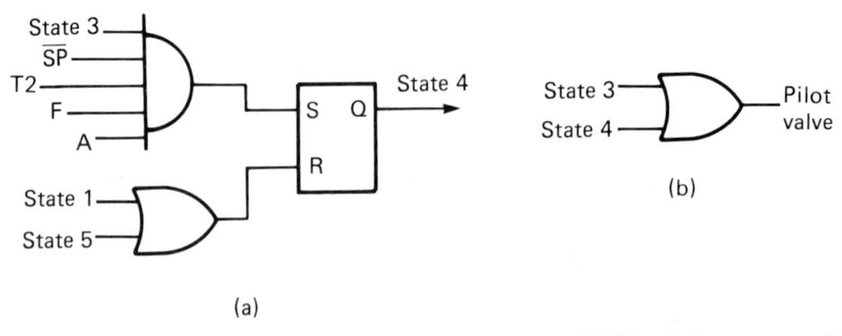

Fig.3.54 Circuit implementation of state diagram. (a) Circuit for state 4. (b) Circuit for pilot valve output.

It is possible to minimise event driven circuits to use fewer flip flops, but such an approach is usually not required in industrial applications. A straightforward state diagram similar to fig.3.53 is easy to design, understand and modify, and simplifies fault finding for maintainance personnel. State diagrams can also be used to write programs for programmable controllers (see section 4.5.5).

3.9. Analog interfacing

3.9.1. Digital-to-analog converters (DACs)

A binary number can represent an analog voltage. An 8 bit number, for example, represents a decimal number from 0 to 255 (or -128 to $+127$ if twos complement representation is used). An 8 bit number could therefore represent a voltage from 0 to 2.55 volts, say, with a resolution of 10 mV. A device which converts a digital number to an analog voltage is called a digital-to-analog converter, or DAC.

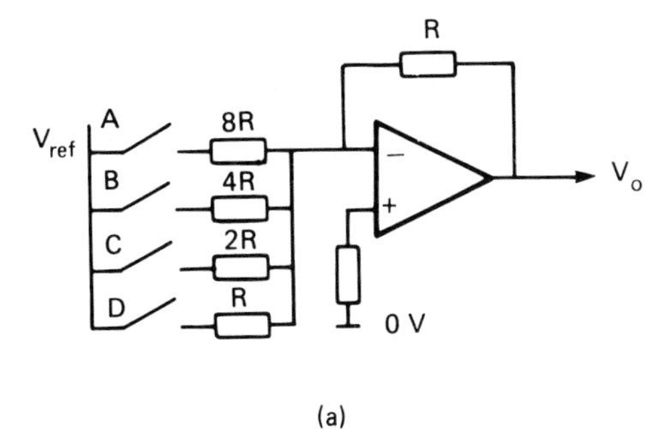

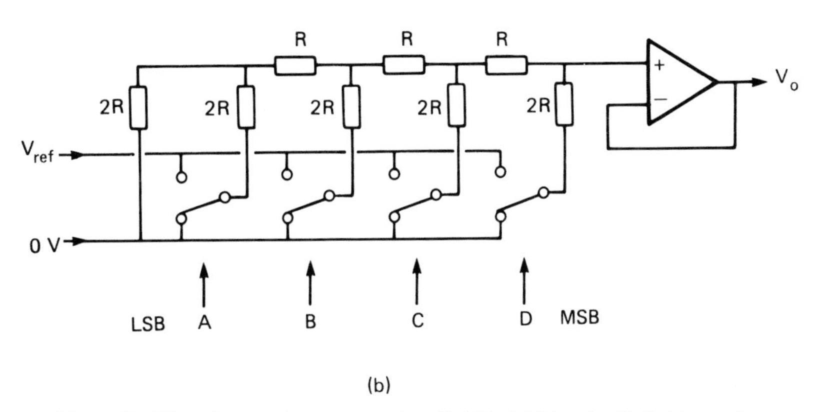

Fig.3.55 Digital to analog conversion (DAC). (a) Simple digital to analog converter. (b) R-2R ladder DAC.

Common DAC circuits are shown in fig.3.55; in each case the output voltage is related to the binary pattern on the switches. In practice, FETs are used for the switches, and usually an IC DAC is used. The R–2R ladder circuit is particularly well suited to IC construction.

3.9.2. Analog-to-digital converters (ADCs)

There are several circuits which convert an analog voltage to its binary equivalent. The two commonest are the ramp ADC, shown in fig.3.56, and the successive approximation ADC of fig.3.57. Both of these compare the output voltage from a DAC with the input voltage.

The operation of the ramp ADC commences with a start command which sets FF1 and resets the counter to zero. FF1

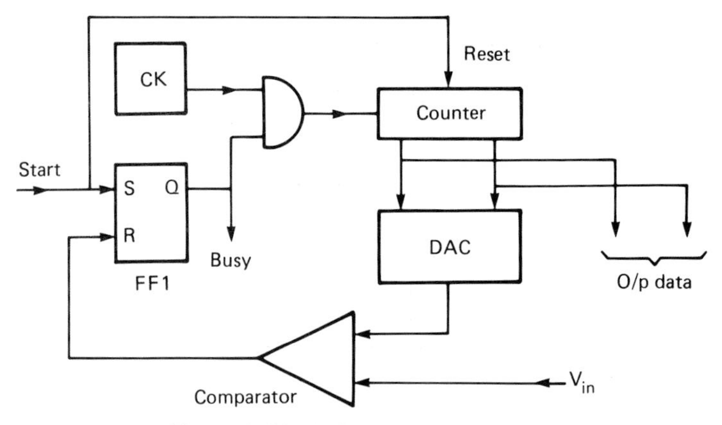

Fig.3.56 Block diagram of ramp ADC.

gates pulses to a counter. The counter output is connected to a DAC whose output ramps up as the counter counts up. The DAC output is compared with the input voltage, and when the two are equal FF1 is reset, blocking further pulses and indicating that the conversion is complete. The binary number in the counter now represents the input voltage. A variation of the ramp ADC, known as a tracking ADC, uses an up/down counter that follows the input voltage.

The ramp ADC is simple and cheap, but relatively slow (typical conversion time >1 mS). Where high speed or high accuracy is required, a successive approximation ADC is used. The circuit shown in fig.3.57 uses an ordered trial and error process.

The sequence starts with the register cleared. The MSB is set, and the comparator output examined. If the comparator shows the DAC output is less than, or equal to, V_{in}, the bit is left set. If the DAC output is greater than V_{in}, the bit is reset. Each bit is similarly tested, in order from MSB to LSB, causing the DAC output to home in on V_{in} as shown.

Successive approximation ADCs are fast (conversion times of a few μS) and accurate (0.01%). Unlike the ramp ADC, the conversion time is constant. They are, however, more complex and expensive than the simpler ramp ADC.

3.10. Practical details

Real-life digital systems have to connect to the outside world, and this can often bring problems when noise and effects such as

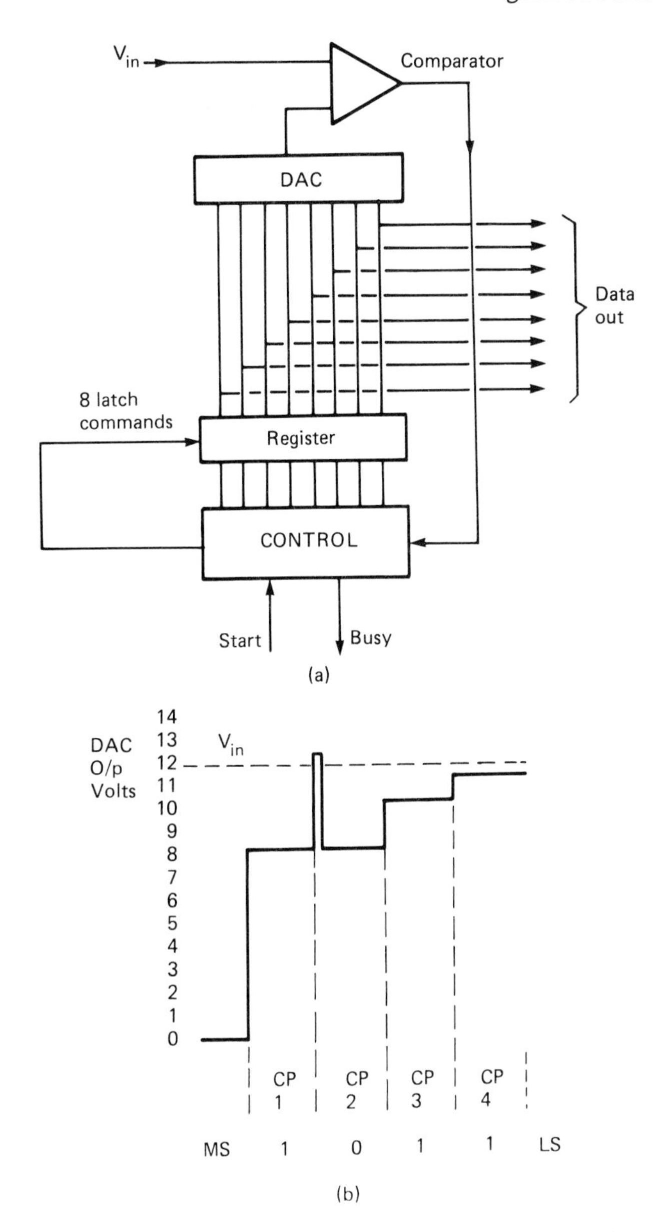

Fig.3.57 Successive approximation ADC. (a) Block diagram. (b) Operation.

contact bounce are encountered. Precautions also need to be taken against inadvertent introduction of high voltages into logic systems via inter-cable faults on the plant.

All signals between a logic system and the outside world should use a technique called opto isolation when cable lengths are

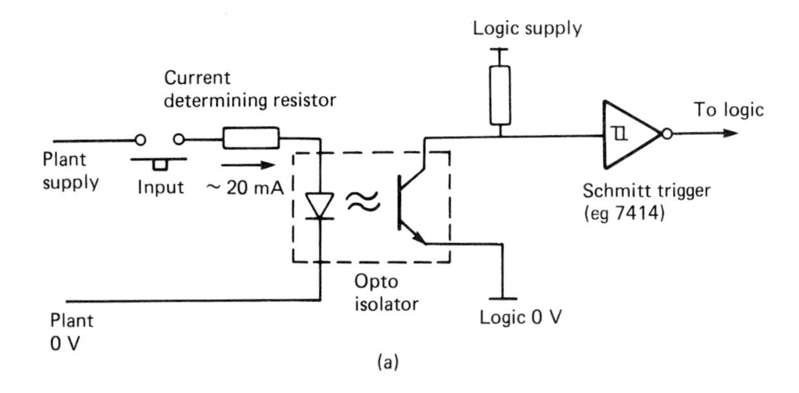

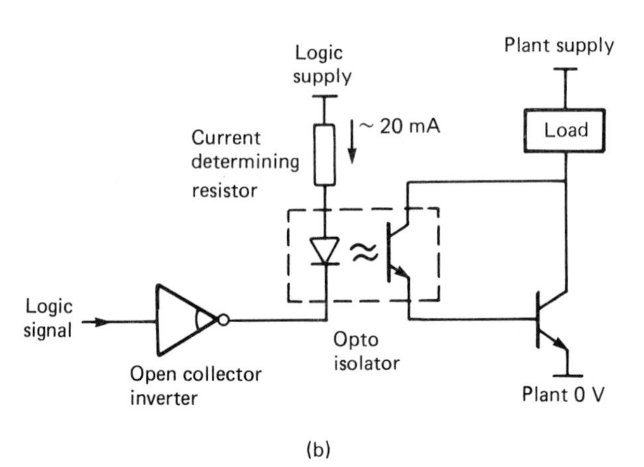

Fig.3.58 Optical isolation of inputs and outputs. (a) Input circuit. (b) Output circuit.

longer than a few metres. Figure 3.58 shows typical input and output circuits. In both, the signal is electrically isolated by using a coupled LED and phototransistor. Because the plant side power supply and digital power supply are totally separate, the system will withstand voltages of up to 1 kV without damage to the digital equipment (although such voltages would probably damage the plant side components, of course). The absence of ground loops and relatively high current levels (around 20 mA) also give excellent noise immunity.

Opto isolators (such as the TIL 107) are usually constructed in a six pin IC, and characterised by a current transfer ratio. This is defined as the ratio between the phototransistor collector current to the LED current. A typical value is 0.3, so 20 mA input current will give 6 mA output current. If Darlington phototransistors are used, transfer ratios as high as 1:2 can be obtained.

Noise can also enter digital systems via the power supply rails so filtering is necessary, both on the DC side and (with LC filter) on the AC supply side. It is particularly important to adopt a sensible segregation of 0 V rails such that digital logic, relays/lamps and analog circuits have separate 0 V returns to some common earth points. Under no circumstances should high currents flow along logic 0 V lines, or the logic 0 V be taken outside its own cubicle.

A digital IC can also generate its own noise on power supplies (TTL is particularly troublesome). It is therefore highly desirable to provide each IC with its own local 0.01 μF capacitor. A single large value electrolytic has no effect as the noise is caused by rapid di/dt and the PCB track inductance.

Mechanical contacts from switches, relays, etc., do not make instantly but 'bounce' rapidly for 1 to 4 mS due to dirt and the uneven contact surfaces. In many combinational logic systems this does not matter, but where counting, sequencing or arithmetic circuits are used, trouble can ensue.

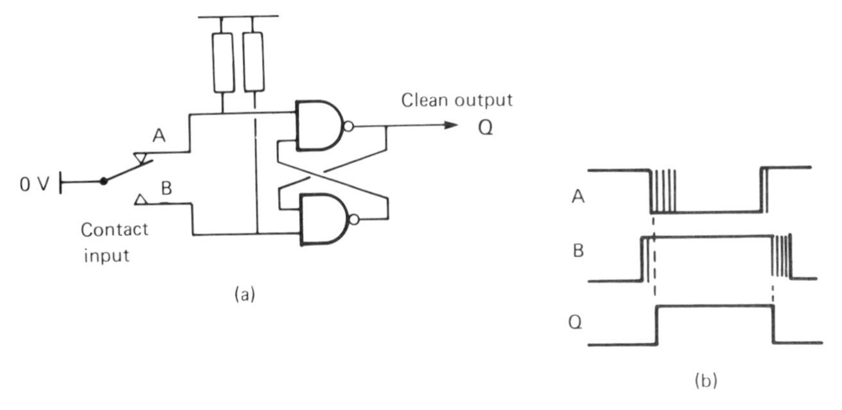

Fig.3.59 Bounce removing flip flop. (a) Circuit diagram. (b) Typical waveform.

Contact bounce can be removed by RC filters, but the best solution is to use a bounce removing flip flop, as in fig.3.59. Provided break before make contacts are used, the circuit gives totally bounce-free true and complement outputs. If the contacts are some distance from the digital system, opto isolation should, of course, be used before the flip flop.

Chapter 4
Computers in control

4.1. Introduction

Until quite recently, a process control system would probably consist of an ad hoc collection of controllers, amplifiers, relays, indicators and recorders. These would be engineered into a total control package, and inevitably each and every system would be a unique design.

The digital computer provides the control engineer with a universal device that can perform most control operations with minimal effort. The same computer can log oil flow on a pipeline, control a steel furnace or manufacture chocolates. The only major difference between these systems will be the instructions written for the computer. The advantages of this are many: standard systems are cheap to manufacture and quick to build, spares holding costs are reduced, and commissioning time is reduced, to name just a few.

Commercial digital computers have been available since about 1950, but early machines were large, expensive and too temperamental for industrial plants operating 24 hours a day. The first industrial application was not until 1959 when a computer was used for logging purposes in an American oil refinery.

The evolution of the transistor, integrated circuits and the microprocessor has brought about a dramatic fall in the cost of computers, and an equally rapid (and important) rise in reliability. The cost of a small computer is now similar to a few conventional controllers, making computers cost effective even for simple control systems.

4.2. Fundamentals

4.2.1. Computer architecture

It is not necessary to have extensive knowledge of electronics to appreciate how a computer operates. For most purposes, the computer can be considered as a collection of functional boxes. Surprisingly, all computers have the same basic 'architecture', shown in its simplest form in fig.4.1.

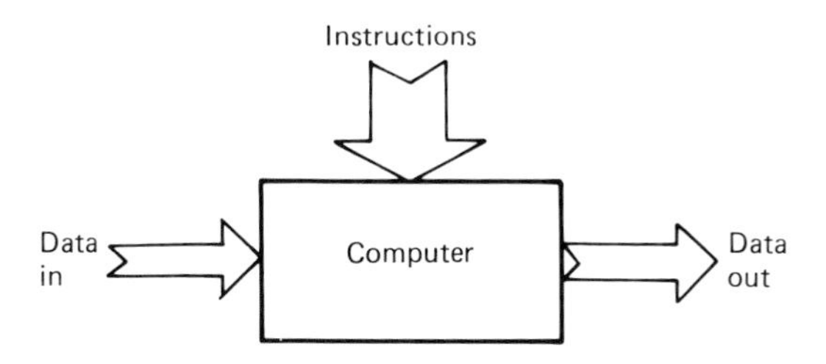

Fig.4.1 Simple representation of a computer.

A computer is a manipulator of 'data', a term which covers all forms of information. A computer accepts input data, performs operations on this data according to a set of instructions predefined by the user, and produces new output data. In a salary calculation, for example, the input data would be employees' hours worked and the output data salary slips.

In industrial control, the input data is information about the plant from transducers, limit switches and similar devices. It should be noted that the operator's controls are input data, not instructions. The instructions are provided by the designer and relate the output data to input data. Typical output data will be settings for modulating valves, on/off signals for contactors and solenoid valves, control panel mimic displays, visual display units (VDUs) and printers.

Figure 4.2 is an expanded version of the simple block diagram of fig.4.1 and is as detailed as most computer users need. Essentially, a computer can be considered as six interconnected blocks. It is interesting to note that an English engineer, Charles Babbage, identified these fundamental computer components in the early nineteenth century, but sadly lacked the technology to turn the theory into a practical machine.

Internally a computer works solely with binary numbers, although the user does not need to be aware of this. The input

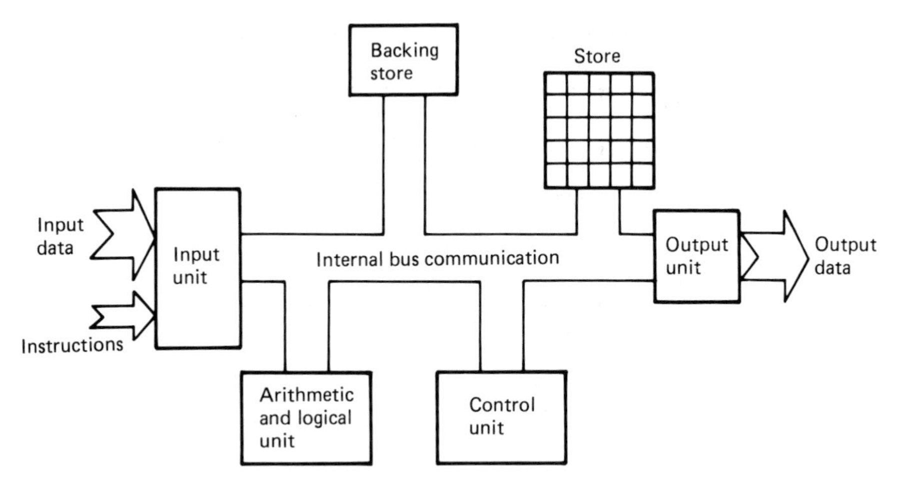

Fig.4.2 Computer architecture.

unit takes data from the outside world (currents, voltages, on/off signals) and converts it to a binary form suitable for use by the computer. It therefore consists of analog scanners and optical isolators similar to those described in sections 3.9 and 3.10.

Input (and output) signals are usually identified by an address, so instructions can be written in the form:

> 'Take the analog signal on input 27, find the square root and put the result to output 56.'

Input 27 and output 56 uniquely identify the signals.

Obviously the output unit takes binary data from the computer and converts it to a form suitable for use by the plant. It will consist of DACs and output optical isolators. The input and output units are collectively known as the I/O of the computer. Additional common I/O devices are VDUs, printers and keyboards. These are known usually as peripheral devices.

The store of the computer (often called the memory) can be considered as an array of pigeonholes similar to fig.4.3. Each pigeonhole is identified by a unique address and can store one number (or one character). Data can be written to, and subsequently retrieved from, a store location. The store is used to hold the computer's program (i.e. its instructions) and for temporary storage of values during calculations. The store does not differentiate between instructions and data; both are held in the same form. The machine's control unit, described below,

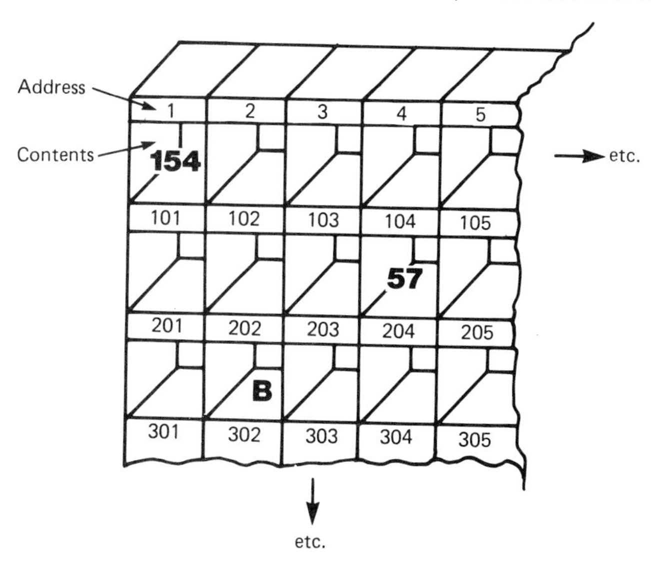

Fig.4.3 The store represented as pigeon holes. Location 104 contains 57 and location 202 contains 'B'.

retrieves instructions to be obeyed and keeps track of where in the store they are held.

There are several 'jargon' words concerned with computer storage. The number of store locations determines the size of program and the number of values that can be used by a computer. Store size is usually measured in 'K' which represents 1024, rather than the usual 1000. This odd value arises because computers work in binary, and 1024 is ten binary bits. A small microprocessor based system could have as little as 4K memory. A minicomputer could have more than 128K.

Most modern computer stores lose their contents if the voltage supply is removed. They are said to be volatile. This is clearly undesirable for commercial applications, particularly in heavy industries where unpredictable supply interruptions are common. One solution is the use of batteries to keep power on the memory boards during power loss. Such memories are said to be battery backed, or battery supported.

Permanent storage of programs can be achieved by the use of memory locations in which the contents are 'fixed' at the manufacturing stage, and cannot be subsequently altered. These memories are known as read only memories, or ROMs. As far as the computer is concerned, a ROM is simply part of the store.

ROMs can also be used to hold data that is not subject to change (e.g. conversion tables for linearisation of thermocouple signals).

The store of fig.4.3 is called a random access memory, or RAM, a term which often causes confusion. The computer can access each and every store location with the same speed. Compare this with data stored on, say a cassette tape. The time taken to access data on the tape is not constant, but depends both on the current position and the data position on the tape. Storage systems such as a magnetic tape, where the access time is variable, are called serially accessed stores. Storage systems such as fig.4.3, where the access time is constant, are called random access stores.

A computer is required to perform arithmetical and logical (AND/OR) operations on data. This is achieved by the block called the arithmetic and logic unit, or ALU. Operating under the instructions of the program, this will take data from a specified store location, perform a specified arithmetic operation and place the result back in some specified store location. An instruction sequence, for example, could be:

> 'Take the number in store location 3220, add the number in location 4057 and put the result in locations 6633.'

The amount of data that can be stored in the RAM is rather limited, even with the 128K plus memory of a minicomputer. Industrial control computers often require large amounts of storage for, say, storing history records or standard applications. This need is met by the backing storage of fig.4.2. This can provide vast storage, equivalent to over five million store locations in some systems, and is usually based on magnetic disks or tape.

The drawback of backing storage is a relatively slow speed. It is possible to store, or retrieve, data from any RAM location in under 1 μS. Typical figures for magnetic disks are around 10 mS and possibly several seconds for magnetic tape. This is not often a problem because access to the backing storage is only required at infrequent intervals.

Backing storage in industrial systems is usually provided by floppy disks or Winchester disks. Both record data magnetically on to concentric tracks on a disk which is spun under a movable read/write head. Floppy disks use a thin flexible disk (hence the name), Winchesters a rigid disk in a totally enclosed air tight enclosure. A floppy disk physically contacts the read/write head, which can lead to reliability problems in dusty atmospheres. A Winchester, being totally sealed, is immune to dust and can hold far more data than a floppy. Winchesters are, however, considerably more expensive.

The final block of fig.4.2 is labelled 'control unit'. This takes the instructions from the store in order, and implements them by controlling the other blocks in the computer. All the other blocks work under the direction of the control unit, which itself works under the direction of the instructions in the user's program.

4.2.2. Digital representation of numbers

Almost without exception, computers work in binary for the reasons given in section 3.6. Internally a computer works with a fixed number of bits; 8 or 16 bits for microcomputers; 16 or 32 bits for minicomputers.

Conventionally, 8 bits are called a 'byte', and can represent any number from 0 to 255 or a single alpha character (i.e. a letter). Sixteen bits are called a 'word' and can represent any number from 0 to 65536 (which is 64K). Usually twos complement representation is used in practice, so 8 bits, for example, represents −128 to +127.

One store location in an 8 bit machine can hold any number in the range 0 to 255; a resolution of about 0.5%. This is not the restriction it might appear. By using two adjacent store locations as in fig.4.4a, 16 bit resolution can be achieved.

Very large, or very small, numbers can be stored using 'floating point' representation. This is analagous to representing decimal numbers as powers of ten. For example:

74057 can be written	7.4057×10^4
2.5412	2.5412×10^0
9285700000	9.2857×10^9
0.003962	3.962×10^{-3}

The numerical portion is called the mantissa and the power of ten the exponent. Floating point numbers are held as in fig.4.4b, with one (or more) locations holding the mantissa and one (or more) the exponent. The exponent represents a power of two in twos complement to allow for fractions.

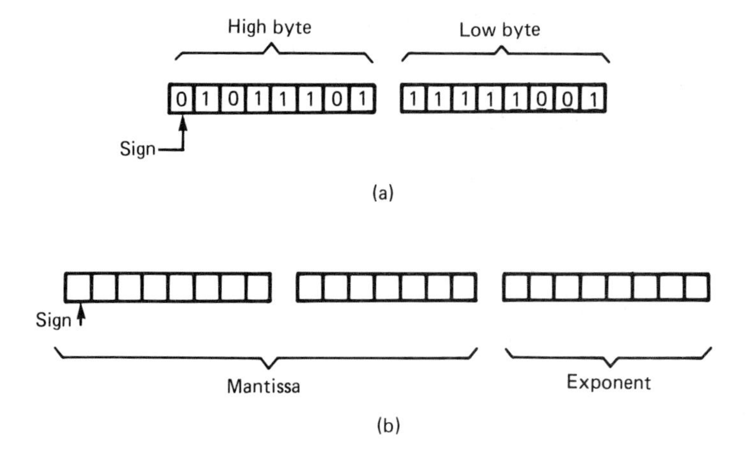

Fig.4.4 Digital representation of numbers. (a) 16 bit representation of +24057. (b) Floating point number.

4.2.3. Microprocessors

The heart of fig.4.2 is the arithmetic and logic unit (ALU) and the control unit. These two units are collectively known as the central processing unit, or CPU, a term which dates from the early valve computers.

Large scale integration (LSI) allows very complex digital ICs to be manufactured. Although it is technically feasible to make a small computer on one IC, it is not economically viable as each computer requires different RAM, I/O, etc. The microprocessor–the CPU of a small computer–was developed instead.

It is very important to realise that a microprocessor is *not* a computer. It requires a power supply, ROM/RAM and I/O to become useful. A computer using a microprocessor is called a microcomputer.

Early microprocessors worked with 8 bit data (one byte) and could address 64K of ROM/RAM. The commonest of these are:

Z80 manufactured by Zilog
6502 manufactured by MOS Technology
6800 manufactured by Motorola (later improved to 6809)
8080 manufactured by Intel (upgraded to 8085)

Second generation microprocessors are based on 16 or 32 bit data, and can address upwards of 8 Mbytes of memory. Microcomputers based on these devices have a power similar to minicomputers. Common 16 bit microprocessors are:

Z-8000 manufactured by Zilog
68000 manufactured by Motorola
8086 manufactured by Intel
9900 manufactured by Texas Instruments

Figure 4.5 shows typical block diagrams for a microcomputer and is, in effect, a rearranged version of fig.4.2 with more detail added. The first thing that is apparent is that all the items are interconnected by three highways (sometimes called buses) which are controlled by the microprocessor.

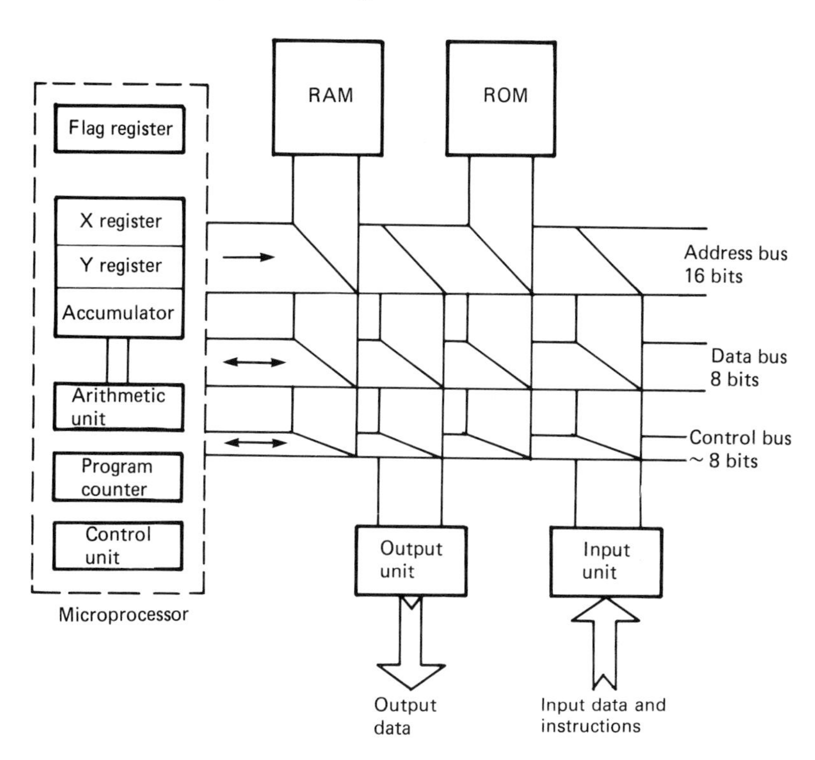

Fig.4.5 Microcomputer architecture.

The store locations and I/O all have addresses (get the number from store location 6633 and sent it to the output device with address 27). The address bus carries the address of the store location or I/O device being used. A typical 8 bit microprocessor can address 64K locations, for which 16 bits are needed.

The data bus carries data between the store, I/O, and the microprocessor itself. Data can pass between all units so a bidirectional bus is required (unlike the address bus which operates solely *from* the microprocessor *to* the store and I/O). Obviously the data bus for an 8 bit microprocessor will have eight lines.

The final bus is the control bus. This carries timing signals, and commands to show the direction of signals on the data bus and to say whether an address refers to a store location or I/O device. There are typically eight control signals.

The microprocesor shown in fig.4.5 is a simplified version of the 6502. A, X, Y are three storage locations called registers each of which can hold one 8 bit number. These can be accessed faster than a store location, however, and A (for accumulator) is used in all arithmetic operations. Most arithmetic takes the form:

> 'ADD contents of store location to contents of A; result going back to A.'

or

> 'SUBTRACT contents of store location from contents of A, result going back to A.'

F, for flag register, indicates whether the last instruction left a positive, negative or zero number in the accumulator. This is used by conditional jump instructions described in the next section.

The PC, for program counter, indicates the address of the next instruction. Normally instructions are held sequentially in the store, so on completion of an instruction the PC is simply incremented. A jump instruction, described in the next section, overules this incrementing and puts a new value in the PC.

The control circuit sequences the operation of the machine. For each instruction this takes four steps:

(1) Bring the next instruction from the store to the control unit.
(2) Decode it (i.e. decide what it is and what store location and/or I/O is to be used).
(3) Move data to execute the instruction.
(4) Set up PC for the next instruction.

The control unit is responsible for the signals in the control bus, and for determining whether the address bus contains an address for data or an address for the next instruction.

4.2.4. Computer instructions

There are surprisingly few instructions that a computer recognises. Most instructions are variations on:

(1) FETCH a number from a specified store location to a register.
(2) STORE a number from a register to a specified store location.
(3) ADD a number from a specified store location or register to A; result to A.
(4) SUBTRACT a number held in a specified store location or register from A; result to A.
(5) INPUT a number from a specified input device to a register.
(6) OUTPUT a number from a register to a specified output device.
(7) JUMP–the next instruction is not sequentially after the current instruction, but at some specified location; instructions continue from the new location.
(8) CONDITIONAL JUMP (positive/zero/negative)–this operates as a jump instruction IF the last instruction gave the specified result. The conditional jump, often called a branch, allows the program to take different routes. The flag register in fig.4.5, is used by the control unit for the tests.

These are called the machine code instructions of a computer. Programming with these is tedious and error prone. Most computers are programmed in high level languages, a topic discussed further in section 4.5.2.

4.2.5. Types of computer

Although all computers are based on the block diagram of fig.4.2, different applications require slightly different capabilities. Computers are generally classified as follows.

The largest machines, used by government departments, banks, insurance companies, etc., are called mainframes. These are generally used to process large amounts of data and consequently require large backing storage. They are rarely encountered in industrial control.

Minicomputers, such as the DEC PDP 11 of fig.4.6, are generally the largest machines used for control purposes. These are usually 16 or 32 bit machines with about 128K of RAM and 5 Mbytes of disk storage. Typical applications would be supervision of a large process where history recording is required.

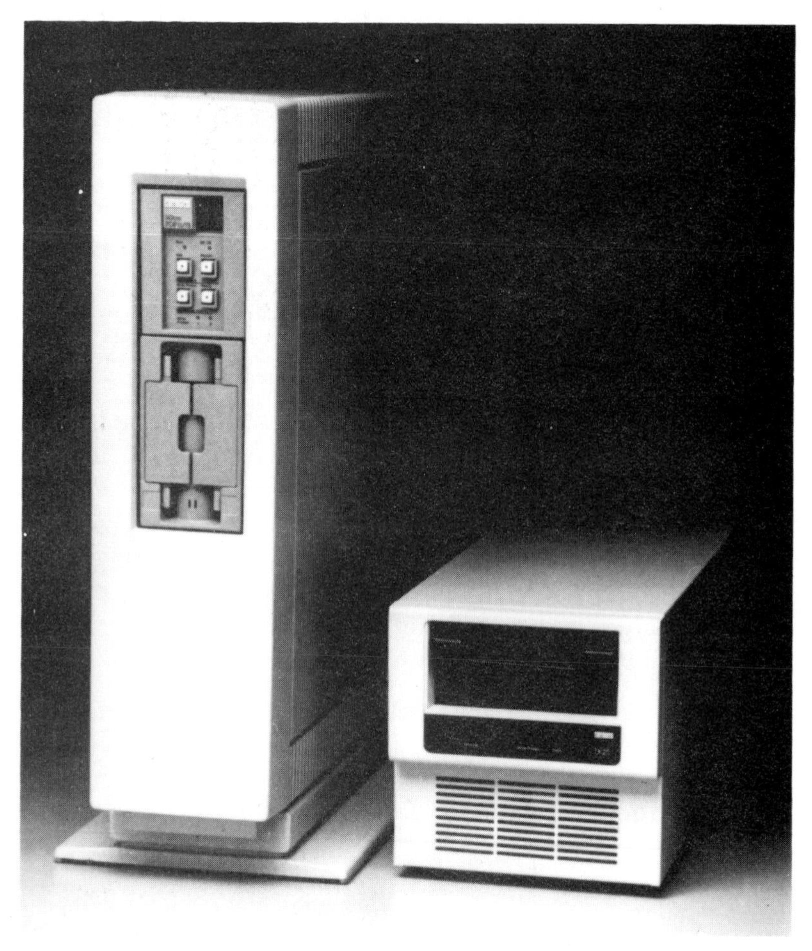

(a)

(b)

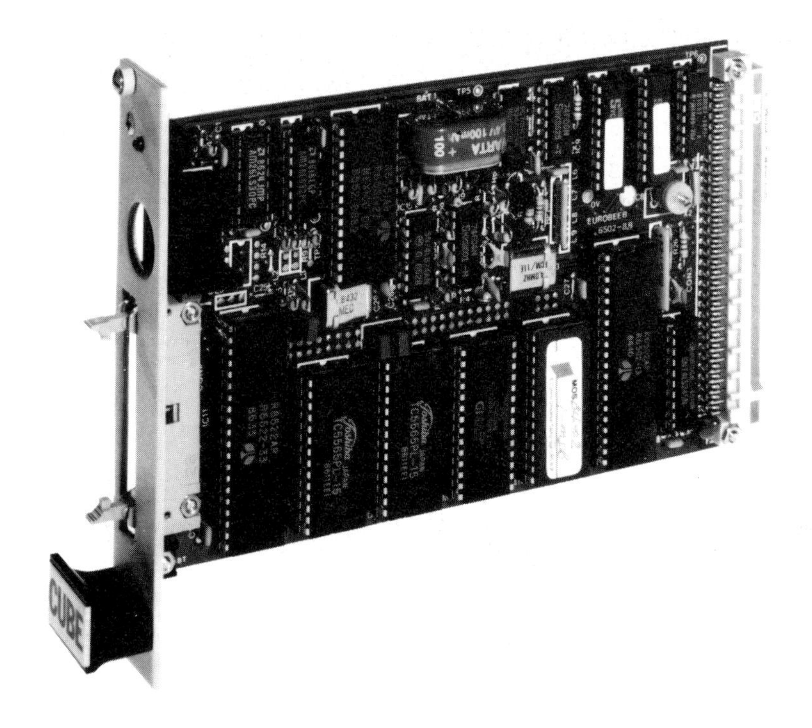

(c)

Fig.4.6 Various computers for industrial control. (a) The DEC PDP 11/73 computer from Digital Equipment with TK25 cartridge tape streamer for disk backup. An early DEC machine can also be seen in fig.4.31. (Photo courtesy of Digital Equipment.) (b) Portable personal computer ruggedised for industrial use. (Photo courtesy of Rochester Instrument Systems.) (c) A typical single board computer, the Eurobeeb from Control Universal of Cambridge. This is based on the popular BBC microcomputer and programmed in BASIC with extensions to the language for control functions. (Photo courtesy of Control Universal.)

The commonest control computer is based on the microprocessor, and is consequently called a microcomputer. This is physically smaller than the minicomputer, often occupying no more than a single 19 inch rack. Microcomputers are specifically designed for industrial control, one example of which is the GEC GEM-80 of fig.4.7. Desk top computers, sometimes called personal computers, are invariably microcomputers. A typical microcomputer will be an 8 or 16 bit machine with between 4K and 64K RAM and possibly 0.5 Mbytes of disk storage. Single board computers are cheap and easy to incorporate into other equipment (e.g. machine tools, vending machines).

(a)

(b)

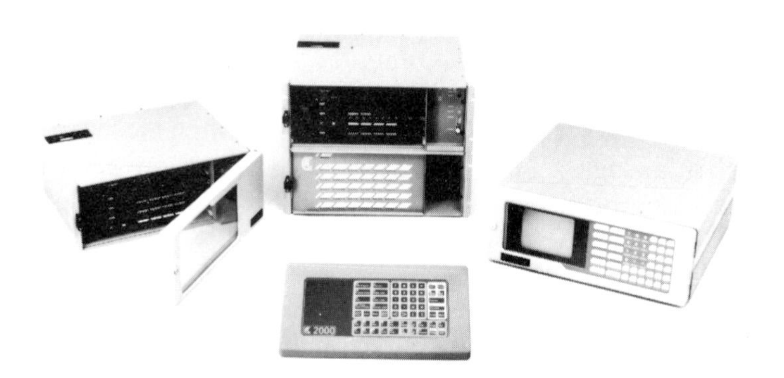

(c)

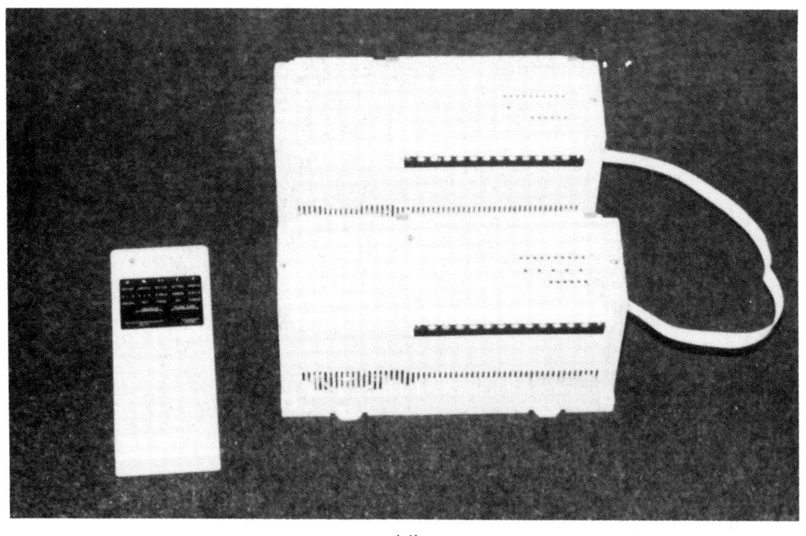

(d)

Fig.4.7 Various programmable controllers. (a) The GEC GEM-80 family of programmable controllers, with programmer, membrane keyboard, VDU monitor, decode switch input and LCD unit. (Photo courtesy of GEC Industrial Controls.) (b) Allen Bradley PLC 2/30. (c) Control & Readout's CRL 1000 and 2000 controllers mounted in IP55 enclosures with membrane keyboard for operator interface in difficult environments. The CRL range is particularly well suited for analog control. (Photo courtesy of Control & Readout.) (d) A small cheap micro PLC; the Allen Bradley SLC 100 with expansion unit and hand held programmer. The set up shown can handle 20 inputs and 10 outputs.

The smallest computers are known as dedicated computers, and are only to be found in control applications. It is possible to manufacture an entire computer on one IC, complete with CPU, ROM program, RAM and I/O, but it is only economically viable when many identical devices are manufactured. Sufficiently large production runs exist for intelligent instruments, robots, and control units for domestic washing machines and similar devices. These are commonly constructed around a dedicated computer IC made specifically for the application.

The following section of this chapter looks at the use of computers for industrial control.

4.3. Classification of industrial computers

4.3.1. Introduction

It is possible to identify five distinct industrial applications of computers: logging, sequencing, supervisory systems, direct digital control (DDC), and dedicated computers. Most real-life systems, of course, cannot be neatly labelled and will combine two or more of these functions. It is, however, useful to consider each classification separately.

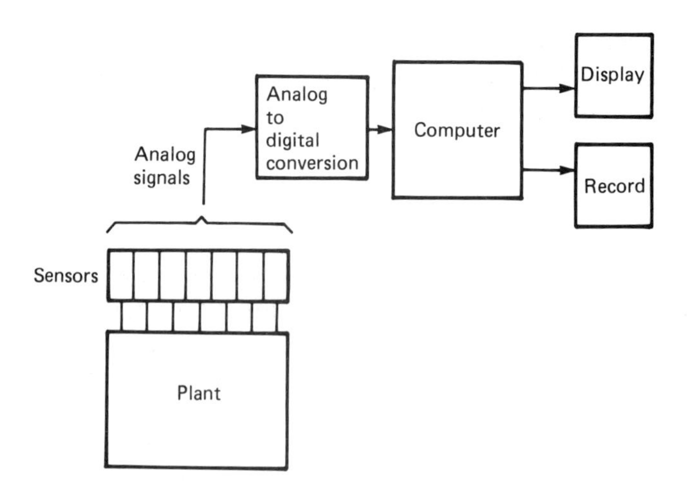

Fig.4.8 Computer logging system.

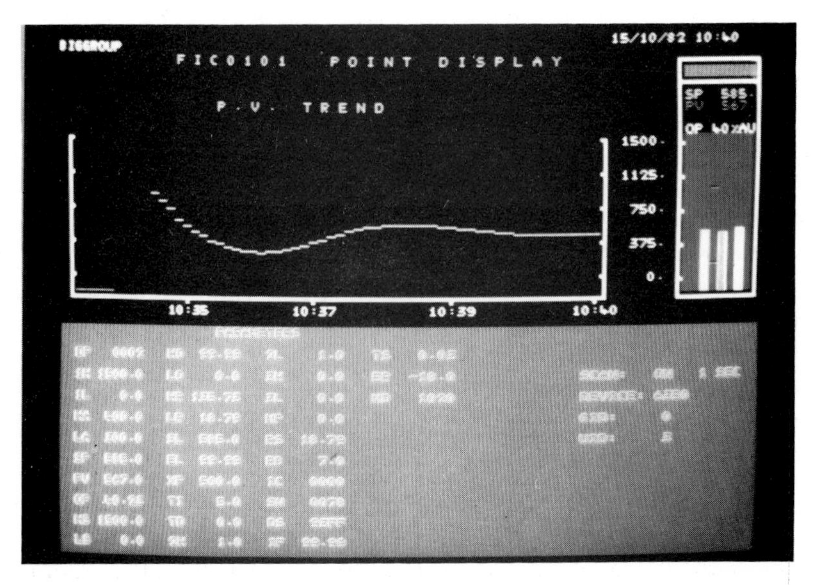

Fig.4.9 A VDU trend display, showing the history of a process variable. The block at the right simulates a controller front panel, with PV, SP and controller output. The lower half of the screen is the controller tuning constants. (Photo courtesy of TCS Ltd, Worthing.)

4.3.2. Logging systems

The simplest computer systems are used for data logging, shown schematically in fig.4.8. A logging system has no control over the plant and is used purely to obtain information from the plant and display it to the operator via displays similar to fig.4.9, or record it for historical purposes. Alarm logging is similar but has the additional facility of checking plant conditions against preset limits and alarming any deviations.

4.3.3. Sequencing systems

Many computer systems are not concerned with analog values but are used to sequence on/off operation with devices such as solenoids, contactors and limit switches. A typical example could be controlling the purge/ignition/run sequence of a gas burner, or starting a series of interconnected conveyors.

Such systems would be conventionally controlled by a cubicle of relays or a hardwired logic system based on the elements described in chapter 3. A computer becomes cost effective for a surprisingly small system; with a simple programmable controller the breakeven point can be as low as six to eight relays. A computer or programmable controller will bring other advantages such as simplicity of installation and ease of maintenance.

A sequencing computer is used to replace relay or TTL/CMOS logic systems. A sequencing computer does not usually work with analog values, but with on/off signals such as limit switches, push buttons, motor contactors and similar devices. A sequencing system is usually implemented via a programmable controller (or PC–not to be confused with the personal computer, which uses the same initials).

4.3.4. Supervisory systems

A common concern about the use of computers for industrial control is the fact that a simple computer failure can lead to a total plant shutdown or an unsafe condition. The effects of failure can be reduced by the use of the computer as a supervisory system similar to fig.4.10. Closed loop control on the plant is achieved by conventional three term controllers, but the set

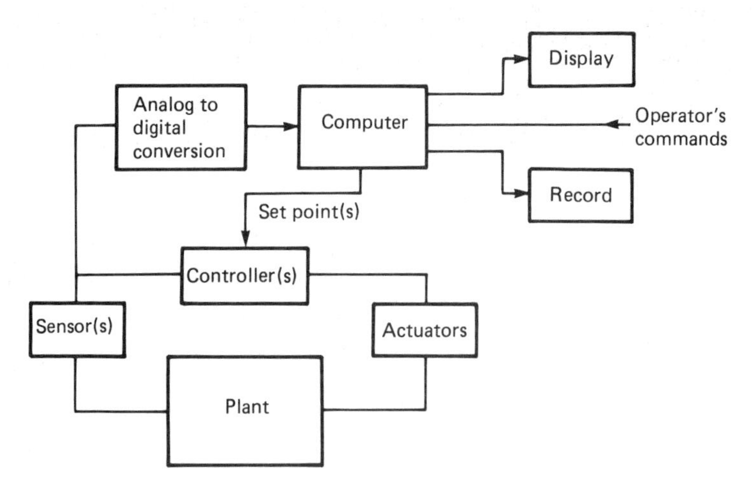

Fig.4.10 Supervisory computer system.

points are set automatically by the supervisory computer which also performs data and alarm logging functions.

There are effectively three possible control modes with a supervisory system: full computer control, semiautomatic with the operator directly setting the controller set points, and fully manual. The last two modes do not require the computer to be operational. Early computers were rather unreliable so the first industrial computer systems were of this type.

There are many applications for supervisory systems. A typical petrochemical plant could have over 500 loops, and it would be tedious for an operator to monitor 500 controllers. In addition many of these loops will interact and be interdependent, so a simple change in plant throughout would require changes in many controller setpoints.

A supervisory computer would provide the operator with centralised control and display plant conditions via VDUs similar to fig.4.11. Setpoints can be changed from the keyboard, and the operator can call for a change in, say, plant throughput and leave the computer to modify all the setpoints accordingly.

Supervisory systems are also used to provide feedforward and take setpoints through some predetermined pattern for batch processes. Many chemical batch processes, for example, require a temperature cycle with controlled rate of rise and controlled rate of fall. This can easily be achieved with a supervisory computer system.

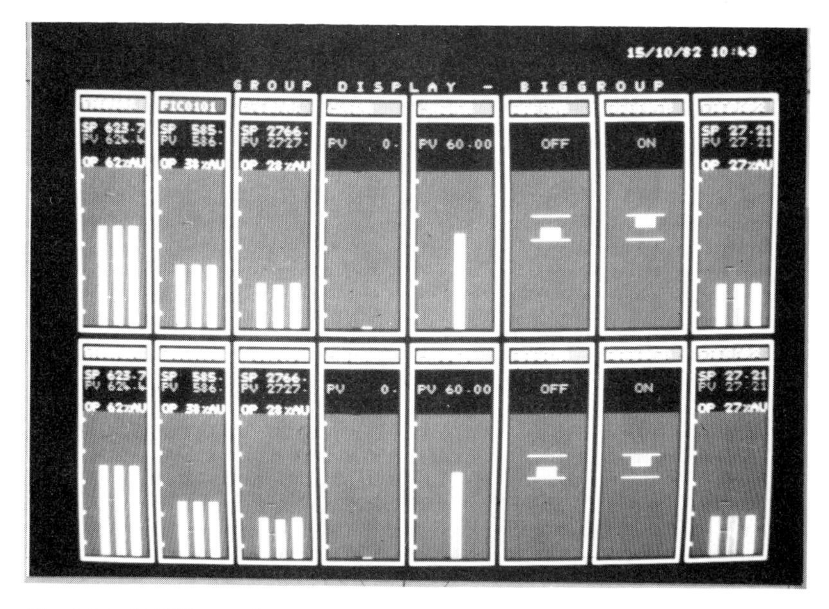

Fig.4.11 A VDU overview page simulating a bank of conventional controllers. (Photo courtesy of TCS Ltd, Worthing.)

4.3.5. Direct digital control

Modern computers are very reliable, so it is now common to dispense with controllers entirely and use the computer itself to perform the three term control algorithm. This is known as direct digital control, or DDC, and is shown diagrammatically in fig.4.12.

Plant variables are converted to a digital form by an analog scanner and ADC (see section 3.9).

Calculations of the error and the three term control algorithm are then performed by the computer program and a digital value produced for the actuator setting. This is converted to an analog voltage or current by a DAC to control the actuator.

Although figs.4.10 and 4.12 fulfil the same purpose, it is important to appreciate the fundamental difference between them. In a supervisory system, the plant can be operated without the computer, albeit with difficulty and at reduced efficiency. With direct digital control, the computer is an essential part of the control scheme and the plant cannot be operated without it. DDC also has safety implications, as a computer failure can leave the plant uncontrolled or with actuators suddenly fully open or shut.

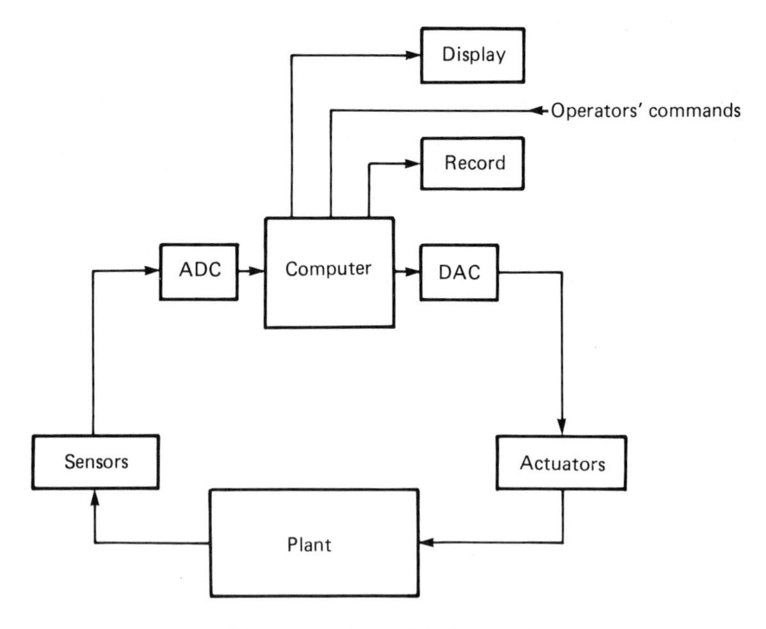

Fig.4.12 Direct digital control.

Such failures are rare, but safety should be carefully considered at the design stage.

4.3.6. Dedicated computers

The final class of computer is often not recognised for what it is. It is possible to produce a complete computer (complete with CPU, RAM, I/O and program held in ROM) in a single IC, but this approach is only economically viable for large production runs as the computer will only be suitable for one specific application.

The previous types of industrial control computer can be modified and reprogrammed after installation, and are known as general purpose machines. The computer on an IC is known as a dedicated computer because it can only perform one task which cannot be changed after manufacture.

Dedicated computers are to be found domestically in games and 'white goods' such as washing machines or cookers. In industry, they are appearing in applications as diverse as intelligent transducers, instruments and simple devices such as automatic grease dispensers for bearings.

4.3.7. Distributed systems

Early process control systems were for reasons of cost usually built around a single large computer. The advent of cheap mini and microcomputers has lead to the development of distributed multicomputer systems comprising many small machines.

A typical example, fig.4.13, has four 'levels'. The lowest level is concerned with direct plant interfacing and consists of

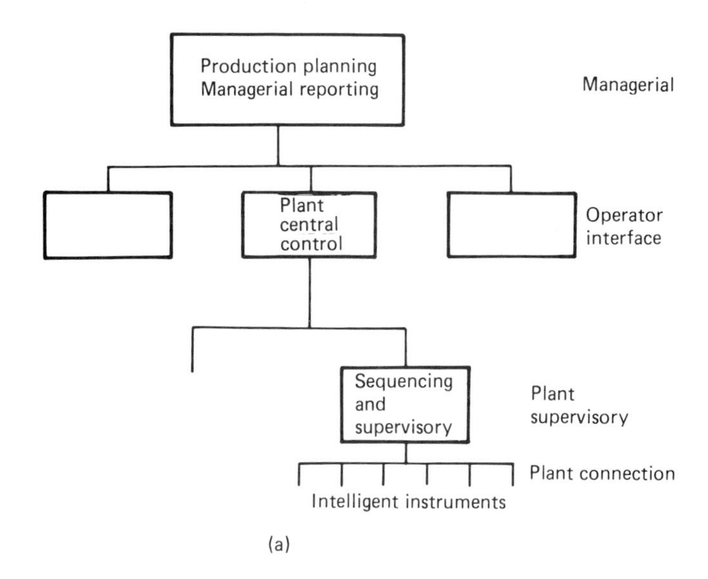

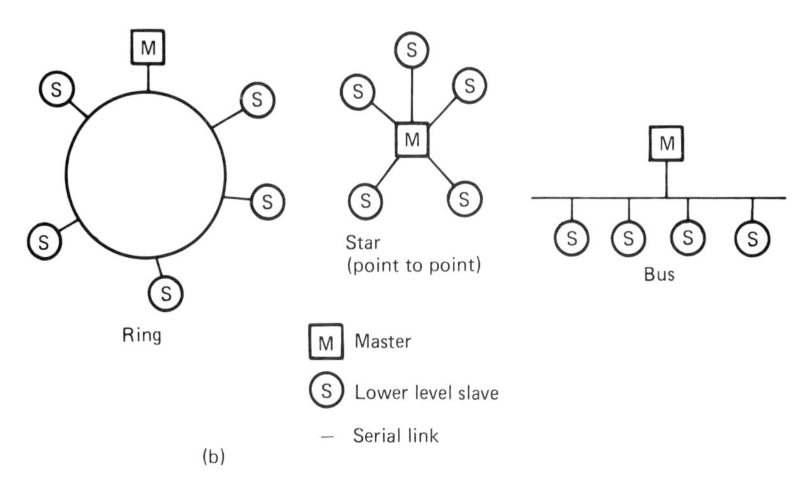

Fig.4.13 Distributed computer system. (a) Four level computer hierarchy. (b) Types of network.

dedicated machines–intelligent instruments, single loop controllers and similar devices. The next level is concerned with supervising control sequencing and multiloop DDC. Operator interfacing and displays come at the third level, along with history recording. The final level in our example is a production planning and control machine, but higher levels to a site mainframe, or even other sites, are obviously possible.

There are many advantages to distributed systems. The more notable are:

(1) The system is conceptually simpler, and as such is easier to design, commission and maintain.
(2) A correctly designed distributed system can tolerate, for short periods, failure of individual machines, and is unlikely to suffer a total failure. Reliability is therefore increased.
(3) Small computers are cheap; cabling and installation are expensive. A distributed system is linked by cheap two or four core serial link cables and can show distinct cost savings where a plant is physically large.
(4) Distributed systems often bring an increase in performance, because lower level machines take work off higher machines. On a single computer system, polling of inputs can be a very time consuming exercise. In fig.4.14, for example, part of a plant consists of four conveyors which have to be started (and stopped) in sequence with timed audible warnings, and be monitored whilst running for belt skew, blockage or spillage.

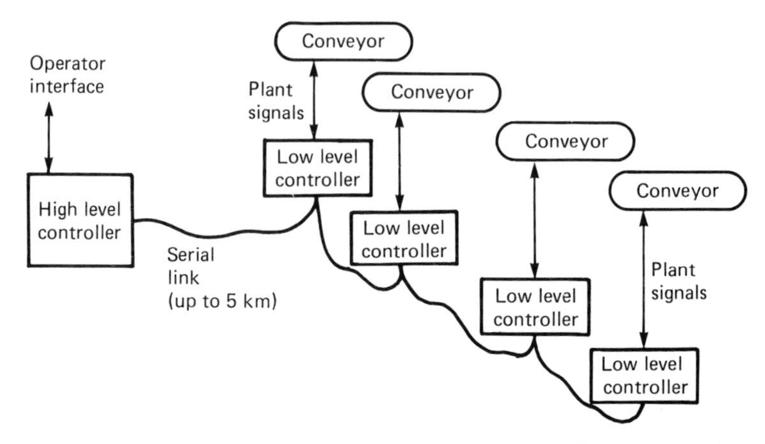

Fig.4.14 A typical distributed system. Low level controllers accept a run command from high level controller, and handle sequence without further high level action. Successful completion of faults or reported back to higher level.

The conveyor task is controlled by a small machine (probably a programmable controller) which accepts a RUN command from a higher machine, and signals 'Sequence OK' or 'Fault' back.

4.4. Advantages of computer control

4.4.1. Introduction

Any control system goes through four distinct stages from conception to completion. These stages are effectively sequential and do not overlap, so the duration of the project is the total time taken for the stages. It is interesting to compare a conventional non-computer system and a computer system at each stage.

4.4.2. Design

The first stage in any project is obviously design, during which the plant is studied and the various control strategies identified. With conventional schemes, the design has to be taken to near completion before construction can commence, as individual controllers need to be ordered.

With a computer based system, all that needs to be specified prior to construction is the size and type of computer and the I/O requirements (so many inputs, so many outputs). The design of the system is, in reality, the writing of the computer program. This can proceed in parallel with the construction of the 'hardware' of the computer, thereby reducing the overall project time. A useful rule of thumb is to double the postulated I/O capacity for a well understood, well documented project, and quadruple it if the specifications are rather hazy. I/O is cheap!

4.4.3. Construction

With conventional control schemes, every job has to be individually constructed and wired. This is inevitably labour intensive, time consuming and expensive. Apart from special control desks, assembling a computer system is largely a matter of bolting together standard items.

There are obvious cost savings in using VDUs in place of purpose built mimic panels, and in sequencing and control by program rather than relays and controllers. A computer system is invariably smaller and tidier than a comparable conventional system.

4.4.4. Installation

One of the most expensive items in control schemes is plant cabling. This can be reduced by the use of serial links (see section 4.3.7.) to convey data between plant areas. By using remote 'data gathering' units (often called remote I/O) as in fig.4.15, the majority of long (and expensive) multicore cable runs can be replaced by relatively cheap twisted pairs.

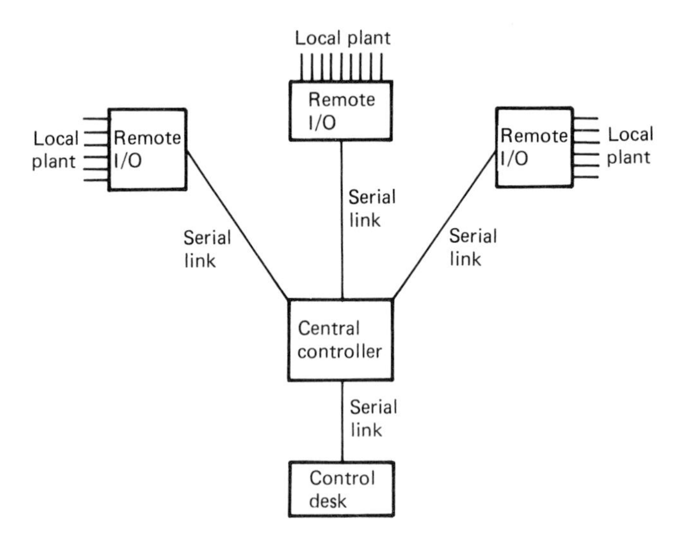

Fig.4.15 Reduction of cabling costs with distributed system.

Control desks can also be connected via serial links which allow complete control consoles to be totally tested prior to despatch from the manufacturer. All that is required for on site installation is a supply for the remote I/O unit, serial link cable and video cables for the VDUs. Installation becomes a matter of days rather than weeks.

4.4.5. Commissioning

It is very rare for any complex system to work correctly the first time. Inevitably there will be design errors, and these will be found and corrected at the commissioning stage.

Correcting errors in a non-computer system can be difficult and messy. Supose during commissioning it is found that cascade control is required to give fast response on a flow control valve. An additional controller will be required, the control panel must be modified to take the controller, and cabling and instrument air piping must be installed.

With a computer system, all that would be required is a simple program change to incorporate an additional PID routine, and minor alterations to a VDU format. Commissioning is therefore easier and takes less time.

Documentation often suffers during commissioning, as design errors are corrected but not recorded. Often the commissioning engineer's modifications are hidden for years, and are usually rediscovered by a harassed maintenance engineer working on a fault in the early hours of the morning. Computer programs are inherently self-documenting, so the documentation should always be up to date.

4.4.6. Maintenance

Computer systems are surprisingly easy to maintain, and an intimate knowledge of computers is not necessary for the fault finding engineer. Computers, being constructed of solid state devices, rarely fail, so well over 90% of faults are caused by devices external to the machine itself–plant mounted equipment and peripherals such as VDUs and printers.

Almost all industrial computer systems can be represented by fig.4.16, and incorporate extensive facilities for monitoring input and output signals. Digital I/O, for example, is easily monitored by LEDs similar to those in fig.4.17. Such features allow the system to be observed and checked by relatively non-technical personnel.

Self-diagnostic facilities are also often provided. At regular intervals (typically 0.5 seconds) the computer checks its own operation by performing tests on its CPU, I/O and store. Failures

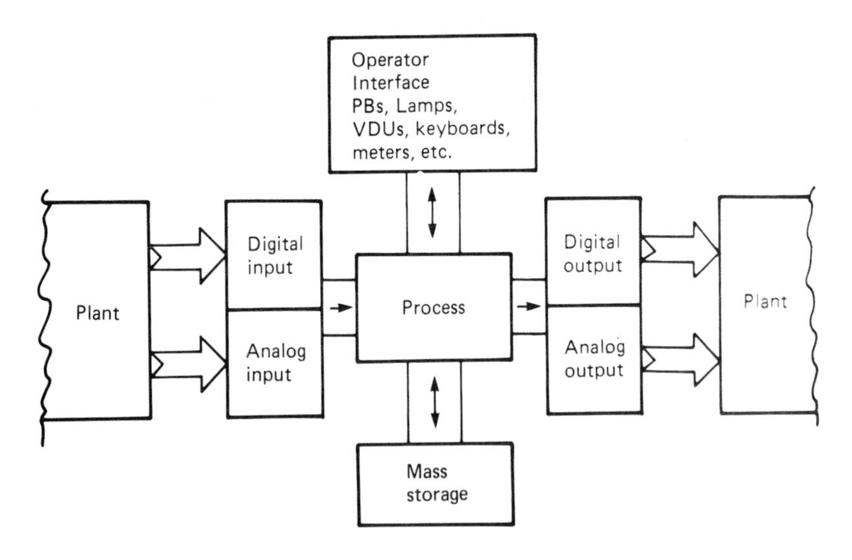

Fig.4.16 Block diagram of a process control computer.

Fig.4.17 Allen Bradley PLC I/O cards. Note the monitoring indication showing the status of plant signals.

produce error messages indicating the fault and also, in many cases, the board to be changed to rectify the fault.

An often overlooked requirement for fault finding on computer systems is the need for an understanding of the plant operation (rather than a detailed electronics knowledge of the computer). If the maintenance engineer does not appreciate the purpose of, say, a pressure regulating damper in the 'system' and what transducers and actuators are related to it, he cannot be expected to have much success in rectifying a fault which causes the damper to suddenly drive fully open.

4.5. Programming for industrial control

4.5.1. Introduction

The design of a computer system is largely concerned with writing the instructions for the computer to follow–usually called a program.

In section 4.2.4 it was shown that there are actually very few individual instructions that a computer recognises. It is therefore very tedious and time consuming to write programs in the actual 'machine code' that the computer uses.

Intuitively we wish to give the computer instructions in an English-like form (e.g. 'open valve 27 until pressure reaches 150 psi') or via process type drawings. This can be achieved by the use of a high level language, which allows the user to write programs in a much more convenient form than machine code.

The key to high level language programming is a special program written (in machine code) by the computer manufacturer which translates the user's program to the equivalent machine code which can be followed by the computer. This can be achieved in two ways.

A *compiler* converts the entire user program from high level to machine code off line (i.e. whilst the computer is not controlling the plant). The resulting machine code program is then loaded and obeyed. This sequence of events is shown in fig.4.18.

An *interpreter* converts the user program to machine code one instruction at a time on line (i.e. whilst the computer is controlling the plant). The computer at all times holds the user's high level language program.

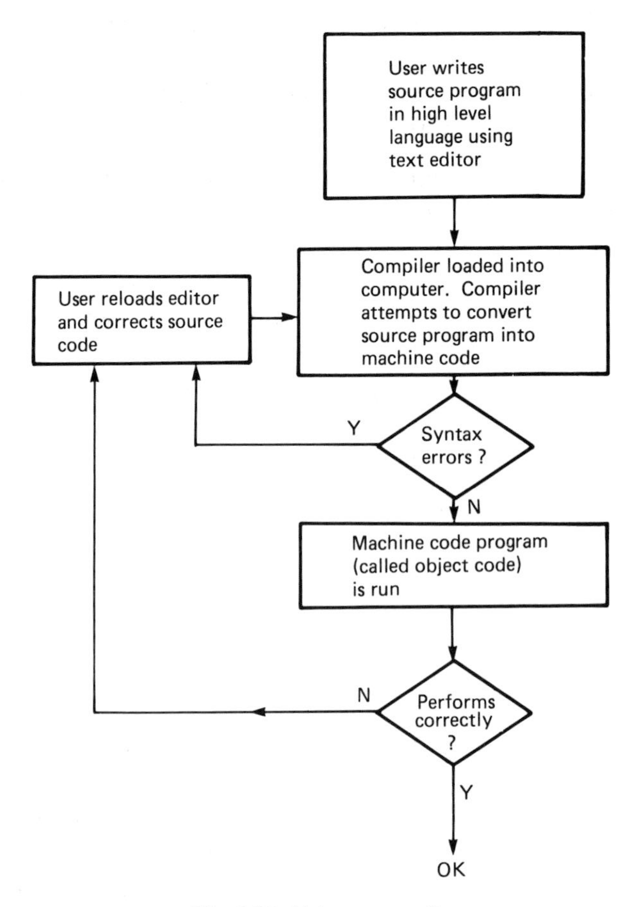

Fig.4.18 Using a compiler.

There are advantages and disadvantages to each approach. A compiler produces a machine code program which is fast and uses the minimum amount of storage. Modifications to the program are tedious and time consuming, however, and it is difficult to incorporate on line monitoring of the program operation.

An interpreted program is slower than a compiled program (by a factor of between 5 and 10) because the conversion to machine code is performed whilst the program is being obeyed. It also uses considerably more storage, as the user's program is stored in its entirety at all times (a compiler does not require the user's program after the conversion). An interpreter is, however, far easier to use, as should be apparent from fig.4.18. It is particularly easy to modify programs. Ladder diagram languages (see 4.5.5.), which are invariably interpreted, incorporate extensive facilities for monitoring program operation whilst the plant is running.

The requirements of a process control high level language can be summarised as follows:

(1) The program should be easy to write and modify and be capable of being understood by technicians with minimal computer experience.
(2) It should be sufficiently fast for real time control.
(3) It should be capable of handling a large amount of input/output; individual bits (on/off) for sequencing, and 16 bit words replace analog signals for analog ADCs and DACs.
(4) It should include extensive facilities for plant monitoring to facilitate fault finding.

The following subsections show various approaches to these ideals.

4.5.2. Conventional high level languages

The traditional high level language, BASIC, FORTRAN, Pascal, etc. were designed for general purpose computing rather than control applications, and consequently have some shortcomings (notably speed and handling of input/output operations). They are, however, widely used in other branches of computing, close to 'English' and easy to understand.

The example below illustrates how a three term control function can be implemented using a high level language. The example is based on MACBASIC, a variant of the home computer language modified for control purposes.

The three term control algorithm is:

$$OP = K\left(E + \frac{1}{T_i}\int Edt + T_d\frac{dE}{dt}\right)$$

where

- K is the gain
- E is the error (actual value − desired value)
- T_i is the integral time
- T_d is the derivative time
- OP is the output value.

This is achieved by a program which executes in a continuous loop which operates repetitively every T seconds (typically T should be in the range 0.1 to 10 seconds, according to the process being controlled).

The integral term is simulated by summing the error on each loop, since:

$$\frac{1}{T_i} \int E dt \doteq \frac{T}{T_i} \Sigma E$$

Similarly, the derivative term is approximated by:

$$\frac{\Delta E}{T} T_d$$

where ΔE is the change in error from the last scan T seconds previously.

The MACBASIC program below follows closely the BASIC language found on home computers, with a few additions. TIME is a 'built in' variable which increments in 0.01 seconds. AIN (X,Y) gets an analog value input from channel Y in card X, and AOT (L,M) similarly sends an analog output to channel M on card L. Obviously, this PID program would be part of a larger program, with K, TI, TD, T set elsewhere, and variables such as OE and SUM initialised before the program starts.

```
1000 OT = TIME: REM OT is used at line 1100
1010 SP = AIN(5,4) : REM Setpoint from operator's
          control
1020 PV = AIN(5,5) : REM Actual value of process
          variable
1030 E = PV - SP : REM Error value
1040 SUM = SUM+E : REM Integral Calculation
1050 IT = T*SUM/TI : REM Integral term. T is loop time
1060 DT = (E-OE)*TD/T : REM Derivative term
1070 OP = K*(E+IT+DT) : REM All 3 terms
1080 AOT (7,1) = OP : REM Analog output
1090 OE = E : REM Old error for next loop at line 1060
1100 WAIT UNTIL TIME - OT > = T : loop timer
1110 GOTO 1000 : Repeat loop indefinitely
```

There are a few modifications to the above program that would be necessary before it could be used in a real application– protection against indefinite increase of the integral term, for example–but it suffices to show how conventional high level languages can be used for control.

At best, though, the use of conventional languages is something of a compromise and there are specialised languages for industrial applications. RTL/2 (for real time language) is an example that was developed by the UK chemical firm ICI for their own use, but it is now generally available. Other approaches are Forth, block structures and ladder diagrams.

4.5.3. Forth

Forth is unique in that it originated as a language for control of an astronomical telescope at Kitts Peak in the USA. It is an unconventional language in many respects, but once its peculiarities are learned it is ideally suited for industrial use.

Forth uses the store as a stack to hold numbers. This is best viewed as being similar to the spring loaded piles of plates seen in cafeterias; adding a plate pushes the others down. Similarly numbers in Forth are placed at the top of the stack with previous numbers being pushed down. Figure 4.19a shows the numbers 3,5,27,2 being placed on the stack.

Arithmetic is performed on the top two numbers on the stack using the usual symbols +, −, *, /. The arithmetic symbols are written after the number that they are to operate on (being the top two numbers on the stack) so the addition 273 + 28 is written:

273 28 +

and operates as in fig.4.19b. The more involved calculation:

(412 + 27 − 16)*3

is performed as in fig.4.19c and is entered:

412 27 + 16 − 3 *

Forth is also unique in that it allows the user to define his own language, because Forth is expandable by the user. When programming any operation, it is always easier to consider it as a collection of tasks. These tasks can then be broken down into smaller subtasks, and so on until the problem is broken down into a series of small, easy-to-program jobs. In the jargon, this is known as 'top down programming'.

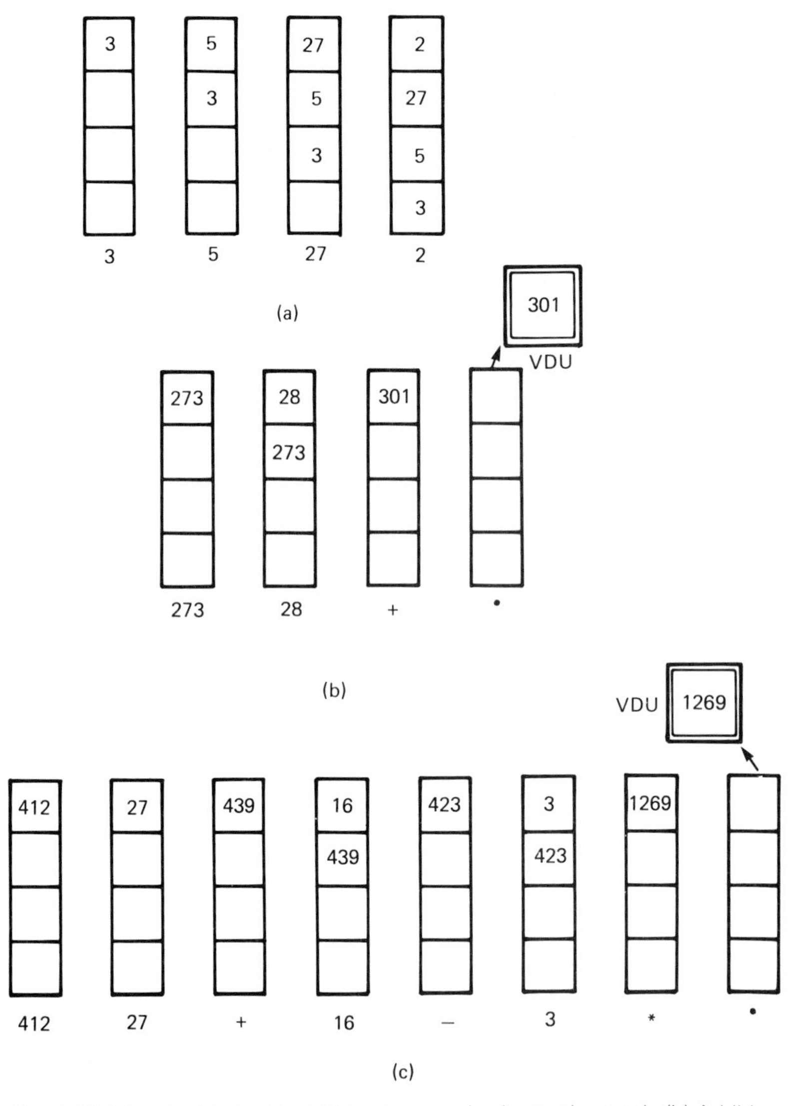

Fig.4.19 The stack in Forth. (a) Numbers pushed onto the stack. (b) Addition and printing a result. (c) An arithmetical sequence. Full stop . prints top number.

Suppose we want to control the batch process of fig.4.20, where two chemicals are added to a vat, mixed, heated to some preset temperature, mixed for some time and then drained. The whole program could be represented by the (new) Forth word BATCH, which is defined as:

```
: BATCH
ADD1 ADD2 MIX HEAT MIX2 DRAIN ;
```

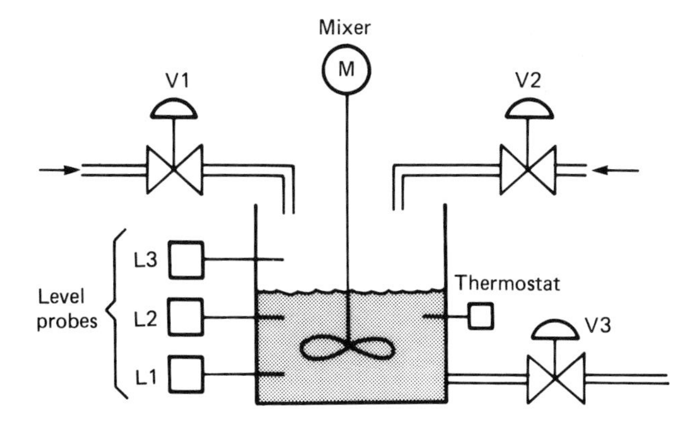

Fig.4.20 Forth control of batch process.

Each of the words ADD1, etc., are new Forth words which are themselves defined, for example:

```
: ADD1
OPENV1
BEGIN ?LI UNTIL
SHUTV1 ;
```

ADD2 is similar. MIX is defined as:

```
: MIX
MOTORON
BEGIN ?TIME UNTIL
MOTOROFF ;
```

In each definition, new words are introduced (OPENV1, ?L1 (for test float switch LI)). The breaking down process continues until each definition is done in simple Forth words. At this point, the whole program comes together, and can be run with the single word BATCH.

The example below is based on part of the control for the boiler of fig.4.21a. Exhaust gases from the boiler are used to preheat the incoming air, thereby saving fuel. A discrete temperature controller controls the air flow, and a flow controller the gas flow. The air temperature will vary, however, which will affect its density. The correct gas/air ratio will therefore need calculating to compensate for temperature changes in the combustion air. The Forth program below is part of a supervisory system to provide a setpoint for the gas flow controller.

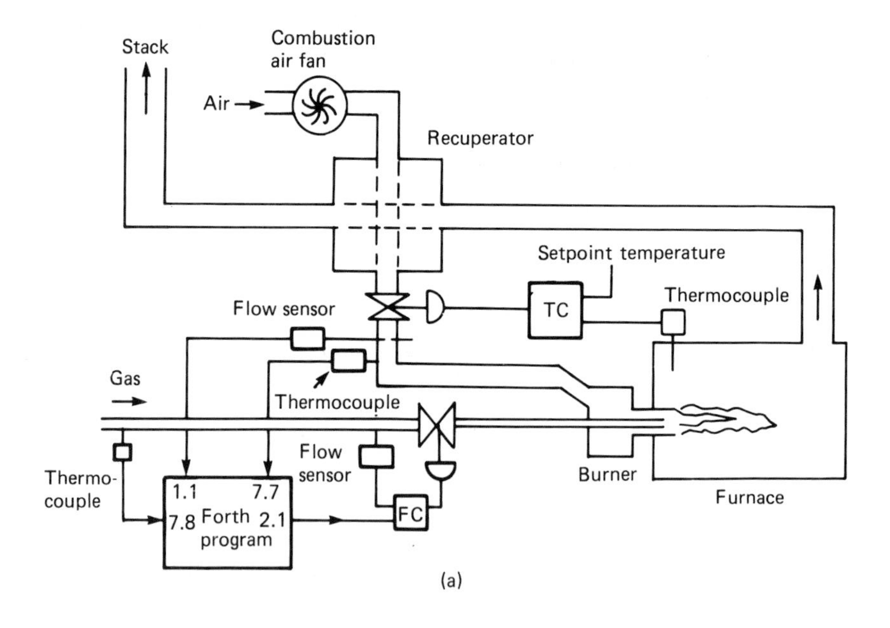

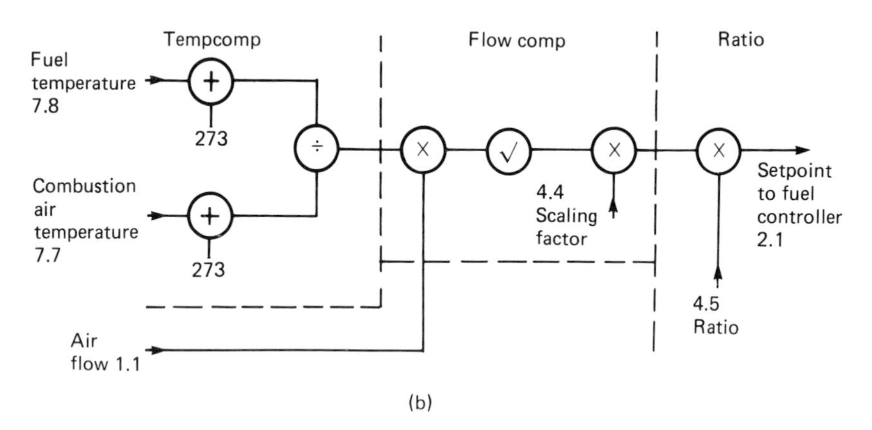

Fig.4.21 Forth program example. (a) Plant schematic. (b) Forth program.

The structure for the program is shown in fig.4.21b. There are three analog inputs on channels 1.1, 7.7, 7.8 (the first digit denoting a card, the second the input on the card–card 1 is, say, for 4 to 20 mA signals, card 7 for thermocouples). One analog output, the setpoint, is on channel 2.1. Internal variables 44, 45 (preset elsewhere) hold the stoichiometric ratio and scaling factors to correct for the different full scale ranges of the air and gas orifice plates/delta-p transmitters.

The program falls naturally into the three areas shown so our program, which we will call SETPT, is simply:

```
: SETPT
      BEGIN
            TEMPCOMP
            FLOWCOMP
            RATIO
      REPEAT ;
```

where BEGIN/REPEAT mark the start and finish of an endless loop, and TEMPCOMP, etc., are words we define as follows:

```
: TEMPCOMP
        7 8 GETAN (ambient temperature, input 78)
        273 +     (convert to absolute)
        7 7 GETAN (air temperature, input 77)
        273 +     (convert to absolute)
        / ;       (leave correction factor on stack)
: FLOWCOMP
        1 1 GETAN   (air flow, input 11)
        *           (multiply by correction factor)
        SQRT        (linearise by square root)
        4 4 GETVAR  (scaling factor for gas and air
                    orifice plates)
        * ;         (leave corrected air flow on stack)
: RATIO
        4 5 GETVAR  (ratio)
        *           (setpoint gas flow)
        2 1 ANOUT ; (output to setpoint for gas
                    flow controller)
```

This is not quite the bottom level of the program. GETAN, GETVAR, SQRT, ANOUT are not 'standard' Forth words and would need definition, probably once for the whole supervisory program on the plant of which the boiler forms a part. The program is also written 'back to front'. A Forth program starts at the bottom definition and works up, so :SETPT should follow, not precede, :TEMPCOMP, etc., which should follow :GETAN, etc. The program was written back to front to simplify explanation.

4.5.4. Block structure

High level languages such as Forth, or even Basic, require a certain amount of programming expertise. The block structure

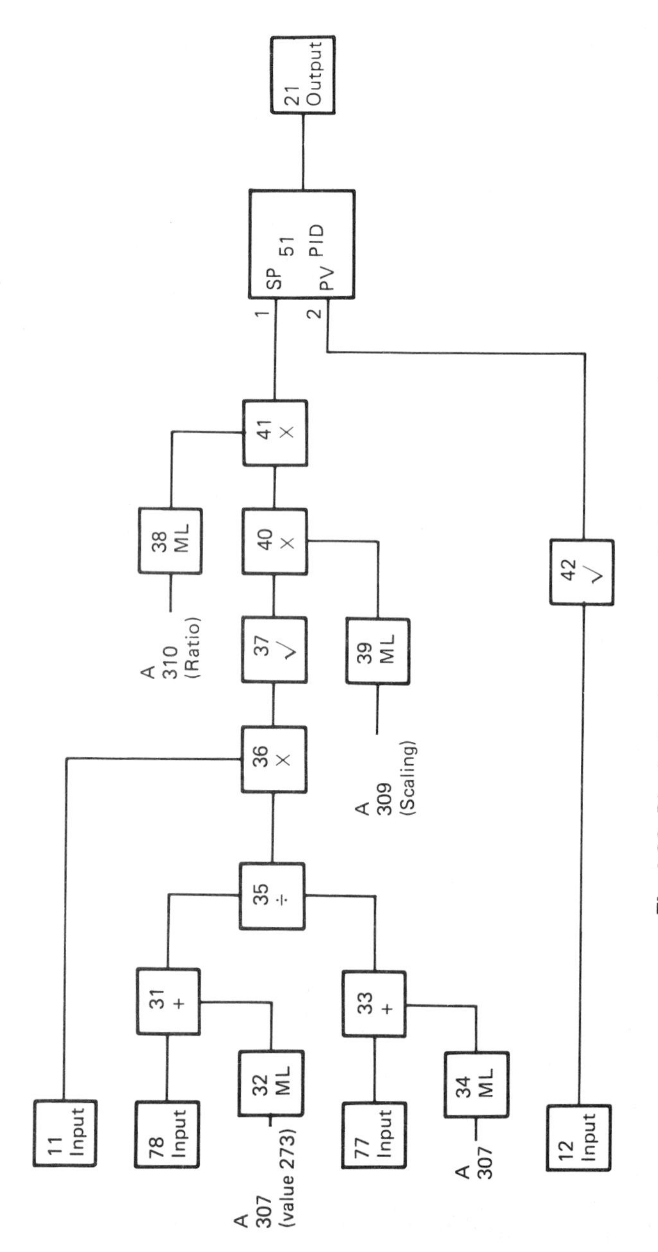

Fig.4.22 Block structure program for burner control.

approach was designed to allow a process control engineer to design a control strategy using 'building blocks' provided by the computer manufacturer.

Figure 4.22 shows the control scheme for the recuperative burner of fig.4.21a arranged in a block structure. For illustrative purposes the three-term controller for gas flow has been included in the structure, giving DDC of gas flow. The block structure described is a somewhat simplified version of that used by Control and Readout in their CRL1000 computer system.

The layout of fig.4.22 is fairly obvious, the only oddity being possibly ML for memory links (blocks 32, 34, 38, 39). These obtain values from store locations–store location A307, for example, contains 273 for the conversion of the combustion air temperature to absolute degrees.

The block structure is constructed on a question-and-answer basis. It would be tedious to list all the blocks of fig.4.22, but the examples below should show the technique. Computer outputs are underlined thus: LINK TO? The user's inputs are not underlined.

INPUT 7 8
THERMOCOUPLE TYPE R
LINK TO ? BLOCK 31
INPUT? 1

BLOCK 31
TYPE? MATHS
FUNCTION? +
LINK TO? BLOCK 35
INPUT? 1

BLOCK 32
TYPE? ML
LOCATION? A307
LINK TO? BLOCK 31
INPUT? 2

INPUT 12
LINEAR 4-2OMA
LINK TO? BLOCK 42
INPUT? 1

BLOCK 42
TYPE? MATHS

FUNCTION? SQRT
LINK TO? CONTROLLER 51
INPUT? 2

and so on for the rest of the blocks. The controller is set up in a similar question-and-answer fashion for gain, T_i, T_d, remote auto/manual changeover, etc.

Usually a computer using a block structure approach for control also incorporates a high level language to allow sequence programs to be written and to obtain values from the block structure for display on VDU screens. This is often a variation on BASIC. Suppose the operator is allowed to change the ratio held in location A310. A button on his desk, labelled 'new ratio' is connected to logic input L500. Part of the sequence program for the control system could be:

```
1000 IF L500 OFF GOTO 1030
1010 PRINT @ (10, 8) "New Ratio?"
1020 INPUT A310
1030 (REST OF PROGRAM)
1040 WAIT 0.55
1050 GOTO 1000
```

This would obviously need some refinement–error checking to prevent ridiculous values of the ratio, for example–but illustrates the principles.

4.5.5. Ladder diagrams

Ladder diagrams originated in programmable controllers to allow electricians and technicians to write, and understand, sequence programs. They aim to emulate relay wiring diagrams. These are drawn similar to fig.4.23a and resemble the steps of a ladder, hence the name.

Figure 4.23a is the relay circuit for a simple motor starter, and its operation is straightforward, and obvious. To implement this on a PC, the components are wired as in fig.4.23b. The notation is based on the GEC GEM-80. A denotes inputs and B outputs. B2.8, for example, is output 8 on card 2.

The ladder diagram program for the motor starter is shown in fig.4.23c. This is directly equivalent to fig.4.23a, as should be obvious; the only difference is that the emergency stop button is

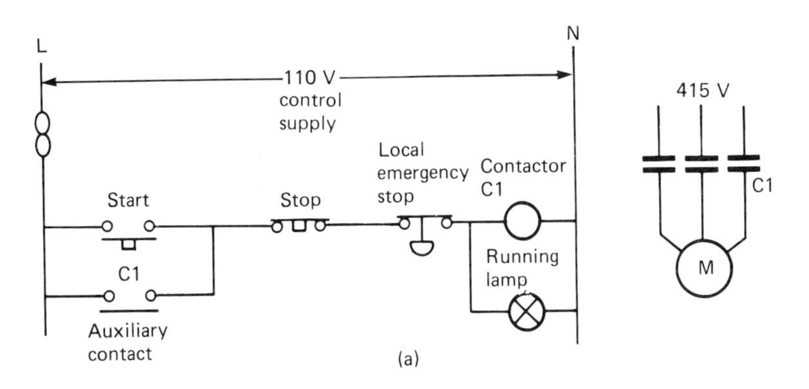

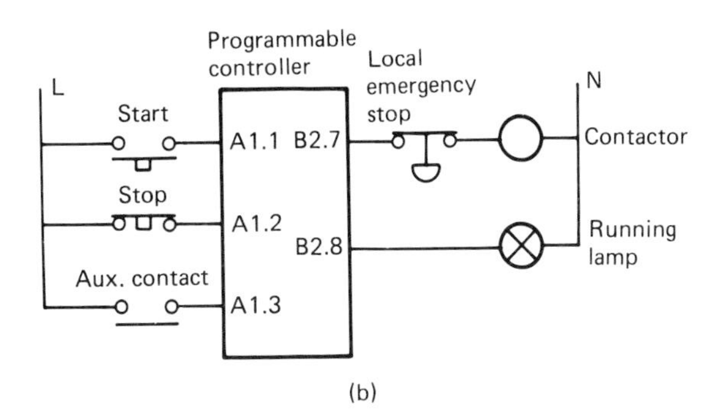

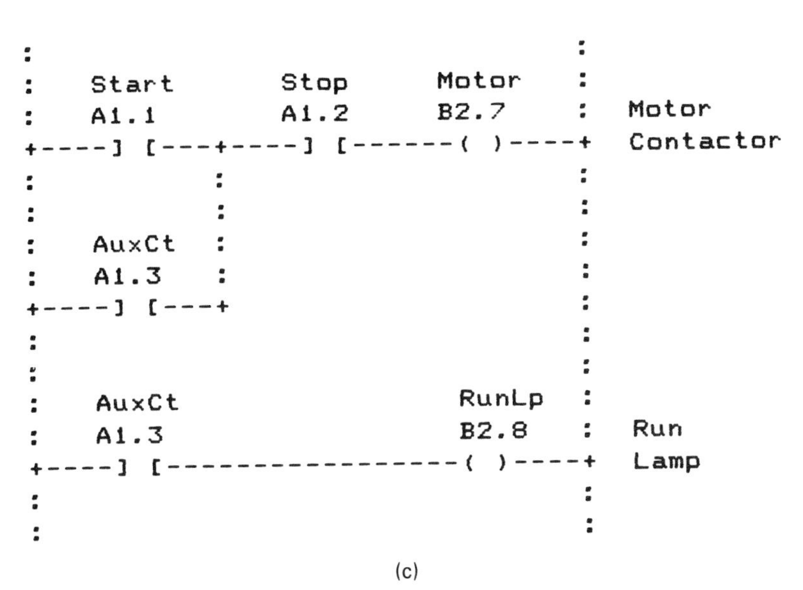

Fig.4.23 Ladder diagram programs. (a) Relay circuit for motor starter. (b) Wiring items to programmable controller. (c) Sequencing ladder diagram.

(a)

(b)

Fig.4.24 Various programming devices. The GEM-80 programmer can also be seen in fig.4.7a, and a small hand held unit in fig.4.7d. (a) Programming an Allen Bradley PLC. (b) Programming an ASEA Master controller. (Photo courtesy of ASEA & Sheerness Steel.)

directly wired in the coil line for safety reasons, a topic discussed later in section 4.11.

The 'program' is loaded via a portable programmer shown in fig.4.24. Contacts, branches, coils and other functions such as delays, sequences, etc., are entered via labelled keys. Extensive monitoring facilities are also included to allow the operation of the plant to be observed for fault finding purposes.

The example of fig.4.23 is very simple, but the ladder diagram concept can obviously be used to construct very complex sequence operations. These will, however, be easy to design, easy to modify and easy to understand and maintain.

Ladder diagrams can also be used to write control programs. Figure 4.25 is a ladder diagram equivalent of the block structure of fig.4.22. It is based on a combination of the features of the GEM-80 and Allen Bradley range of controllers for ease of

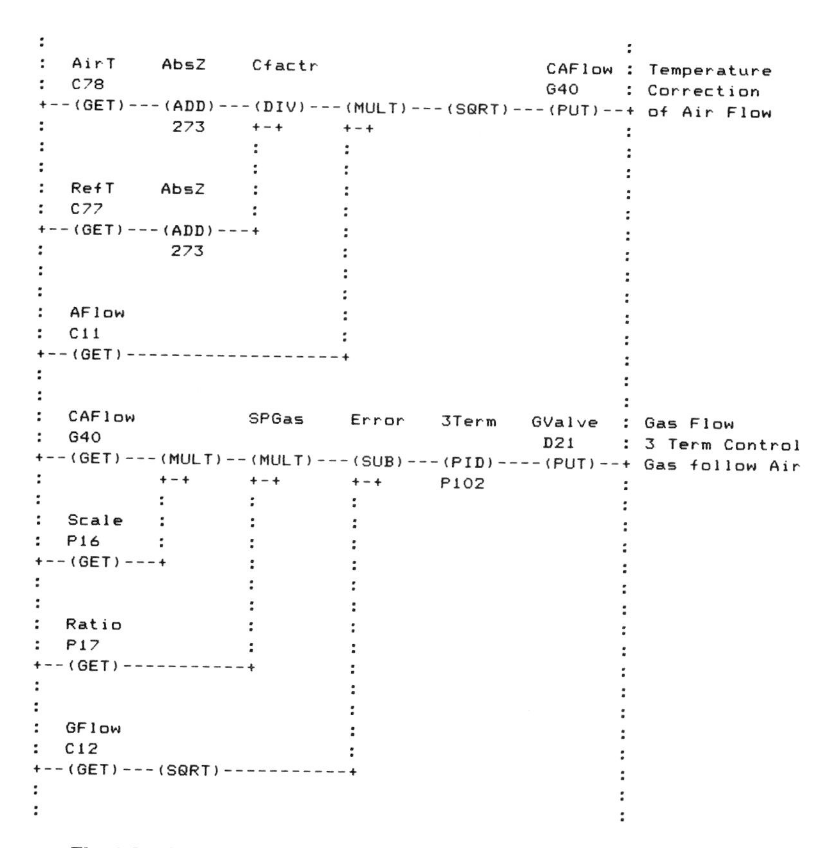

Fig.4.25 Data handling in ladder diagram format, based on burner of Fig.4.21(a). Analog process control ladder diagram.

explanation. C denotes analog inputs, D analog outputs. G40 is an internal store, and P denotes preset values. GET obtains a value, and PUT sends it to the designated address. PID performs the three term control algorithm on the error signal generated by the preceding SUB. Preset P102 holds the gain and other adjustments for the three term control algorithm.

4.5.6. Logic symbols

Digital circuits, described in chapter 3, are designed using logic elements: AND gates, OR gates, memories, etc. There is a close resemblance between the design of digital circuits and the writing of computer programs for industrial control. It is therefore not surprising that languages have evolved which are based on the interconnections of logical and functional elements.

Figure 4.26 is based on the language used by the Swedish company ASEA in their Master series of controllers. This is very hierarchical, with the program being broken down into a tree-like structure, the bottom level (or leaves!) of which are the elements themselves. In some aspects, Master has the same relationships to other controller languages as Pascal has to, say, BASIC, because it imposes a useful discipline on the user.

Elements are defined by the user; element 5.1.3 could be defined as a memory with three set inputs and two reset inputs, for example. The inputs and outputs of the defined elements are then interconnected. Pin 6 of element 5.1.3, for example, could be connected to Pin 8 of element 5.1.7.

Elements are not limited to logical elements. The language also includes a wide range of elements for analog signal processing such as filters, arithmetic units, PID controllers and specialised blocks for specific applications such as weighing.

4.5.7. Assembly language

Any form of high level language, including block structures and ladder diagrams, brings constraints of speed or storage requirements. In cases where speed is essential, storage is limited or some form of non-standard operation is required, there is little option other than to write programs in the actual form used by the computer itself. This is called machine code programming.

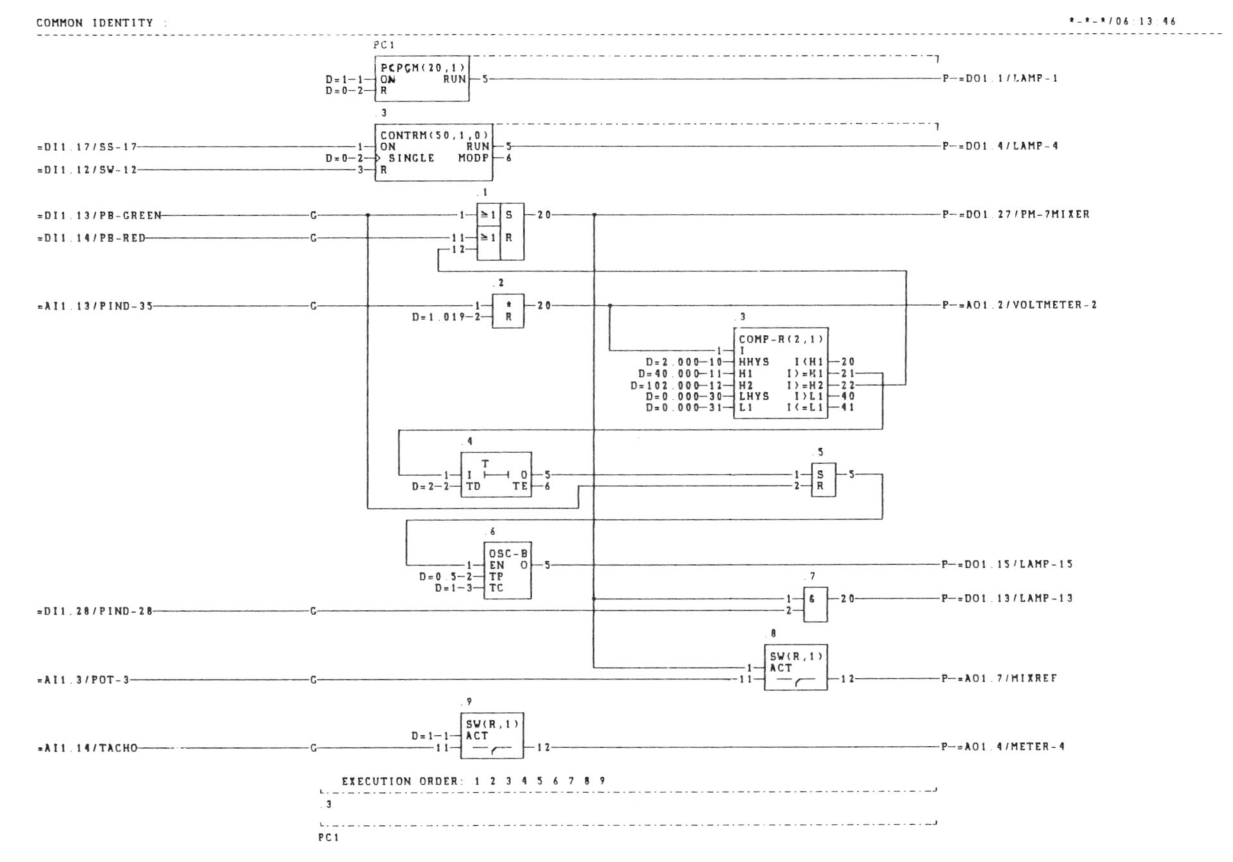

Fig.4.26 Block/logic symbols used by ASEA for their master series of controllers.

It is an almost impossible task to remember all the instructions for even one microprocessor, as they are apparently meaningless, hexadecimal numbers. For example, 87 is the code for ADD in a Z80 microprocessor.

Machine code programming is invariably written in assembler. This uses mnemonics for each machine code instruction. Some of those available with the Z80 are:

CP	operand	Compare operand with accumulator
ADD	operand	Add operand to accumulator
INC	operand	Increment (add one) to operand
JP	destination	Unconditional jump
LD	destination source	Load; used for general data movement

A typical extract from an assembler source program could be:

```
LD B,C            ; SET UP COUNTER FOR TESTS
DEC B             ; ONE OFF COUNTER
LD IX<(FIRST)     ; IX NOW CONTAINS FIRST ITEM
CP E              ; REGISTER E HOLDS REFERENCE
JR NC ALARM       ; JUMP IF NOT EQUAL TO ALARM
```

and so on.

This source program is converted to the machine code equivalent (called the object program) by a manufacturer-supplied program called an assembler. The above extract, starting at location 5000, say, would be converted as below:

Location	Object	Source
5000	41	LD B,C
5001	05	DEC B
5002	DD2A2850	LD IX,(FIRST)
5006	BB	CP E
5007	3008	JR NC ALARM

The reader should not attempt to follow the above, which is purely for illustrative purposes. The important points are that the instructions are written in mnemonic form and converted to object code by the machine itself.

Assembler programming is not easy, and requires careful logical thought. Inevitably errors in initial source programs require a rerun of the assembler operation, so using an assembler is superficially similar to using a compiler, described earlier in section 4.5.1. and fig.4.18. It is, however, the only way of getting the ultimate performance out of a machine.

4.5.8. Interrupts

Suppose it is required to monitor the state of an overpressure limit switch as part of a control scheme, and it is desired to take some form of remedial action within 0.5 seconds of the switch changing state. One possible way of achieving this would be to examine the state of the switch every 0.5 seconds. Whilst feasible for small systems, this would be impossibly time consuming where many such events need to be monitored.

Most computers incorporate an interrupt system. This allows the user to designate priorities to input conditions. When an output occurs, the computer executes a routine to deal with it. If a higher priority input occurs, the computer temporarily interrupts its current routine to deal with the new conditions.

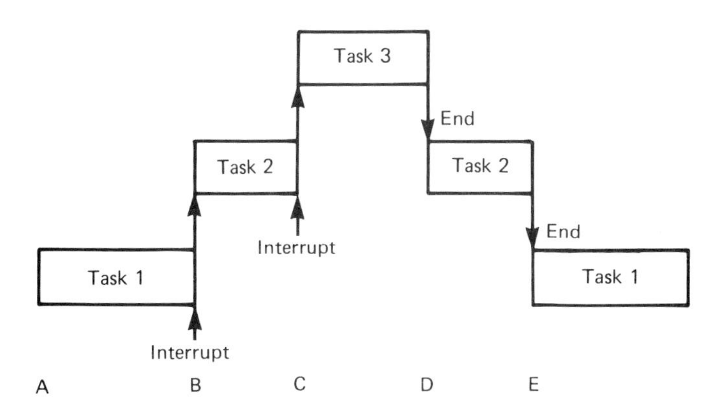

Fig.4.27 Interrupts.

In fig.4.27, at time A the computer is running a low priority routine task 1. At time B, a higher priority interrupt occurs and the computer suspends task 1 and starts task 2 to service the interrupt. At time C an even higher priority interrupt is received, and task 2 is suspended to allow the higher priority task 3 to run. This terminates at time D when task 2 recommences. This in turn finishes at time E when the initial task 1 continues.

Interrupts can also be generated from timing signals if some part of a program needs to be obeyed at fixed regular time intervals; a three term control algorithm or digital filters are typical examples.

Interrupts are best handled via an assembler or, to a lesser extent, high level languages. They are not usually directly accessible from block structures or ladder diagrams. The use of interrupts requires intimate knowledge of a computer and its construction.

4.6. Mass storage

The amount of fast access storage on most computers is limited by both cost and the addressing capability of the central processor. A typical 8 bit microprocessor (such as the 6502, 8080, Z80, etc.) can address a mere 64Kbytes (enough to store about twenty A4 pages of text). The more modern 16 bit microprocessors can address several Mbytes of memory, but even so it is unusual to find machines with more than 640K of RAM.

Computer storage is required for the programs, scratchpad (short term) memory, medium term data storage, and long term storage for archiving infrequently accessed data, files and possibly different programs which can be loaded for different operating conditions. In general, the first three requirements are met by RAM, the rest by relatively slow mass storage. Typical uses for mass storage are long term storage of event logs and production records, production planning schedules, plant usage and cost records, and so on.

Almost all common mass storage systems are based on magnetic storage systems although optical storage methods are under development. Computer storage uses non-modulated digital techniques, described further in fig.7.44. The cheapest and simplest systems use floppy disks, various types of which are shown in fig.4.28. The larger 20 cm disk is used on some minicomputers, the 13 cm disk on some microcomputers and personal computers. Minidisks have not yet been widely used on industrial computers, but their sealed construction should make them attractive.

A disk is organised into concentric tracks; each track is further divided into sectors. A typical example is shown in fig.4.29, which has 80 tracks and 10 sectors. Each sector can store 256 bytes of data, giving a total capacity of 204,800 bytes.

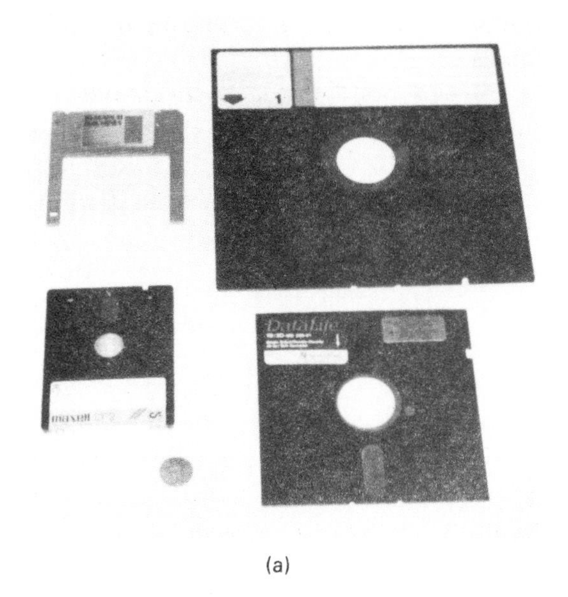

(a)

(b)

Fig.4.28 (a) A selection of floppy disks, 20 cm and 13 cm conventional disks, plus two styles of minidisks. (b) Construction of a 13 cm disk drive. The spindledrive, head positioner stepper motor and screw, plus the index photocell can be clearly seen.

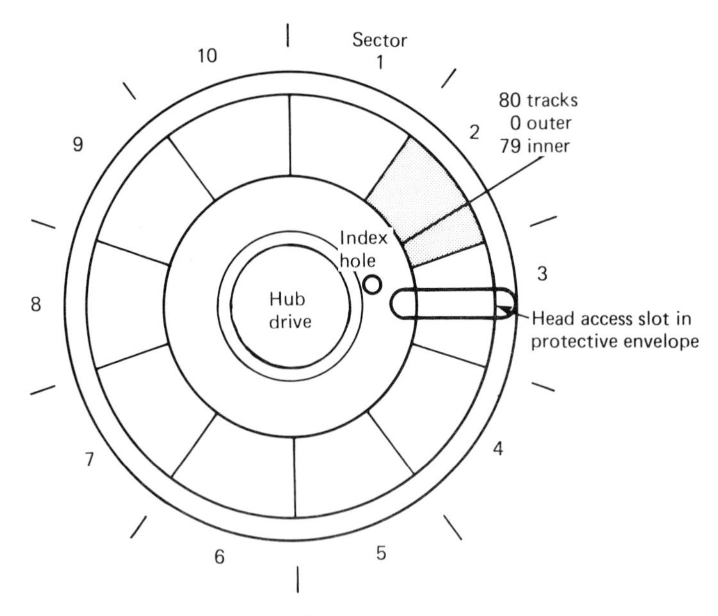

Fig.4.29 Disk tracks and sectors.

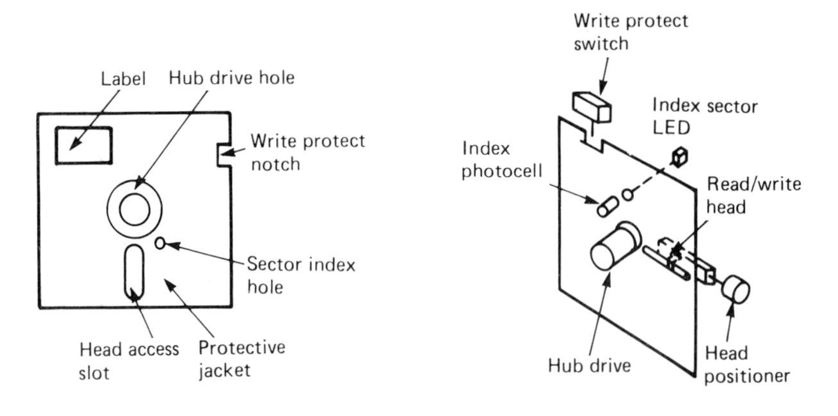

Fig.4.30 Components of a disk and disk drive.

Data is recorded/retrieved by a movable read/write head. The construction of a typical disk drive is shown in fig.4.30. The track is selected by linear positioning of the head (usually achieved with a stepper motor, described in section 2.11). Sectors are selected by an angular position measuring device which measures angular rotation from an index hole. A photocell is used to identify the index position.

The head is in contact with the surface of the floppy disk at all times. To reduce wear, the disk is only rotated when data is to be read or written. There is therefore an inherent delay of about 0.5 seconds as the disk spins up to its operating speed of 300 rpm. There is an additional delay of about 0.75 seconds as the head is positioned to the correct track and the correct sector to come round. Data transfer takes place at around 20 kbytes per second.

Dust is the enemy of floppy disk systems in industry. Because the head is in contact with the disk, dust acts as an abrasive powder, wearing away both the head and the disk surface. Systems which read from or write to floppy disks continually in a typical industrial atmosphere will have a high failure rate. Even systems which perform infrequent read/writes but operate with a disk permanently in a drive will have trouble, as the dust will settle either side of the head and be ground in when the drives start. In the author's experience, 'wet' head cleaners do not seem to help as they turn abrasive dust into abrasive sludge.

Floppy disks in process control applications are therefore best suited to clean air-conditioned control rooms, or systems where the disk is only on the machine intermittently (e.g. an end-of-shift

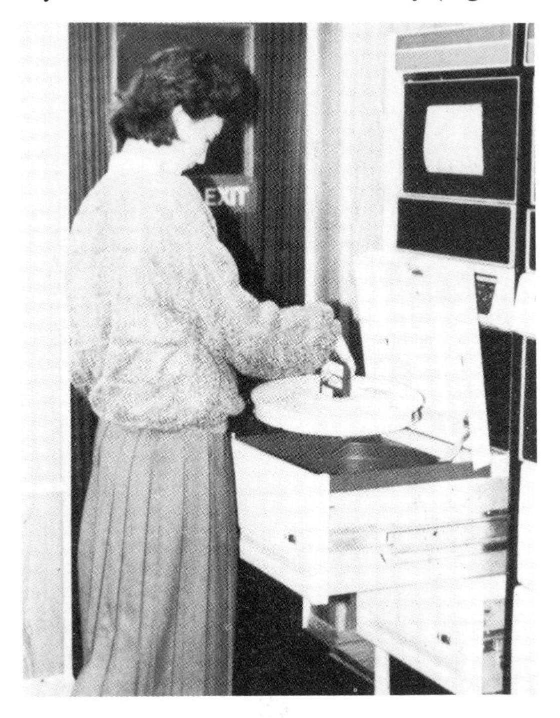

Fig.4.31 Loading a hard disk into a DEC PDP11 computer.

dump of shift records or the occasional saving and reloading of programs).

Floppy disks are inherently portable, but care must be taken to avoid damage in transit. Typical problems are surface contamination from finger grease and accidental erasure by stray magnetic fields (motors, electromagnetics, solenoids, etc.). Regular backup copies should be made of all important disks.

Large amounts of data (upwards of 20 Mbytes), can be stored on changeable hard disks similar to fig.4.31. These are precision aluminium disks which are spun continuously at high speed (typically 2400 rpm) to reduce access time. To prevent wear, the read/write head 'flys' above the surface of the disk on the air carried round as the disk spins. Dust can, however, be an even

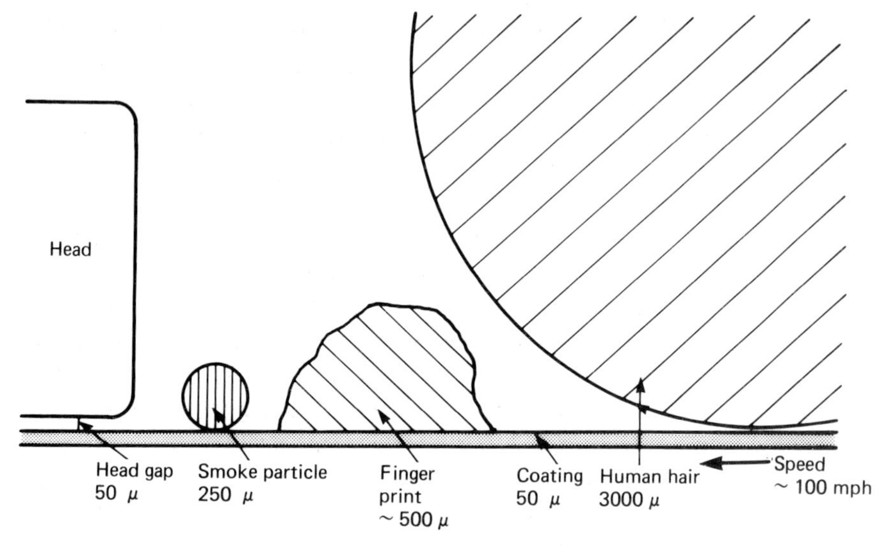

Fig.4.32 The need for cleanliness.

bigger enemy, as fig.4.32 demonstrates. A single smoke or dust particle passing between the head and disk surface will damage both and possibly cause the head to contact the disk on the rebound. A 'disk crash' invariably destroys the head and the disk. Hard disk systems therefore incorporate complex filters and closed loop air flows. Routines for the inspection and changing of filters should be established as part of routine plant maintenance. Hard disks are also susceptible to mechanical vibration and shock which can again cause the head to contact the disk surface, resulting in a costly head crash.

Fig.4.33 A 20 Mbyte Winchester disk drive. This has a similar capacity to the hard disk of fig.4.31.

The so-called Winchester disks are probably best suited to industrial applications. These can store up to 20 Mbytes in a small compact unit of which fig.4.33 is a typical example. These are manufactured with one or more solid disk platters and heads in an air tight container filled with a clean inert gas. Because the unit is totally sealed, dust is not a problem although Winchester disks are still vulnerable to shock and vibration. The disk units are not easily changed, so a backup mass storage unit (usually floppy disks or magnetic tape) is usually employed along with the Winchester disk. Regular backups or archiving are done to prevent the Winchester unit running out of disk space. With 20 Mbytes of storage, however, the need for disk purges is rare on most systems. In computer networks, Winchesters are widely used to provide buffer storage between levels. A low level computer can follow a production schedule written to a disk by a higher level computer. At the end of each shift (or day, or week) the higher level computer reads back production records written to the disk by the lower level machine.

4.7. Visual display units

Visual display units, or VDUs, are the commonest computer/operator interface units providing simple mimic displays as in fig.4.34. The use of VDUs is described in section 7.5, but the main advantages are simplicity of installation, ease of modification and the ability to display complex information concisely and clearly. There are disadvantages, however. TV monitors are

Fig.4.34 Plant mimic drawn with the Intercolor 8820. (Photo courtesy Techex Ltd UK.)

probably the most unreliable part of a computer system, and duplication or redundancy is advisable to avoid the 'eggs in one basket' syndrome. The amount of data that can be displayed on one screen is also limited, necessitating the careful design of a tree structure of displays which lead the operator from overviews down to the level of detail required.

Most VDUs use a raster scan display identical to that used on domestic televisions. A TV picture is constructed from lines–625 in the UK, 525 in the USA. The lines are 'drawn' on to the face of a cathode ray tube (CRT) by an electron beam. The beam is moved by electromagnetic coils and scans the tube face, as shown in fig.4.35b. The whole tube face is scanned in 1/50 of a second (1/60 of a second in the USA).

The brightness of any point on the screen is determined by the intensity of the electron beam at that point. The beam current is therefore controlled by an electrical signal which represents the

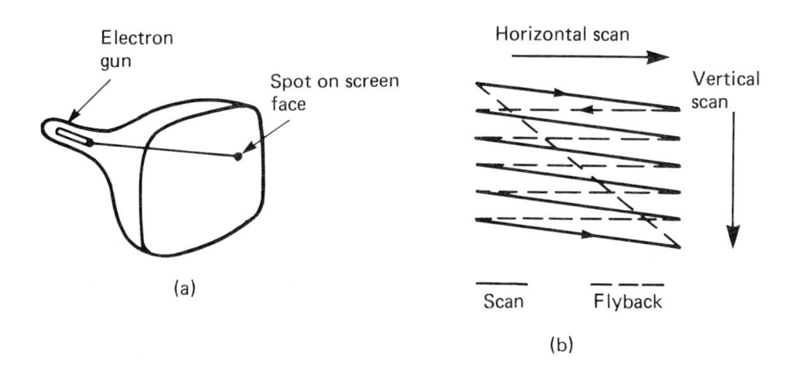

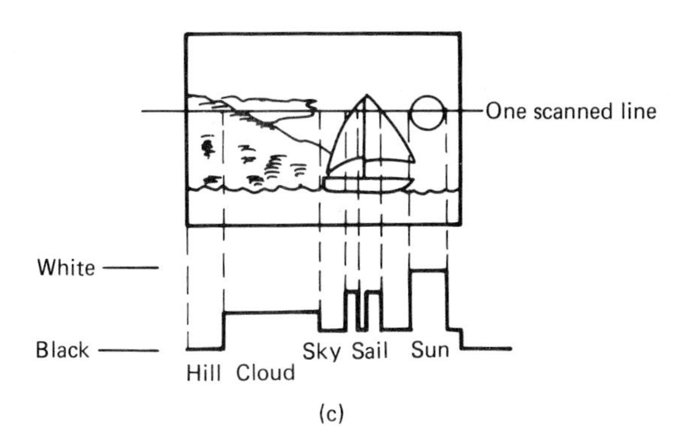

Fig.4.35 Makeup of a TV picture. (a) Electron beam and TV tube. (b) TV screen scanned by electron beam. (c) One line of a TV picture.

required luminous intensity of each line in turn, as shown in fig.4.35c. The scanning of the screen is too fast for the eye to follow, and the screen is perceived as displaying a picture.

The simplest VDUs can display alphanumeric text only, plus the usual punctuation and mathematical symbols. For simplicity, the operation of a simple VDU displaying 24 rows of characters with 72 characters per row will be described. The display effectively consists of 24 rows and 72 columns (the term 'row' being used to avoid confusion with the TV 'line').

The first requirement of the VDU is a store to hold these characters. It is usual to store these characters in ASCII coding (for Americal Standard Code for Information Interchange) as in table 4.1. This requires one byte per character so a 24 × 72 display will require a 1728 location store. Note that there is a direct relationship between the address of a store location and the

Table 4.1 ASCII coding

Decimal	*Hex*	*Character*	*Decimal*	*Hex*	*Character*
32	20		81	51	Q
33	21	!	82	52	R
34	22	"	83	53	S
35	23	#	84	54	T
36	24	$	85	55	U
37	25	%	86	56	V
38	26	&	87	57	W
39	27	'	88	58	X
40	28	(	89	59	Y
41	29	)	90	5A	Z
42	2A	*	91	5B	[
43	2B	+	92	5C	\
44	2C	,	93	5D	]
45	2D	–	94	5E	^
46	2E	.	95	5F	_
47	2F	/	96	60	'
48	30	0	97	61	a
49	31	1	98	62	b
50	32	2	99	63	c
51	33	3	100	64	d
52	34	4	101	65	e
53	35	5	102	66	f
54	36	6	103	67	g
55	37	7	104	68	h
56	38	8	105	69	i
57	39	9	106	6A	j
58	3A	:	107	6B	k
59	3B	;	108	6C	l
60	3C	<	109	6D	m
61	3D	=	110	6E	n
62	3E	>	111	6F	o
63	3F	?	112	70	p
64	40	@	113	71	q
65	41	A	114	72	r
66	42	B	115	73	s
67	43	C	116	74	t
68	44	D	117	75	u
69	45	E	118	76	v
70	46	F	119	77	w
71	47	G	120	78	x
72	48	H	121	79	y
73	49	I	122	7A	z
74	4A	J	123	7B	{
75	4B	K	124	7C	:
76	4C	L	125	7D	}
77	4D	M	126	7E	~
78	4E	N	127	7F	
79	4F	O			
80	50	P			

Codes less than 32 are control codes. Common are (decimal): 07 bell, 10 linefeed, 12 form feed, 13 carriage return

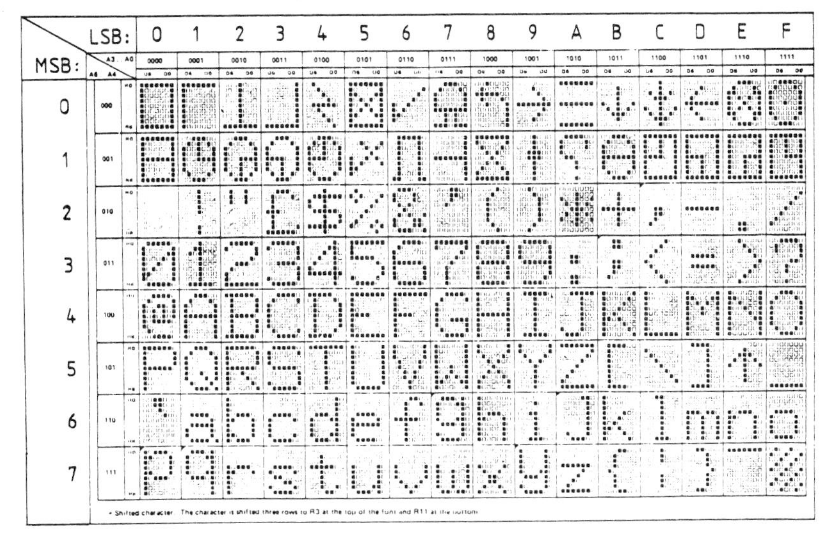

Fig.4.36 Dot matrix table. Character set for the MCM 6576 ROM.

corresponding screen position. If the top left screen position has address 0, and the bottom right has address 1728, to put a letter A in the third column of the second row we would write the ASCII code for A (65) into address 74.

Characters are built up by a dot matrix, a typical example being shown in fig.4.36. It will be seen that this is arranged in ASCII code represented in hexadecimal. The code for R, for example, is 52. These dot matrix patterns are stored in a character generator

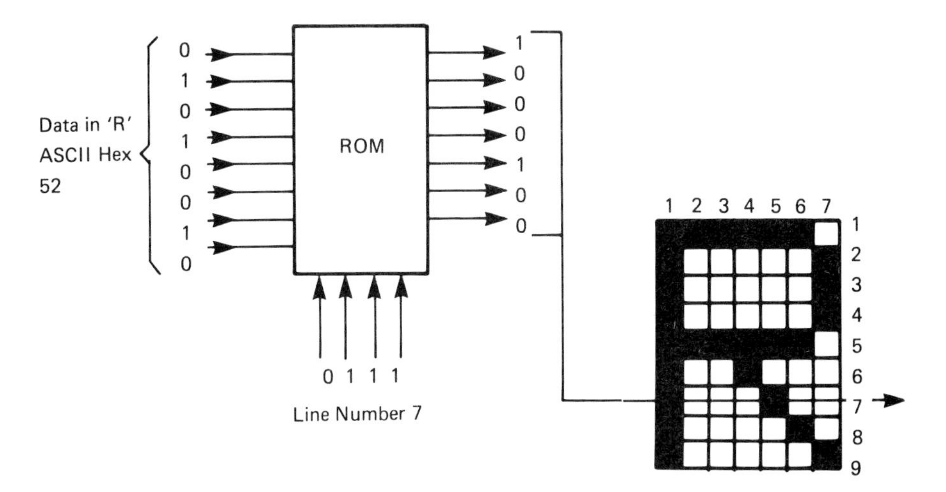

Fig.4.37 Character generator ROM.

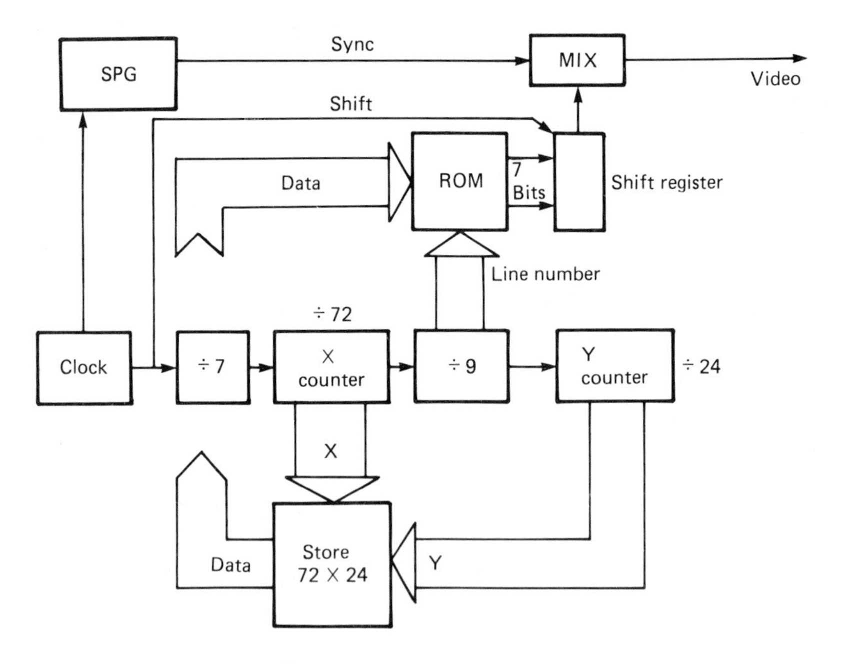

Fig.4.38 VDU block diagram.

ROM. This accepts the ASCII code for a character and the line number within the character, as in fig.4.37, and produces the dot pattern as shown.

Figure 4.38 shows a much simplified block diagram of the display portion of a VDU. For simplicity it is assumed that the store is addressed with XY coordinates corresponding to the column and rows of the display. Two counters X, Y determine the screen position being addressed.

The operation starts with the sync pulse generator (SPG). This is a complex IC which produces all the timing pulses and clock for the system. The SPG pulses are divided by a dot counter corresponding to the number of horizontal dots per character, the X address counter, a counter corresponding to the number of lines per character and the Y address counter.

The X and Y counters bring characters to the character generator ROM, whose parallel output is converted to a serial video signal by a shift register. As there are nine lines per character, each row is repeated nine times with a different line number before the Y row counter is stepped to the next row. When all 24 rows have been displayed, the sequence starts again.

The SPG also produces synchronising pulses which are added to the video signals. These inform the display monitor when to start a new line or a new picture.

Characters are 'written' to the VDU by putting the ASCII code for the character into the required store location. The simplest, and fastest, method of achieving this is by making the VDU store part of the computer RAM store. This technique is used in many home computers, and is called a 'memory mapped' VDU.

Memory mapped VDUs require the VDU to be local to the computer, so it is more common for data to be transmitted via a (usually RS232) serial link. Data is written to the store at a point identified by the cursor, which steps on as characters are received. Control characters can be transmitted to move the cursor to a required position.

The VDU described so far can only display alphanumeric characters. Most industrial VDUs are required to display plant mimics. Examination of table 4.1 will show that ASCII only utilises 128 of the 256 codes available from 8 bits. The remaining 128 codes are commonly allocated to useful graphical symbols, some of which are shown in fig.4.39. Using the graphical symbols, pipe lines, pumps, motors, valves, etc., can be drawn by placing

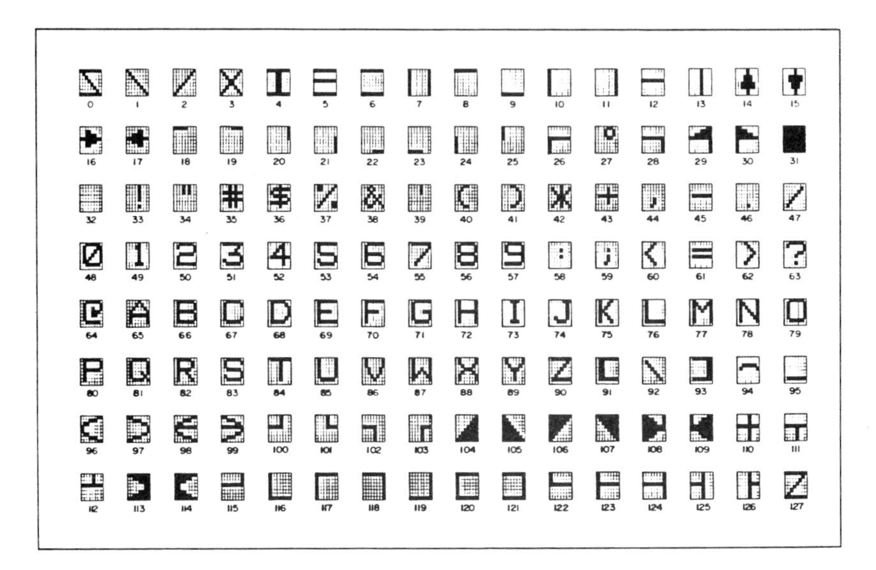

Fig.4.39 Block graphics table. Character and graphic characters used in the Allen Bradley Advisor VDU. Semi graphic characters similar to fig.7.30 are also included for area filling and bar charts. (Courtesy of Allen Bradley.)

the required graphical code into the requisite store location. This approach is sometimes called 'block graphics'.

Block graphics allows the user to construct very effective mimics, but the screen layout is still constrained by the underlying character cells: 24 × 72 in our earlier example. For complete flexibility, each and every point on the screen needs to be individually controllable. This feature is called line, or pixel, graphics–pixel being the name given to an individual screen point.

Pixel graphics allows any picture to be constructed up to the limit of resolution of the displays. Circles can be drawn at any position, and of any required radius, for example. The penalty is a large increase in VDU complexity. A store is required with 1 bit for each and every pixel on the screen. A screen with a resolution of 640 × 320 points (a reasonable definition) will require a video store of 204,800 pixels or 25,600 bytes. The character generator ROM is not accessed sequentially, as in fig.4.38, but places the bit patterns for text direct into the store.

A pixel graphics VDU will also incorporate a large amount of intelligence to allow the user to request, say, 'a circle radius R centre X, Y' or 'a line from X1, Y1 to X2, Y2', leaving the VDU to calculate which pixels need illuminating. Usually, skeleton screens are developed off line by the user with a specialist graphics language, and stored in the VDU. Screens are called by the host computer, and dynamic data written into the skeleton.

Colour displays increase the readability of a VDU display. A valve, for example, can be displayed in green for open or red for shut. Colour displays are generally built around the three primary colours red, green and blue, which can be combined together to give eight colours as below, with 1 denoting present and 0 absent:

Red	Green	Blue	Resultant
0	0	0	black
0	0	1	blue
0	1	0	green
0	1	1	cyan
1	0	0	red
1	0	1	magenta
1	1	0	yellow
1	1	1	white

Full colour displays therefore require 3 bits per pixel (for pixel graphics) or 6 bits per character cell (for block graphics with independent foreground and background colours). Colour increases the store requirements considerably. A common technique to save memory with block graphics is the use of control characters to change foreground and background colours. Subsequent characters are displayed in the new colours until a different colour combination is selected by new control commands.

4.8. Plant interfacing

To be useful, a computer must communicate with the outside world. Figure 4.40 shows the various forms this communication can take. Essentially these are:

(a) Digital inputs.
(b) Digital outputs.
(c) Analog inputs.
(d) Analog outputs.
(e) Operator interface (keyboards, VDUs plus I/O above).
(f) Peripherals (e.g. printers).
(g) Other computers.

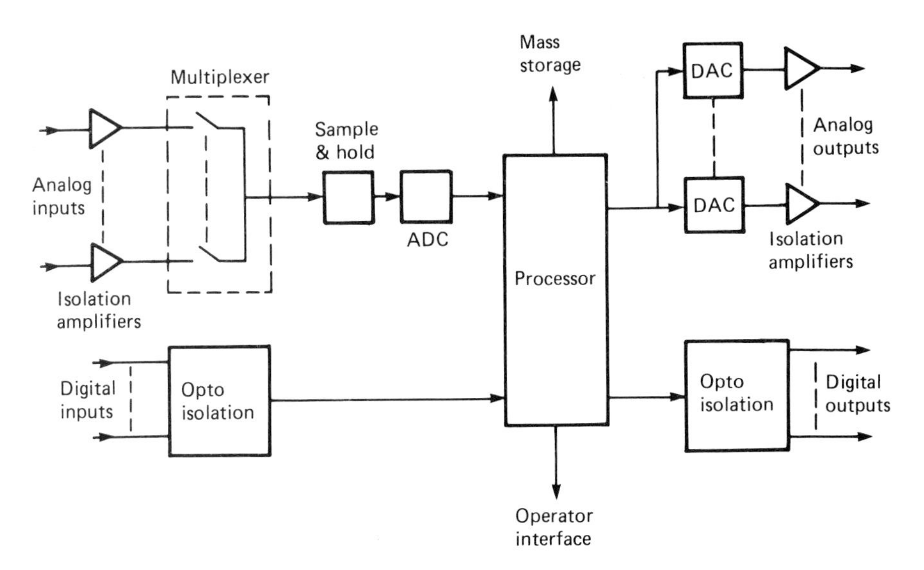

Fig.4.40 Computer to plant interfacing.

Computers operate with low voltage, low power signals (typically 5 V). Electrical noise can easily be introduced into a computer via the external cabling, causing program and data corruption with potentially dangerous consequences. Particular care must therefore be taken with power supply and grounding (topics discussed further in section 8.5).

Optical isolation should be used on all digital inputs and outputs to separate external and internal supplies and grounds, and also to limit damage from plant faults such as intercable shorts. Typical circuits are given in section 3.10.

Analog inputs need to be converted to digital form before being processed by a computer. This is performed by analog-to-digital converters (ADCs). Common ADC circuits are given in section 3.9. Because an ADC takes a finite time to perform a

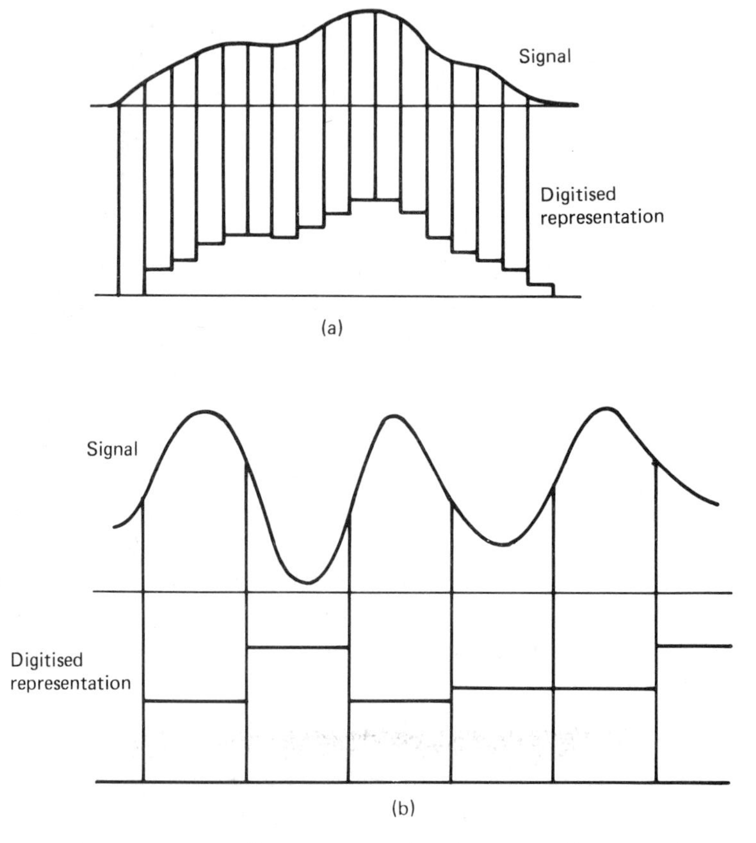

Fig.4.41 Sample rate and Shannons sampling theorem. (a) Reasonable sample rate. (b) Slow sample rate.

conversion, analog inputs are sampled at regular intervals and not scanned continuously. If this sampling is to give a faithful representation of an analog signal, the sample rate must be chosen with care. In fig.4.41a, the sample rate is adequate. In fig.4.41b, too low a sample rate gives a false representation.

Sample systems are governed by Shannon's sampling theorem which, somewhat simplified, says that the sample rate should be at least twice the maximum frequency of interest. In practice, very few real-life systems require sampling more than ten times per second; for most, two times per second is adequate. Slow systems (e.g. temperature loops with long time constants) can be sampled once every few seconds. Too high a scan rate can overload the processor and cause timing problems elsewhere.

Digital inputs are also scanned, and the sample rate must again be chosen. For most sequencing applications, ten samples per second is adequate.

ADCs have conversion times of around 1 mS, so analog inputs are usually multiplexed in turn to a single ADC, as in fig.4.40. In old equipment, this multiplexing was performed by reed relays, but CMOS switches are now used. Multiplexing need not be sequential, and can be designed to allow inputs to be scanned at different rates as determined by Shannon's theorem.

ADCs (particularly successive approximation ADCs) can give an incorrect result if the analog value changes during the conversion. The multiplexer is therefore connected to the ADC via a sample and hold circuit which takes a 'snapshot' of the signals to freeze it for the ADC. A typical circuit is described in section 1.6.6.

Analog inputs again need protection against noise and stray fault voltages. This should be at least differential amplifiers (see section 1.3.3) but ideally should be full isolation amplifiers (see section 8.5.7).

Analog outputs are achieved by DACs (see section 3.9.1). Yet again, isolation is required after the DACs for protection and noise immunity unless the DACs are driving displays or meters local to the computer.

Serial communication is used for operator keyboards/VDU, peripherals and inter-computer communication. Computer serial ports commonly operate as an RS232 device. This is based on a protocol designed for modems, and uses a voltage swing of approximately $\pm$ 12 volts. RS232 can be used for distances up to about 20 metres (dependent on the speed of data transmission and the electrical environment). Beyond this distance, current loop transmission (called line drivers), modems or fibre optic transmission must be used.

Noise can easily be introduced into a computer system, and is the cause of many spurious intermittent faults. Sensible grounding, cable segregation (separation of power and signal cables) and screening are essential. Cable routes must be planned with care to avoid parallel runs of power and signal cables and to ensure any unavoidable crosses occur at right angles.

4.9. Computers for design

The past few years have seen a remarkable growth in the power of the desk top personal computer (PC) coupled with an even more remarkable fall in price. Most engineers now have access to a PC, and it is worth considering how these can assist with process control.

The ubiquitous spreadsheet was originally designed for accountancy, but has many uses in industry. It is an amazingly simple idea that produces an 'if only I'd thought of that' response in most people. Essentially it consists of an array of cells similar to fig.4.42a. Cell contents can be an entered as values, text, or an equation involving other cells. The MTBF calculations for table 8.1, for example, were calculated on Lotus 123, a popular spreadsheet for the IBM PC. Figure 4.42b shows the cell contents for part of table 8.1. Spreadsheets have powerful cell copy routines that allow formulas to be copied *relatively* from one cell to a range of cells.

Spreadsheets can be used for all sorts of applications; the author has used Lotus for budgeting, preparation of cable schedules and sequencing simulations, to name but a few.

The planning of complex projects requires the scheduling of many different contractors and the determination of resources needed at each stage. This is generally done via a PERT chart (for program evaluation and review technique, also known as CPA or critical path analysis), a simple example is shown in fig.4.43a. Events (e.g. foundation complete) are represented by boxes, and activities by arrows which are annotated by their planned time. The shortest route through the network is the time for the completion, and the route identifies which activities are 'critical' and which have slack time.

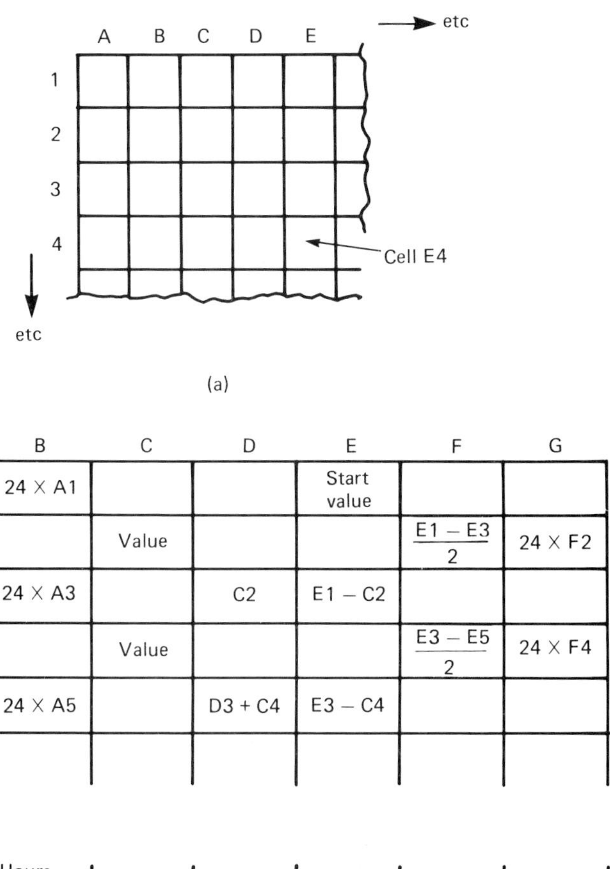

A B C D E
etc
1
2
3
4
Cell E4
etc
(a)
A B C D E F G
1 0 24 X A1 Start value
2 Value E1 – E3 / 2 24 X F2
3 1 24 X A3 C2 E1 – C2
4 Value E3 – E5 / 2 24 X F4
5 A3 + 1 24 X A5 D3 + C4 E3 – C4
6
Day
Hours A column X 24
Entered failures
Cumul fails
Survive
Mean number
Mean hours
(b)

Fig.4.42 The spreadsheet. (a) Representation of a spreadsheet. (b) Part of spreadsheet used for table 8.1. Entries for columns entered using cell replicate routines. Cell B1, for example, copied relatively to B3, B5 etc.

PERT charts are time consuming to produce, and need constant review as a project progresses. Computer packages are available which plot PERT charts quickly, allowing tight control to be exercised over a project. By allocating resources (both manpower and equipment) to activities, requirements can also be planned. Figure 4.43 shows typical printouts based on the simple PERT chart of fig.4.43a.

Computer aided design (CAD) is often thought of as a tool for large drawing offices. Admittedly, CAD packages for PCs are

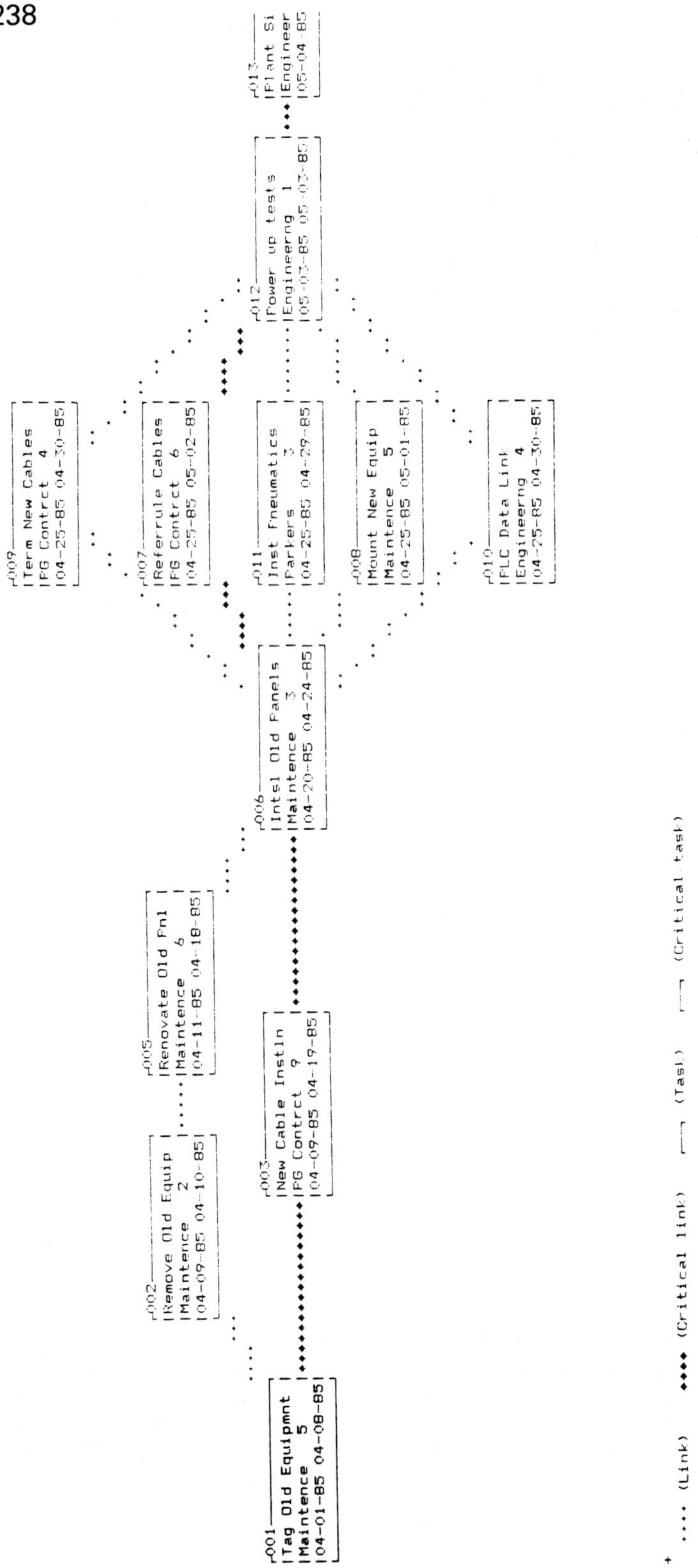

001 Tag Old Equipmnt
Maintence 5
04-01-85 04-08-85

002 Remove Old Equip
Maintence 2
04-09-85 04-10-85

003 New Cable Instln
FG Contrct 9
04-09-85 04-19-85

005 Renovate Old Pnl
Maintence 6
04-11-85 04-18-85

006 Intsl Old Panels
Maintence 3
04-20-85 04-24-85

009 Term New Cables
FG Contrct 4
04-25-85 04-30-85

007 Referrule Cables
FG Contrct 6
04-25-85 05-02-85

011 Inst Pneumatics
Parkers 3
04-25-85 04-29-85

008 Mount New Equip
Maintence 5
04-25-85 05-01-85

010 PLC Data Link
Engineerng 4
04-25-85 04-30-85

012 Power up tests
Engineerng 1
05-03-85 05-03-85

013 Plant Si
Engineer
05-04-85

.... (Link) ♦♦♦♦ (Critical link) (Task) (Critical task)

(a)

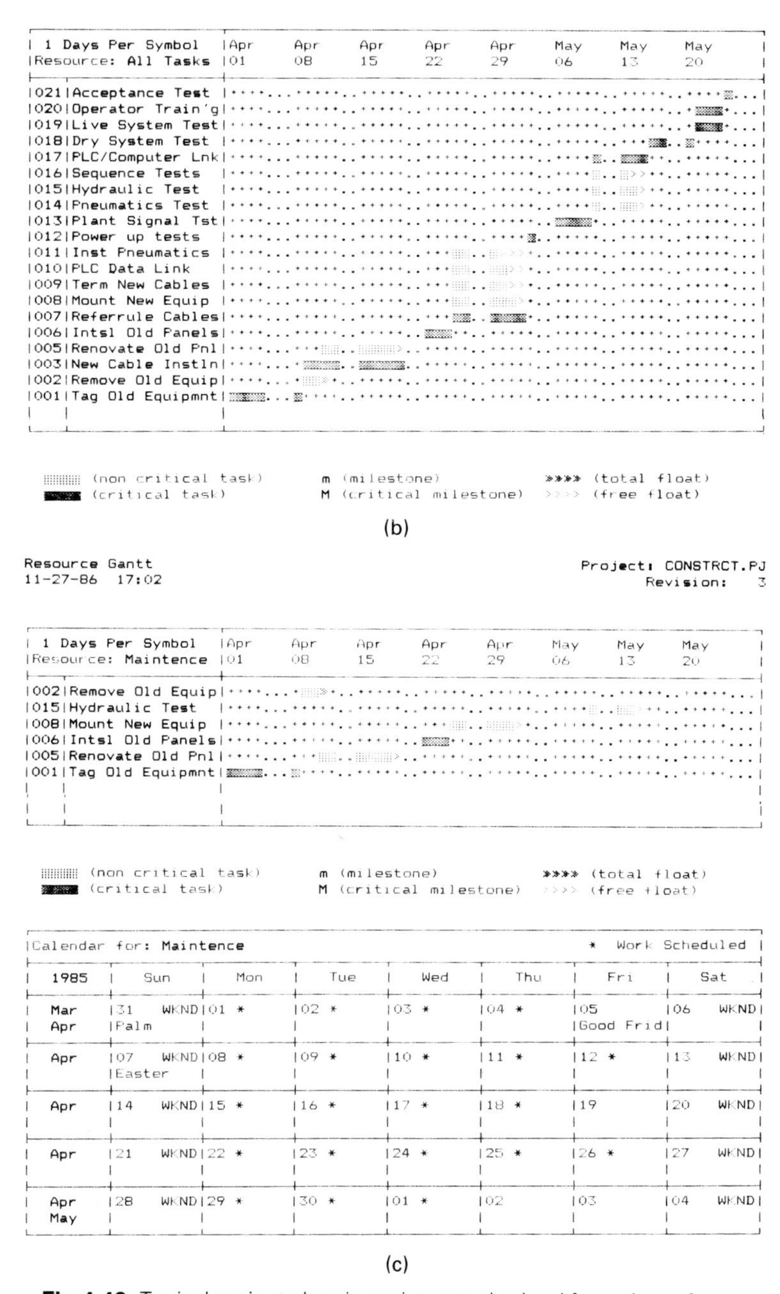

Fig.4.43 Typical project planning printouts obtained from the software package 'Super Project'. (a) PERT chart, critical path goes through tasks 1, 3, 6, 7, 12 etc. Boxes show task name, resource for task; task duration and dates. (b) The Gannt chart is an alternative way of representing a project. Task float can be clearly seen. (c) Gannt chart and calender for one resource, in this case the maintenance department.

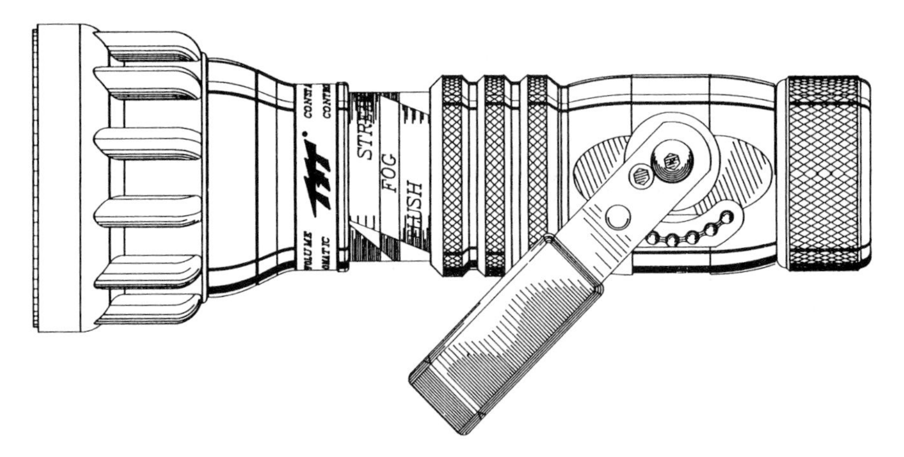

Fig.4.44 Computer aided design (CAD) drawing produced with AUTOCAD program on an IBM-XT personal computer with A3 size plotter.

somewhat restricted but the results can be impressive, as demonstrated by fig.4.44. The minimum requirement is a PC with an A3 flat bed plotter, and a mouse, light pen or graphics tablet. PC based CAD packages tend to be slow unless the PC is equipped with a maths co-processor. The most useful applications are those which require drawings which are variations on a basic theme, where complex symbols (e.g. valves or piping) are repeated. The basic symbols can be drawn once, defined and then repeated as required.

Databases store related data records for easy access (e.g. how many Delta P transducers Honeywell Class 41 have been changed on the acetylene plant between 9.9.74 and 31.3.86). Data is stored in files, which can be subdivided into records relating to one item, each record consisting of fields as in fig.4.45. Data is entered, and subsequently retrieved by matching conditions with the records. The most mundane applications (known to all engineers) are records of stores' spares holding and issues. More interestingly, databases are the basis of plant fault recording and planned maintenance.

If the behaviour of a plant is known, or can be predicted, computers can be used to simulate plant performance and test proposed control strategies. The plant is represented by a series of blocks whose response is defined–usually in the form:

$$(a + bs + cs^2 + ds^3)/(e + fs + gs^3 + hs^4)$$

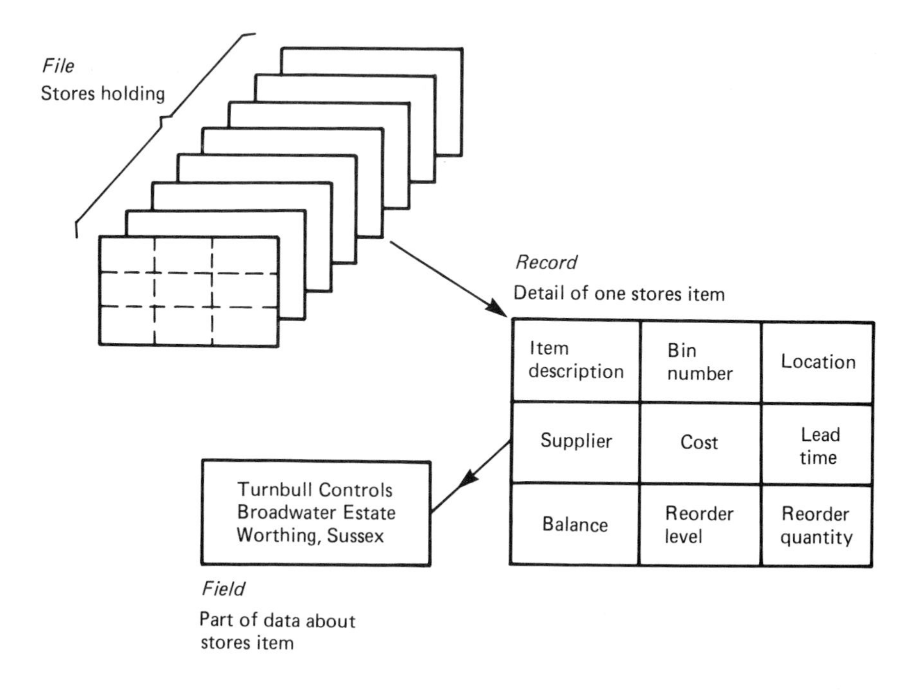

Fig.4.45 Databases and data organisation.

From these blocks the computer can calculate Nicholls, Bode and Nyquist diagrams plus a predicted step response, as shown in fig.4.46. The above analysis techniques are discussed further in Volume 3.

The above is only a sample of the power of the PC. There are many other packages of interest to the engineer: the word processor for report and tender documents (this series of books was written on a BBC computer with Wordwise), expert systems, data loggers, etc., etc. The PC should be as much a part of the process control engineer's tool kit as the screwdriver.

4.10. Robotics

For many people, the terms 'robotics' and 'industrial computing' are synonymous and interchangeable. This view is no doubt fostered by spectacular media reports, but compared with the impact of dedicated computers, programmable controllers,

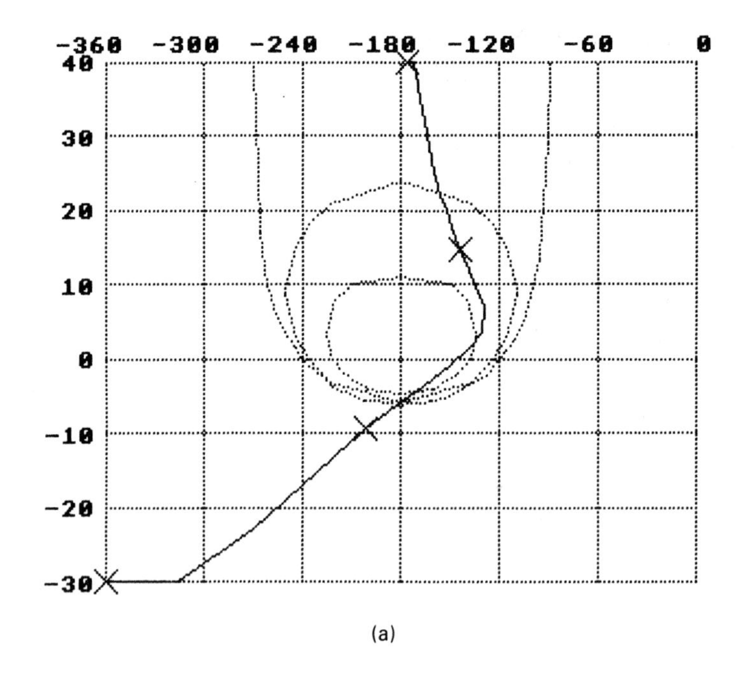
-360
-300
-240
-180
-120
-60
0
40
30
20
10
0
-10
-20
-30

(a)

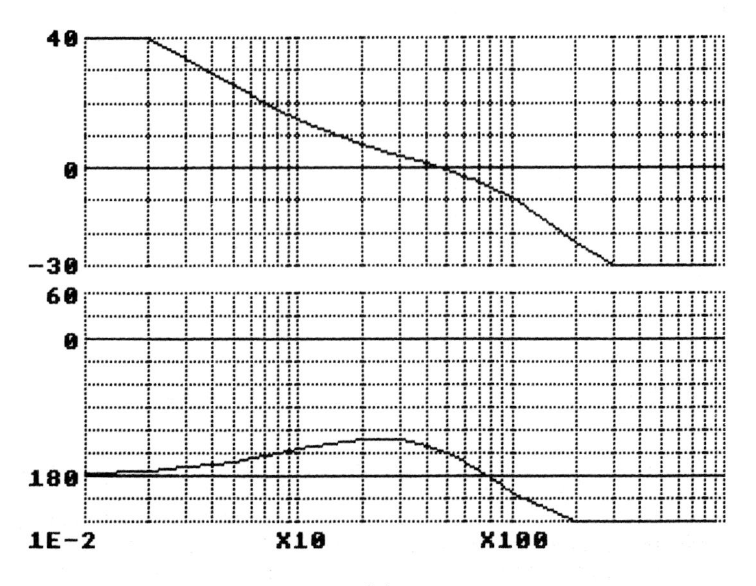
40
0
-30
60
0
180
1E-2
X10
X100

(b)

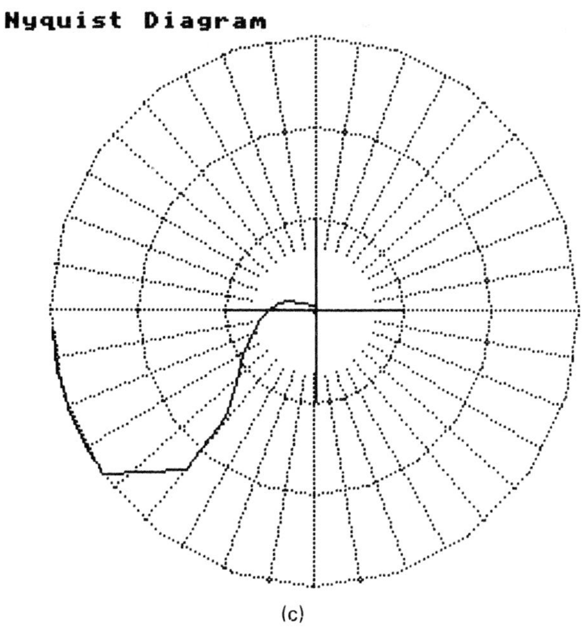

(c)

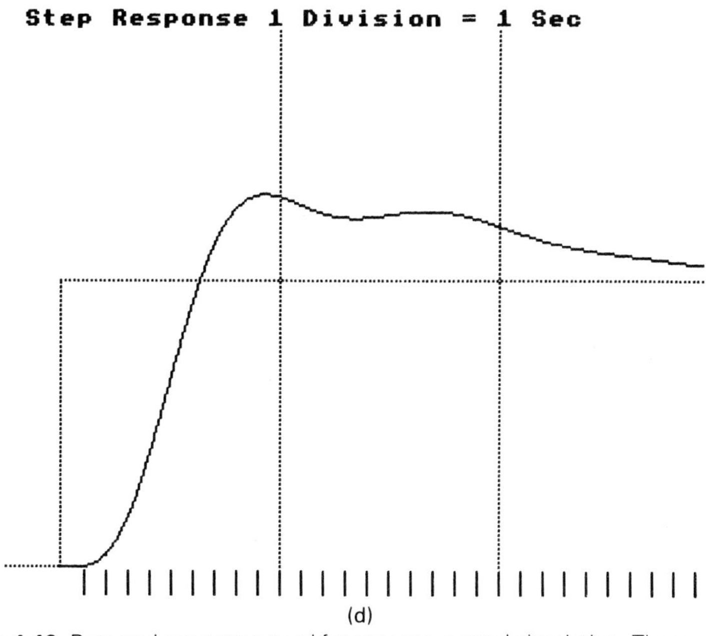

(d)

Fig.4.46 Personal computer used for process control simulation. These were all produced on a BBC model B, and are for a system comprising 3 lags, (2 of 1 second time constant, and one of 2 seconds time constant), a transit delay of 0.5 seconds and integral action; all controlled by a PID controller with T_i 12 seconds, T_d 3 seconds. (a) Nicholls chart. (b) Bode diagram. (c) Nyquist diagram. (d) Predicted step response.

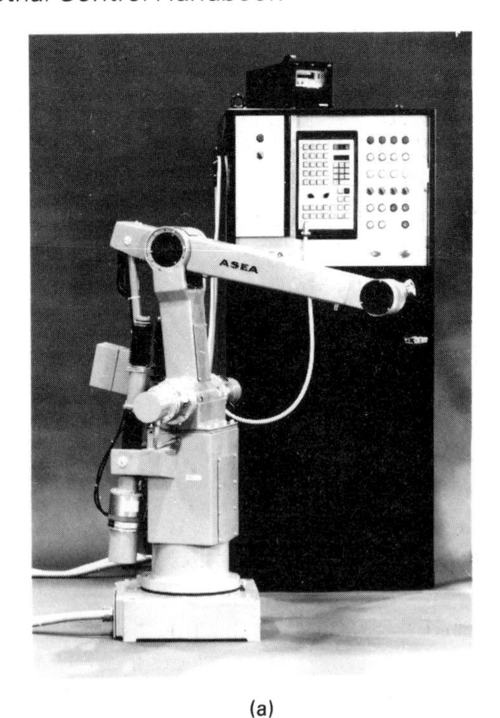

(a)

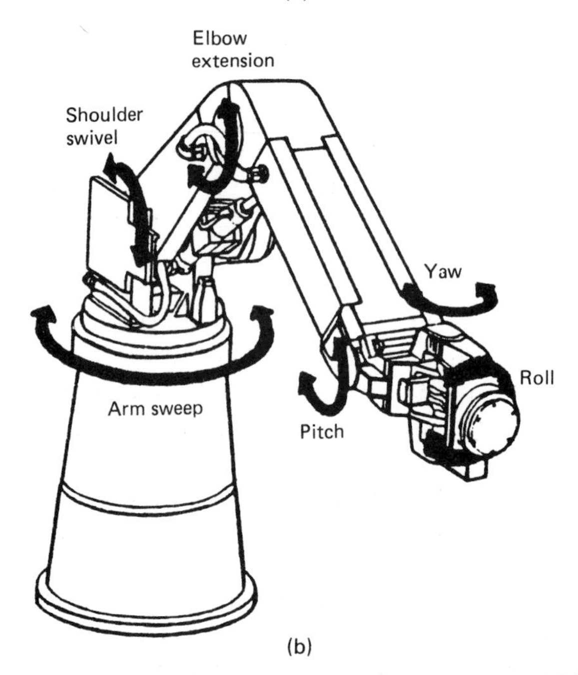

(b)

Fig.4.47 Robotics. (a) An industrial robot. (Photo courtesy of ASEA, Vasteras, Sweden.) (b) A robot with 6 axes (degrees of freedom).

microcomputers and microelectronics based instruments, the impact of robots can, to date, only be described as minimal.

Robots, of which fig.4.47 is typical, are a natural development of the numerically controlled (NC) machine tool. Digitally controlled lathes, millers, grinders, drills, etc., have been around for many years, but are dedicated to one task. A robot is a general purpose machine; the same robot can weld panels on motor cars or spray-paint chairs.

An industrial robot bears little resemblance to the basically humanoid shape of science fiction stories. Most have an articulated arm mechanism which allows the head to be moved to any position and orientation. Robots are described by the number of degrees of freedom possessed by the robot, i.e. the number of axes about which movement can occur. The robot of fig.4.47b, for example, has six axes, which is acceptable for most welding, spraying and assembly applications. (For comparison, the human arm has nine basic axes.)

The robot control system must guide the head through a predetermined path in space. Each axis must therefore have a position transducer (usually a resolver or encoder), an actuator (electric, pneumatic or hydraulic) and a position control servo. The control system must be capable of performing complex trigonometrical calculations.

A robot generally follows a predetermined sequence, and does not usually involve IF-THEN-ELSE decisions. The sequence can be loaded in coordinate form from a programming panel or, more commonly, the robot can 'learn' from a human operator who guides the head through the required operation.

There are generally four applications for robots. The first, and simplest, is the pick-and-place robot which transfers objects from one location to another. A typical operation occurs in metal pressing, where robots are used to take metal sheets from a conveyor to a press, and the resulting formed panels from the press to another conveyor.

Welding robots are common in the automotive industry. Spot welding uses a gripper type head, and continuous welding a continuous wire and flux feed mechanism.

Paint spraying robots use an air driven paint spray head. The head motion required from paint sprayers can be very complex, as the paint must reach every nook and cranny. Many axes of movement are therefore required.

The most complex, and difficult, robot applications involve the assembly of an object from component parts–assembling a gearbox from a collection of frames, shafts, bearings and gears, for example. The difficulties arise because a robot has, at present, no real visual or tactile senses, and consequently cannot see the orientation of objects, or slide objects together by feel. Considerable research is underway to provide these senses, but for the immediate future the simple act of threading a bearing on to a shaft is beyond most robots. Parts must be delivered to an assembly robot with consistent quality, position and orientation. It is usual for the delivery system to cost as much as the robot it is feeding.

Robots perform well where the task is repetitive or dangerous, and the working 'environment' can be tightly controlled. Humans are essential where the work is ill defined and non-repetitive, decisions are required or the work requires visual or tactile skills. Most jobs fall into the latter category, so it would seem that the impact of robots will be small outside of assembly lines until visual and tactile sensors are developed.

4.11. Safety aspects

The safety aspects of any control scheme should be a prime consideration for the designer. The use of computers brings some unique safety problems. In general there are three areas for consideration. The first of these is the failure of plant sensors (or the associated input circuits in the computer interface). The second area is the failure of an output device (or again the associated output circuit). Finally, of course, there is the computer itself, which could suffer some form of memory corruption and start energising or de-energising outputs at random.

The latter possibility is extremely remote, but is surprisingly easy to guard against. Figure 4.5 showed the layout of a typical computer, where all internal communication takes place via three highways. Industrial control computers usually incorporate an independent circuit called a 'watchdog' which is connected to these highways. At regular intervals (typically 0.5 seconds) the computer carries out self-tests, and if these are satisfactory sends a unique code to the watchdog.

If the watchdog fails to receive a correct code *or* receives the right code at the wrong time interval, the computer is assumed faulty and the watchdog forces outputs to some predefined safe state. With careful design, this will give adequate protection against failure of the computer itself.

Input and output failures are, surprisingly, more difficult to guard against. It is not usually possible to predict how an input or output may fail; an output triac, for example, could fail short circuit or open circuit. In simple systems, safety signals should be performed outside the processor. In fig.4.23b, for example, the latching emergency stop push button has been placed directly in series with the computer output. This directly drops out the contactor whose auxiliary contact (input A1.3) removes the output B2.7. Should the computer have failed or output B2.7 stay on, the contactor will stay de-energised as long as the button is pressed.

An alternative, but expensive, approach is to use majority voting. Critical inputs and outputs are triplicated, and the circuit arranged so that at least two (out of three) signals must be present for an input to be present or an output device to react. The system will thus tolerate (and identify) one failure. In exceptionally critical applications, three-out-of-five majority voting may be used.

Majority voting gives very high levels of safety, but its use must be tempered with care. In particular, the probability of a common mode failure affecting all channels simultaneously must be considered. (A trivial example would be having belts and braces whilst suffering a failure of the stitching connecting the waistband to the rest of the trousers!) Similar weak points in a system could be power supplies–instrument air, common hydraulics or even common poor maintenance practice (as happened at Three Mile Island where a triplicated safety system was negated by malpractice).

The ease with which programmable controllers can be reprogrammed makes them vulnerable to overriding of interlocks. This can be overcome, to some extent, by controlling the issue of programming units, but this can cause maintenance problems if the programming unit is also used as a diagnostic tool. One common solution in critical applications is the use of ROMs to hold the program. Reprogramming requires a new ROM, which will deter ill-advised use of 'software frig clips'.

The users of computers and programmable controllers in the UK are strongly advised to study the Health and Safety Executive's consultative document on the subject which covers the above points in more detail.

Chapter 5
Hydraulics

5.1. Introduction

Electric motors and solenoids are widely used where an object is to be moved or a force applied. Electricity is not the only prime mover, however. Hydraulics (the use of liquids) and pneumatics (the use of air) offer a viable alternative in many applications. They are particularly useful where small size is important, precise control is required, a force (as opposed to movement) is needed or explosion hazards preclude the use of electrical machines.

This chapter discusses the use of hydraulics in industrial control. Pneumatic systems are described in chapter 6.

5.2. Fundamentals

5.2.1. Pressure

Hydraulics and pneumatics are largely concerned with pressure in closed systems. It is therefore important to appreciate the precise meaning of the term 'pressure'. It is defined as force per unit area i.e.

$$P = \frac{F}{A} \tag{5.1}$$

where F is the force, A the area acted on, and P the resultant pressure.

In fig.5.1 a ram is supporting a weight of 20 pounds. The ram cross sectional area is 5 square inches, so a pressure of 20/5 or 4 pounds per square inch is transmitted into the fluid.

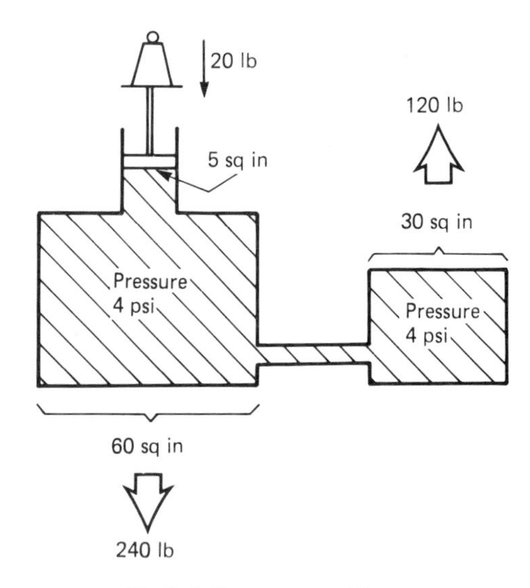

Fig.5.1 Pressure and force.

In the SI system, the unit of pressure is the pascal, defined as 1 newton per square metre. For historical reasons, however, most hydraulic systems in the UK and the USA work in pounds per square inch (psi). The bar (100 kPa; approximately 14.5 psi) is also becoming common. Pressure in hydraulic systems is measured with respect to atmosphere (called gauge pressure). Strictly a g suffix (e.g. psig) should be used. This chapter will mostly follow convention and refer to pressure in psi, and it is hoped the reader will accept this deviation from SI units.

The pressure of 4 psi in fig.5.1 is transmitted uniformly through the confined fluid, and acts with equal force on every unit area of the surface of the system. The base, area 60 square inches, experiences a downward force of $60 \times 4 = 240$ pounds. The top of the side chamber, area 30 square inches, experiences an upward force of $30 \times 4 = 120$ pounds.

Very large pressure can be developed by applying relatively small forces to small areas. A 20 pound force applied to the 0.5 square inch cork in a filled bottle will develop a pressure of 40 psi. A typical bottle has a bottom of area 7 square inches and experiences a force of 280 pounds–which could easily shear the bottom off the bottle.

The equality of pressure in a confined fluid is dignified by the name ‘Pascal’s law’, and is the basis of hydraulics.

In fig.5.2a, a piston of surface area 5 square inches is linked hydraulically to a larger piston of area 50 square inches. If a force

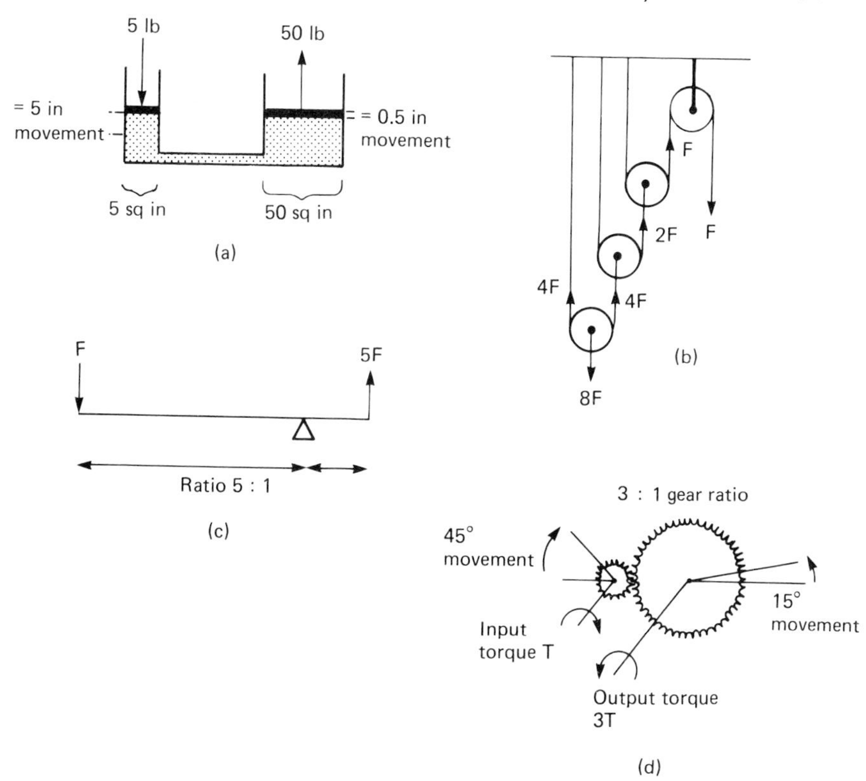

Fig.5.2 Examples of mechanical advantage. (a) Hydraulic. (b) Pulleys. (c) Lever. (d) Gear train.

of 5 pounds is applied to the smaller piston, a pressure of 5/5 = 1 psi will develop in the fluid. This will be transmitted to the larger piston, which will experience an upward force of 50 × 1 = 50 pounds. The input force of 5 pounds has been magnified to an output force of 50 pounds. This force magnification, which is the basis of hydraulic presses and jacks, is the ratio of the areas of the output and input pistons.

Energy must, however, be conserved. Suppose the force causes the smaller piston to move down 5 inches. A volume of fluid 5 × 5 = 25 cubic inches is transferred from left to right. This causes the large piston to rise 25/50 = 0.5 inch. Although there is a force magnification, there is a movement reduction by the same factor.

Work is defined as the product of force and distance moved, and energy as a capacity for performing work (see section 5.2.3). As the force magnification factor and the displacement reduction factor are equal, there is no net system gain or loss of energy (other than that lost by friction). There is an obvious analogy to the pulley of fig.5.2b, the lever of fig.5.2c and the gears of fig.5.2d. In each case a small input force is translated to a larger output force, but displacement is reduced by the same factor, giving conservation of energy.

5.2.2. Volume, pressure, flows and movement

Hydraulic pumps (described further in section 5.4) deliver a constant flow regardless of output pressure (and are accordingly known as positive displacement devices). In electrical engineering terms they are constant current rather than constant voltage sources. A pump specification, for example, could quote a delivery of 7 gallons per minute, with a maximum working pressure of 1000 psi. This does *not* mean that the pump will develop a pressure of 1000 psi in a given system. As we will shortly see, the system pressure is determined by other components. A pump driving fluid into a closed system without some form of pressure regulation will cause a continuing pressure rise until the pump or the piping fails. The pump *creates* the pressure, but does not directly determine the value provided that the pump output meets the system's needs.

Figure 5.3 is a simple hydraulic system, where a hydraulic ram is lifted by a hydraulic pump driven by an electric motor (it could,

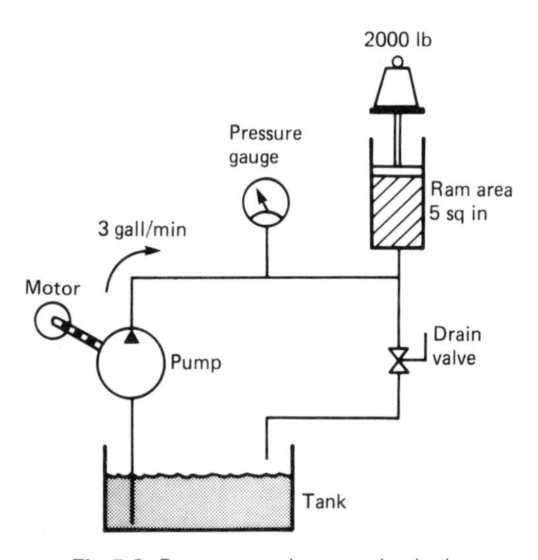

Fig.5.3 Pressure, volume and velocity.

for example, be a car jack in a garage). Suppose there is a 2000 pound load, and the piston area is 5 square inches. With the motor stopped, the pressure indicator will show 2000/5 = 400 psi.

The pump is rated at 3 gallons per minute. If the motor is started, 3 gallons per minute will be transferred to the ram (as a practical aside, it is not desirable to start a hydraulic pump directly on load; this is a 'thought experiment' we are considering). One gallon is 230 cubic inches so 690 cubic inches per minute, or 11.5 cubic inches per second, will be transferred, causing the ram to extend. The speed of advance is simply:

$$\text{velocity} = \frac{\text{volume transferred in unit time}}{\text{ram area}} \tag{5.2}$$

The ram therefore moves at 11.5/5 = 2.3 inches per second.

The ram, however, is still supporting a 2000 pound load, so the system pressure remains at 400 psi. There is no change of pressure when the pump is running (neglecting pressure drops caused by fluid flow in the connecting pipes). Similarly if the pump is stopped and the drain valve opened, the ram and load will fall but the fluid pressure will still be 400 psi.

Let us suppose the drain valve is left open accidentally, and the pump is started. Let us assume also that 2 gpm flows out of the valve. This is a major leak, but the pump will still deliver its 3 gpm, giving a net transfer to the ram of 1 gpm. The system pressure will still, however, be 400 psi, and the ram will still rise, albeit at a reduced speed of about 0.8 inches per second (by a similar calculation to that above).

Fluid flow in hydraulic pipes should be non-turbulent (called laminar flow). Turbulent flow is undesirable on several counts: it creates wear, it is noisy and it causes friction which in turn lowers the system efficiency. The lost energy appears as heat which must be removed by oil coolers.

In general, hydraulic fluid flows should be kept around 10 fps (3 ms^{-1}) to avoid turbulence. Flow velocity is given by flow rate/pipe area. Note that the area is the *inside* area of the pipe, and the area is proportional to the square of the inside pipe radius. Halving a pipe's diameter leads to a fourfold increase in flow.

Flow, particularly through restrictions, causes a pressure reduction as illustrated in fig.5.4a. The pump and associated (unshown) pressure regulating components maintain a constant

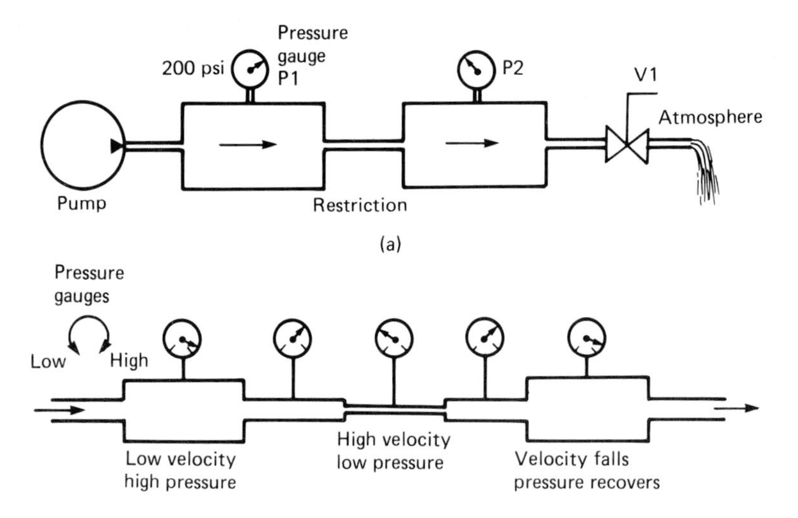

Fig.5.4 Pressure drops in a system. (a) Pressure drop caused by a restriction. (b) Velocity/pressure relationship.

pressure in the left-hand chamber of 200 psi. With the valve V1 closed there is no flow, so by Pascal's law P2 must also be 200 psi, and there is equal pressure at all points.

If V1 is cracked open, fluid will flow from the pump, and a pressure drop–say, 5 psi–will be created across the restriction. P1 will remain at 200 psi, but P2 will fall by 5 psi to 195 psi. As the valve is opened further, the flow and the pressure drop will increase, lowering P2 further. Calculation of pressure drops across restrictions is analogous to orifice plate calculations described in section 5.2 of Volume 1. The calculations are, however, complicated by the laminar flow requirements for hydraulic fluid.

Energy in a hydraulic system is conveyed as a mixture of potential energy (pressure) and kinetic energy (mass flow). As flow velocity increases, pressure falls. In a system not venting directly to atmosphere and with changing diameters as in fig.5.4b, pressure will rise and fall as a complex function of the inverse of flow.

Work is done whenever a force is exerted through a distance, i.e.:

$$\text{work} = \text{force} \times \text{distance moved} \tag{5.3}$$

The British (fps) unit of work is the foot pound; the SI unit is the joule (J).

Mechanical systems are more concerned with the *rate* of doing work (how fast a load is lifted by a crane, for example). Rate of doing work is called power, which is defined as:

$$\text{power} = \frac{\text{work}}{\text{time}} = \frac{\text{force} \times \text{distance}}{\text{time}} \tag{5.4}$$

The British unit is the horsepower (hp), defined as 550 foot pounds per second. The SI unit is the watt (joules per second). One horsepower is 746 watts (0.746 kW).

It follows that power in a hydraulic system is related to flow rate and system pressure, i.e.:

$$\text{power} = K \times \text{flow rate} \times \text{pressure} \tag{5.5}$$

where K is a constant dependent on the units being used. If British units are used, with flow rate in gallons per minute and pressure in psi:

$$\text{power} = \frac{\text{gpm} \times \text{psi}}{1714} \tag{5.6}$$

5.3. Pressure regulation

Figure 5.3 controlled the ram extension by starting and stopping the pump. This is not a particularly convenient (or desirable) control system. Figure 5.5a is one possible attempt at a solution. The pump is kept running all the time, and V1 is opened to raise, and V2 to lower, the piston. With V2 open the piston falls under gravity and fluid is returned to the tank. Unfortunately, when the piston is stationary or being lowered, the pump is driving into a closed system. As explained earlier, this will lead to a theoretically infinite pressure rise; in practice either the pump or the piping will fail.

Figure 5.5b shows a more practical solution. A pressure regulating device is connected from pump output back to the tank. This device will be closed for low pressure and opens when some preset pressure is reached. Figure 5.3 requires a pressure of 400 psi to raise the load. Assuming the same conditions in fig.5.5, a suitable setting for the regulator could be 500 psi.

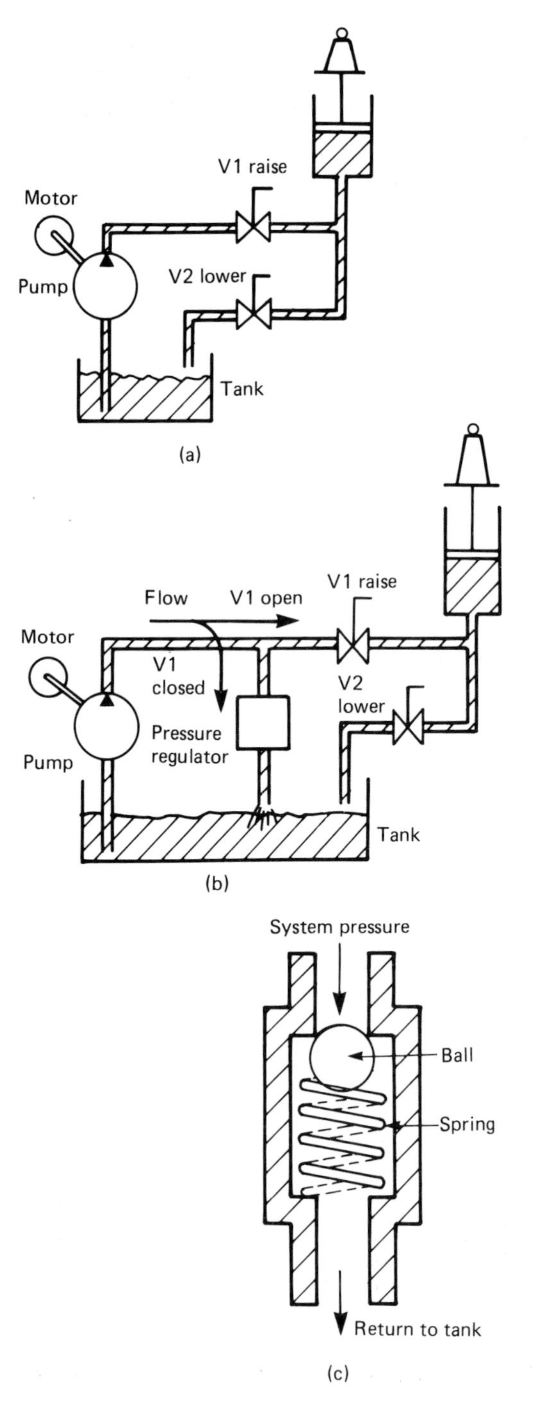

Fig.5.5 Pressure regulation. (a) System without pressure regulation. (b) System with pressure regulation. (c) Simple pressure regulator.

With V1 closed, the regulator will open and the pressure indicator will show 500 psi. With V1 open, the system pressure will fall and the indicator will show the 400 psi necessary to move the load. Note that with V1 closed the *entire* pump flow will go through the pressure regulator. The pressure regulator is called a relief valve. These will be discussed in more detail later in section 6.3.3, but can be considered to act as in fig.5.5c. Under low pressure conditions the spring keeps the ball seated, blocking the flow. When the pressure rises sufficiently high the ball lifts off its seat, allowing fluid to pass. The spring tension (usually adjustable) and the ball cross section determine the pressure at which the valve starts to open.

5.4. Pumps

5.4.1. Introduction

The function of a hydraulic pump is to provide an adequate volume of fluid to the rest of a hydraulic system. The pump is usually driven by a constant speed electric motor. As explained earlier, pumps for hydraulic systems move a fixed volume of fluid per revolution regardless of outlet pressure, and are consequently known as positive displacement, or hydrostatic, pumps. Non-positive displacement devices such as the centrifugal pump of fig.5.6a and the propeller of fig.5.6b are called hydrodynamic pumps and are used purely to shift fluids from one location to another.

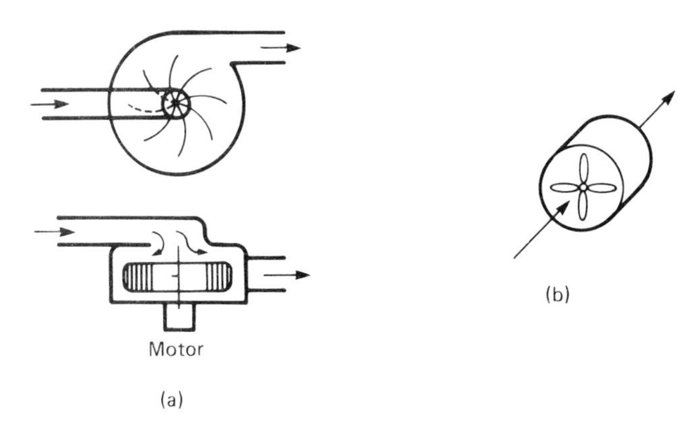

Fig.5.6 Non-positive displacement pumps. (a) Centrifuge blower. (b) Propeller.

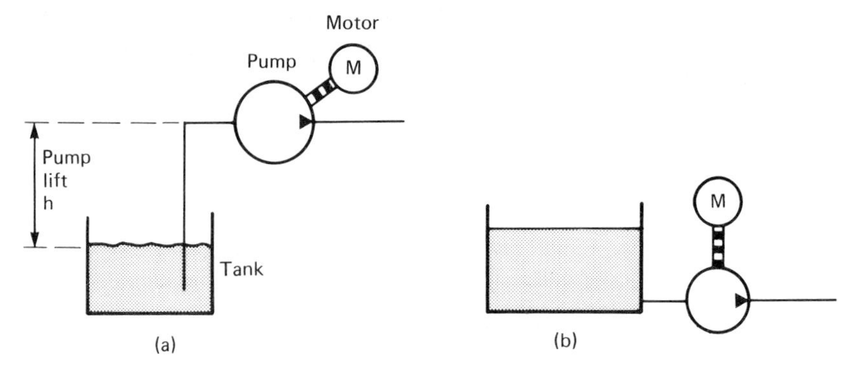

Fig.5.7 Pump to tank connection. (a) Definition of pump lift. (b) Self priming pump.

The symbols for a pump, motor and tank are shown in fig.5.7. The arrow shows the direction of flow (compare this with the rotary motor symbol of fig.5.31). Figure 5.7a also illustrates the fact that in many applications a pump must lift fluid a height 'h' from the tank to the pump inlet. To achieve this, the pump must be capable of creating a partial vacuum at its inlet.

There are, not surprisingly, limits to h. In theory, if the pump could create a perfect vacuum, atmospheric pressure could support a column of oil about 30 feet (9 metres) high. In practice far lower lifts are used. Liquids tend to produce vapour as pressure falls below atmospheric, creating bubbles in the fluid which collapse abruptly at the transition from low to high pressure inside the pump. This is known as cavitation and causes severe wear inside the pump. Similar effects occur from entrapped air and water vapour in the fluid. In general lifts of less than 6 feet (about 2 metres) are recommended, and the self-priming arrangement of fig.5.7b is often used (provided care is taken to avoid passing any sludge from the tank bottom to the pump).

Cavitation is also induced by high flow velocities. Pump inlet lines will always handle total pump flow, and consequently have the maximum flow in any hydraulic system. Inlet lines should be as large as possible to reduce flow velocity and minimise cavitation and frictional losses.

Pumps are specified by their delivery (in gallons per minute, say) and their maximum working pressure at some fixed input shaft speed (usually 1500 rpm to allow use of a simple 3 phase 4 pole induction motor). Delivery rate is proportional to shaft speed, and is sometimes given in terms of the displacement for one rotation of the pump shaft. Pumps are often sold as a pump/electric motor assembly.

5.4.2. The gear pump

The gear pump is the simplest and most robust hydraulic pump, as it has only two moving parts and these are rotating at uniform speed. The principle is shown in fig.5.8.

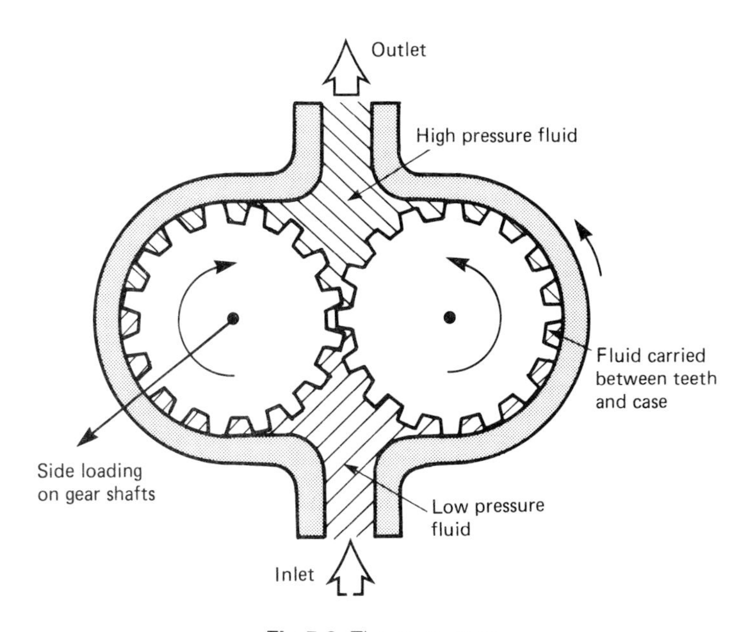

Fig.5.8 The gear pump.

A partial vacuum is created at the inlet as the gear teeth come out of mesh at the centre. This partial vacuum causes fluid to enter the inlet chamber. Fluid is now entrapped between the outer teeth and the housing and carried round to the outlet port where it is discharged. Displacement is determined by the volume between teeth and the number of teeth.

A close machined fit between the teeth ensures no oil leaks back where the teeth mesh at the centre. Close machined side plates are also required to avoid leakage over the gear faces. These are often designed as replaceable wear plates.

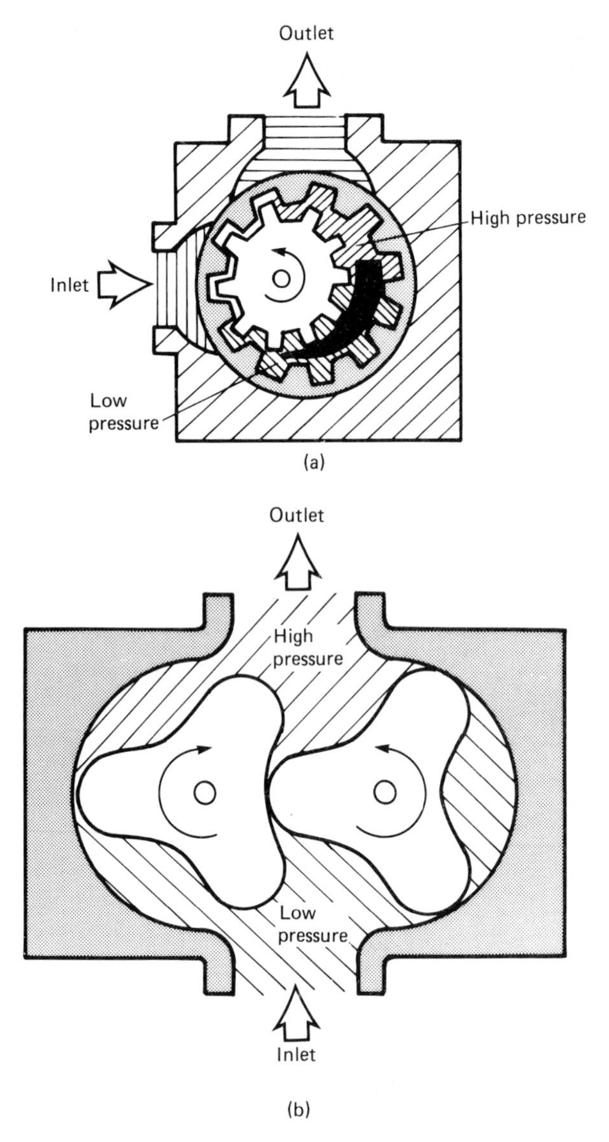

Fig.5.9 Variation on the gear pump. (a) Internal gear pump. (b) Lobe pump. Rotors are linked by external gears.

The difference in pressure between the inlet and outlet port causes a large side load to be applied to the gear shafts (at 45° to the centre line of the pump).

There are many variations on the gear pump principle; fig.5.9a is called an internal gear pump and fig.5.9b a lobe pump. Gear pumps are, in general, simple, robust and reliable. The side loading is usually the limiting factor.

5.4.3. Vane pumps

The basic principle of the vane pump is shown in fig.5.10a. A slotted rotor is fitted with several free moving vanes, and turns

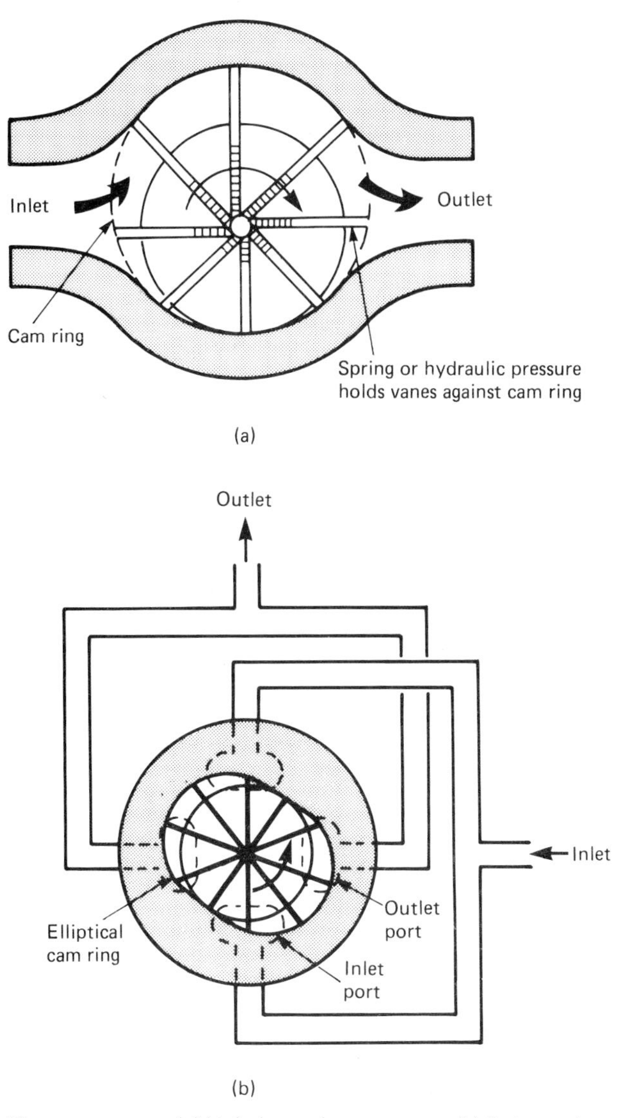

Fig.5.10 The vane pump. (a) Unbalanced vane pump. (b) Balanced vane pump.

inside a cam ring. Hydraulic or spring pressure keeps the vanes in contact with the cam ring at all times.

The rotor and cam ring centres are offset, so oil from the inlet chamber enters the compartments between vanes at the top of the diagram, and is carried to the outlet port. The displacement is determined by the throw of the vanes and the rotor thickness.

As before, the difference in pressure between inlet and outlet ports causes a severe side loading on the shaft. This can be overcome by the balanced arrangement of fig.5.10b. Here two inlet and outlet ports are used with an elliptical cam ring and chamber. Both pairs of ports induce side loads, but they are equal and opposite, and hence cancel.

Vane pumps are obviously more complex than gear pumps, with many moving parts. Vane tips and the cam ring are also prone to wear, but this is mitigated by the hydraulic fluid itself acting as a lubricant. Seals are also required between the face plates and the rotors/vanes.

5.4.4. Piston pumps

Piston pumps are based on the principle of fig.5.11a. As the input shaft rotates the piston oscillates up and down. On the downward stroke, fluid is drawn through the check valve CV1 to fill the space above the piston. On the upward stroke, fluid is discharged from above the piston, through check valve CV2 to the outlet port. This simple arrangement, however, delivers a pulsing flow, and is unbalanced mechanically and hydraulically. Practical piston pumps utilise several pistons to even out the flow and balance the rotating components. Pump designs also eliminate the need for the check valves in fig.5.11b.

Figure 5.11b is called a radial piston pump. The cylinder block carries several pistons and rotates, off centre, inside the pump casing. The pistons are kept in contact with the casing at all times by springs, hydraulic pressure or mechanical linkage. The inlet and outlet ports are at the centre of the pump, and open on to chambers.

As the cylinder block rotates, pistons are moving out in the region of the inlet chamber, and in adjacent to the outlet chamber. Fluid is thus conveyed from inlet to outlet port, without the need for separate check valves. Displacement is dependent on the number of pistons and their bore/stroke.

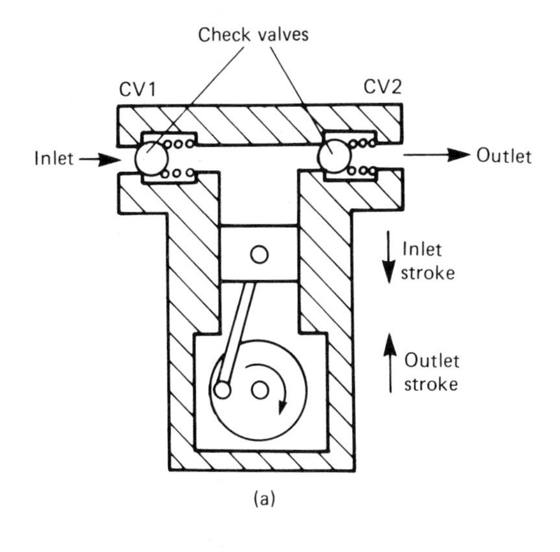

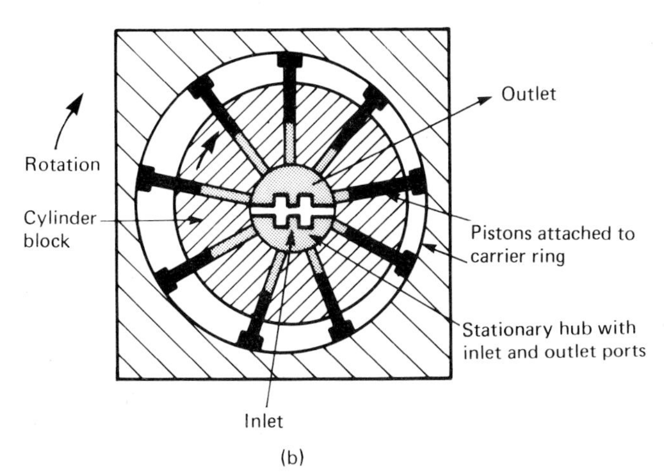

Fig.5.11 Piston pumps. (a) Single cylinder piston pump. (b) Radial piston pump.

An alternative piston pump, called a swash plate or axial pump, is shown in fig.5.12a. The cylinder block is directly connected to the input shaft. The pistons are attached through ball and socket joints to a shoe plate which revolves on a fixed, angled plate (called the swash plate). As the input shaft (and the cylinder block and pistons) rotates, the angled swash plate causes the pistons to reciprocate, transferring fluid from the inlet to the outlet port as before.

The angle of the swash plate determines the piston stroke giving displacement control from zero to maximum flow. If the

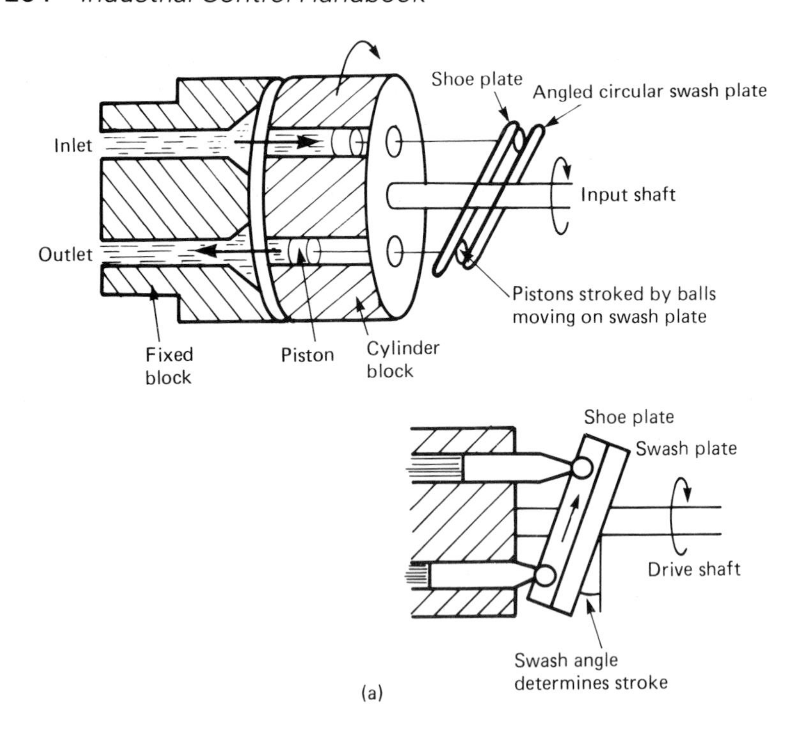

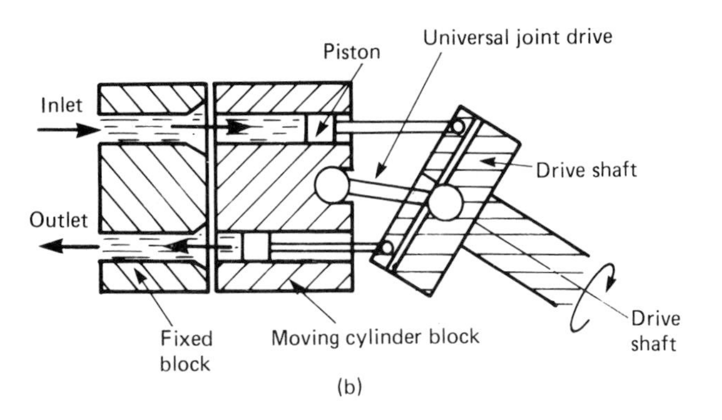

Fig.5.12 Variations on the piston pump. (a) Inline swash plate pump. (b) Bent axis pump.

swash plate goes beyond the vertical (zero flow) position the pump direction is reversed. The swash plate angle can be adjusted manually by a lever, or via electric or hydraulic actuators. Such pumps are said to be variable displacement.

The bent axis pump, shown in fig.5.12b, is a variation on the piston pump. Here reciprocation of pistons is achieved by the angular difference between the drive shaft and the axis of the cylinder block. The cylinder block is driven by a universal joint to avoid side loads on the pistons. Again pump displacement can be controlled by varying the shaft angle.

5.4.5. Pump unloading

Figure 5.5b used a pressure relief valve to protect the system when no flow was required from the pump. This arrangement keeps the pump output at a high pressure, and by equation 5.6 the pump inlet power will remain high, even when no useful work is being done in the system. Apart from consuming electricity unnecessarily, the input power is converted to heat and leads to a rapid rise in fluid temperature.

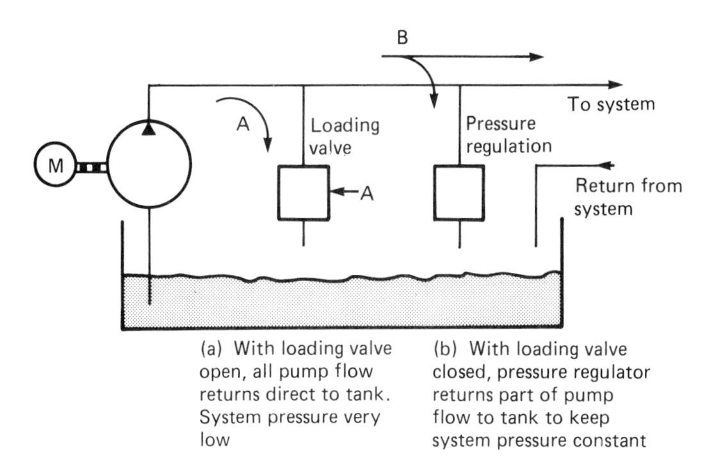
B
To system
A
Loading valve
Pressure regulation
M
A
Return from system
(a) With loading valve open, all pump flow returns direct to tank. System pressure very low
(b) With loading valve closed, pressure regulator returns part of pump flow to tank to keep system pressure constant

Fig.5.13 Pump loading valve.

If the hydraulic pump is required to be kept running at all times, but is only called upon to supply fluid intermittently, the arrangement of fig.5.13 may be used. A separate valve (called a loading or unloading valve, according to the sense of the control signal) goes from pump output back to the tank. When this valve is opened by the control signal (electrical or hydraulic) the entire pump output is returned direct to the tank. The pump output pressure is minimal, and input power to the motor greatly reduced. When the valve is closed, the relief valve operates and sets systems pressure as normal.

Often the relief valve and unloading valve are combined by utilising a relief valve with remotely adjustable setting. Unloading is then obtained by setting a low relief pressure.

5.4.6. Combination pumps

Many hydraulic systems require two separate operating conditions, usually high flow/low pressure and minimal flow/high pressure. An example is a clamping vice. High flow/low pressure will be required to bring the vice jaws from their open position until they touch the object to be clamped. Once in contact, the jaws require high pressure but no flow.

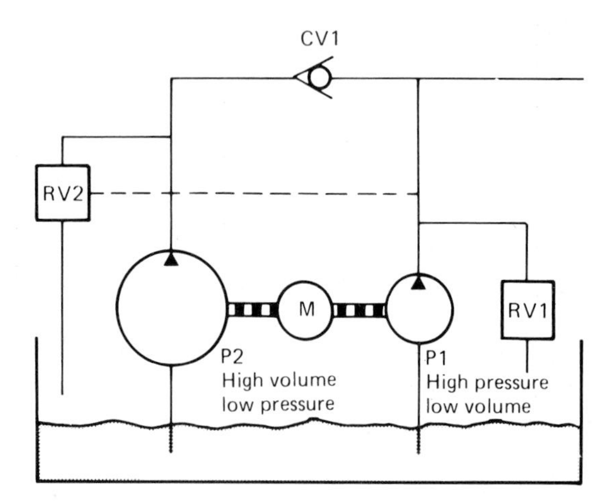

Fig.5.14 Combination pump.

These requirements can be provided by the two pump arrangement of fig.5.14. Pump P1 is a high pressure low volume pump, and P2 a high volume pump. RV1 is a normal relief valve, and RV2 a relief valve operated by a remote pressure (shown dotted and called a pilot line). RV1 is set at, and RV2 set lower than, the high pressure needed by the system.

In the high volume mode system, pressure is low as only friction needs to be overcome. RV1 and RV2 are both shut, and both P1 and P2 deliver fluid. When higher pressure is needed (e.g. when the vice jaws clamp) system pressure rises, causing

RV2 to open fully. Pump P2 now unloads via RV2, and check valve CV1 isolates it from the rest of the system. The system pressure is determined by the relief valve RV1 which is set at the required level.

The arrangement of fig.5.14 is very common, and complete assemblies called combination pumps (complete with two pumps, relief and check valves) are manufactured.

5.5. Hydraulic valves

5.5.1. Graphical symbols

Process control valve symbols have been used so far, but these are inadequate to describe the wide range of hydraulic valves (check valves, proportional valves, changeover valves, relief valves, etc.). A range of graphical symbols for hydraulic valves has evolved.

There are two basic types of valves. Infinite position valves (of which a relief valve is an example) can take any position between fully closed and fully open. Finite position valves (of which a directional valve is an example) can only be fully open or fully closed. Finite position valves generally switch flows between different ports.

The basic valve symbol is a square. Infinite position valves have a single arrow, as shown in fig.5.15a. The arrow shows flow direction and is generally drawn in the non-operated position. Control of the valve is shown by symbols on the side of the square. Figure 5.15b incorporates a spring push to right, and pilot pressure push to left. Pilot pressure increases the flow, and spring pressure reduces the flow as the pilot pressure is reduced.

Figure 5.15c therefore is the symbol for a relief valve (inlet pressure increases flow). The arrow on the spring shows adjustable tension.

Finite position valves generally have four ports, as in fig.5.16a. The pressure port P is connected to the pump, and the tank port T to the tank. The A and B ports are connected to the device being controlled. In fig.5.16b a reversing valve is connected to a ram. In the extend position, P and B are linked as are A and T (to return fluid to the tank from the space above the piston). In the retract position, P and A are linked and B and T.

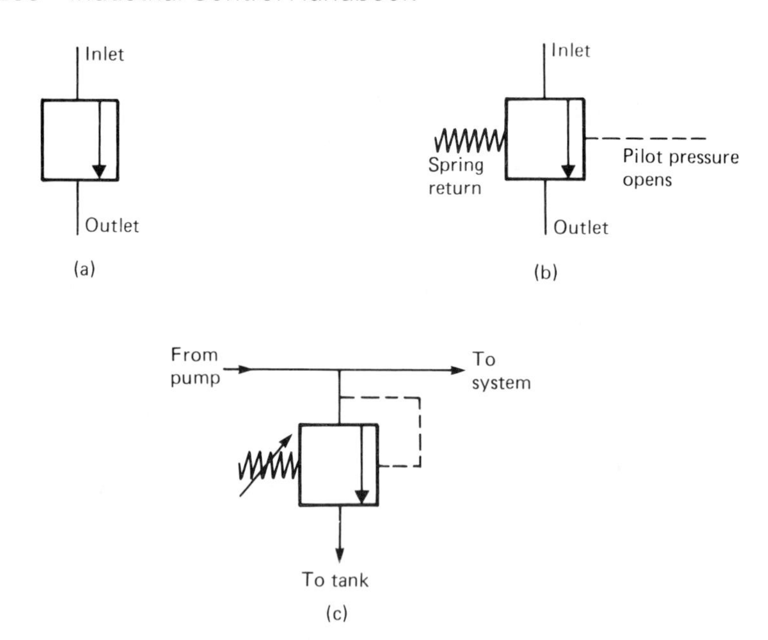

Fig.5.15 Valve graphical symbols. (a) Infinite position valve symbol. (b) Pilot pressure open, spring close. (c) Pressure relief valve.

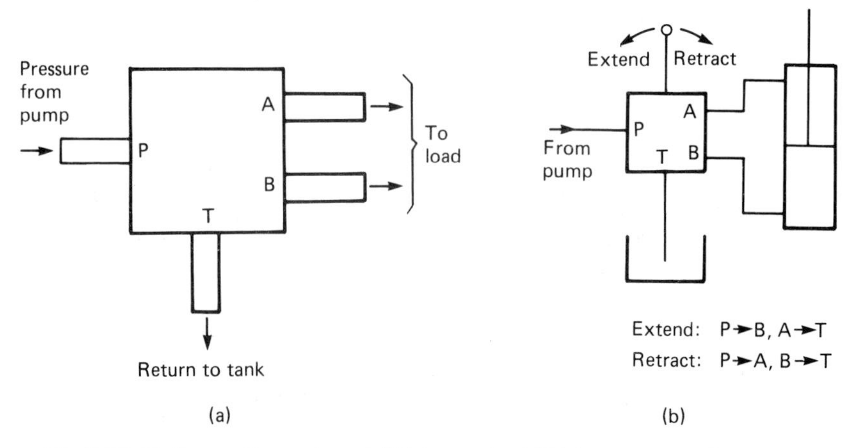

Fig.5.16 Finite position valves. (a) Valve ports. (b) Valve connected to a ram.

Finite position valve symbols are constructed from squares, one for each possible valve position. Lines and arrows inside the squares show the port linking in each position. Figure 5.17 shows various examples: fig.5.17a is a reversing valve with two positions; fig.5.17b is a three position reversing valve with centre off.

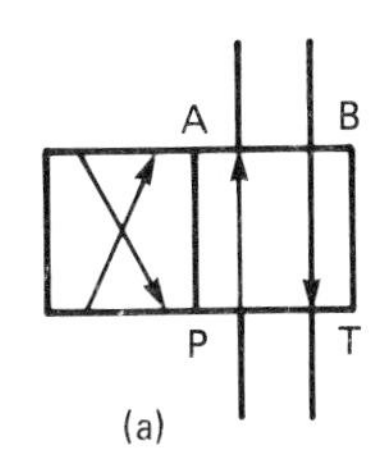

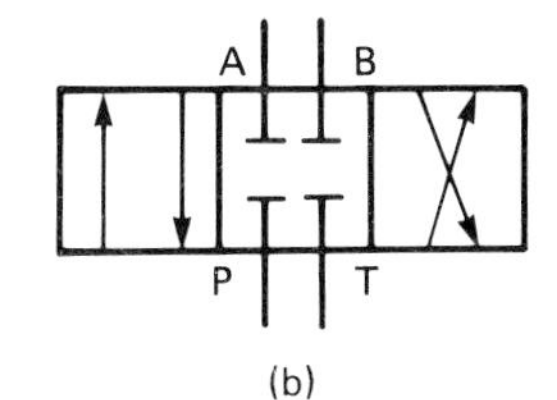

Fig.5.17 Finite position valves graphical symbols. (a) Two position reversing valve. (b) Three position reversing valve, centre off.

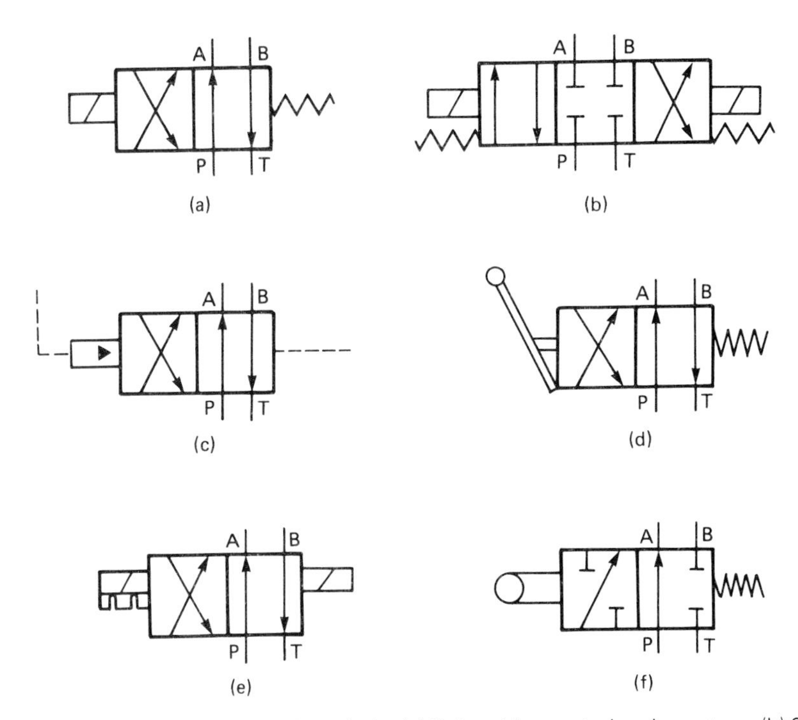

Fig.5.18 Actuator graphical symbols. (a) Solenoid operated spring return. (b) 3 position solenoid operated spring return to centre. (c) Pilot operated direct pressure return. (d) Hand operated spring return. (e) Solenoid operated with detent. (f) Cam operated, spring return, changeover valve.

Actuating control is shown by symbols at the ends of the valve. Figure 5.18 shows various options. These are for the most part self-explanatory, with the exception of fig.5.18e, with detents. Detent valves hold the last position. The solenoid in fig.5.18a must be continually energised to reverse the AB lines. The A solenoid in fig.5.18e need only be pulsed. The AB lines will then be reversed until the B solenoid is pulsed.

5.5.2. Check valves

The check valve only allows flow in one direction and as such is analogous to the electronic diode. It is used to block unwanted flow. In its simplest form it is an inline ball and spring, as in fig.5.19a. Pressure from the left lifts the ball off its seat, and unimpeded flow is permitted. Pressure from the right pushes the ball tight on to the seat, and flow is blocked. Graphically the valve is represented by fig.5.19b, which is rather unnecessarily complex. Usually the representation of fig.5.19c is used. There are many variations of check valves, but most are similar in principle to fig.5.19a.

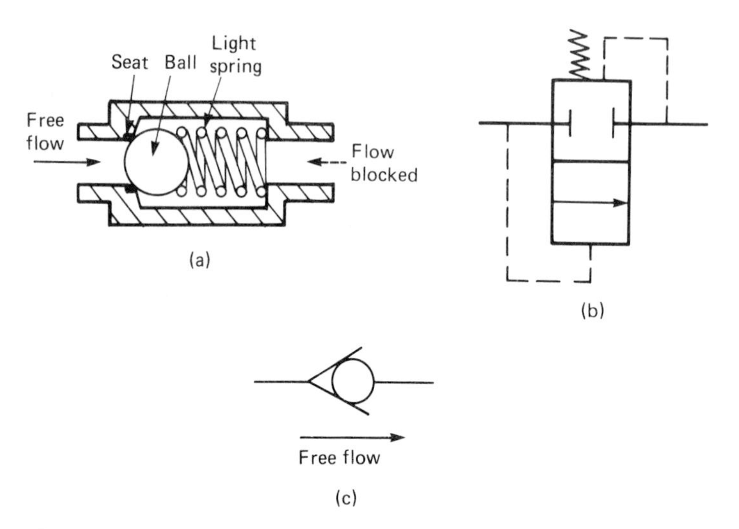

Fig.5.19 The check valve. (a) Construction. (b) Functional symbol. (c) Conventional symbol.

In lifting ram applications similar to fig.5.20a, a changeover valve is used to raise or lower a load. Pump pressure is used to raise the load, and gravity to lower, with oil being returned to the tank. In the valve centre (stop) position, any leakage in the control valve will cause the load to creep down. A check valve in the pipe to the ram will prevent the creep, but prevent lowering when required. In this type of application a pilot operated check valve, shown in fig.5.20b, is required.

With no pilot pressure, the valve functions as a normal check valve, flow from A to B being permitted, and blocked from B. If pilot pressure is applied, the valve is open at all times, and flow is

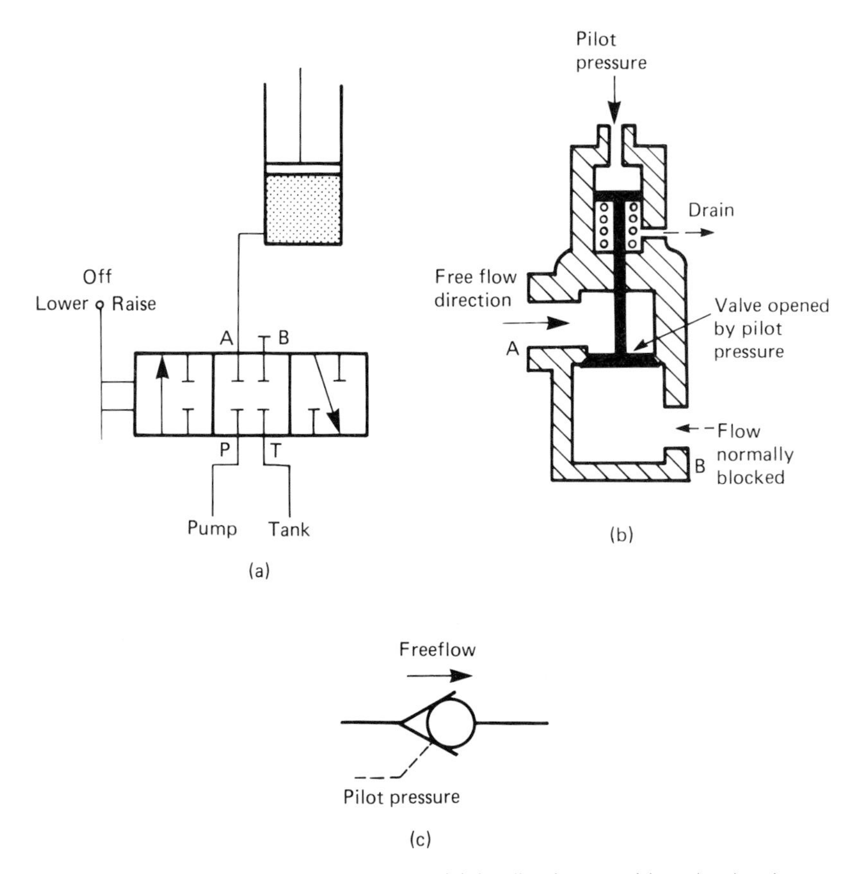

Fig.5.20 Pilot operated check valve. (a) Application requiring check valve. (b) "2C" type check. (c) Symbol.

allowed freely in both directions. The symbol for a pilot operated check valve is shown in fig.5.20c. This device would be inserted in the line from the control valve to the ram. When raised the check valve permits flow to the ram. In the stop position, the check valve prevents creep. To lower, pilot pressure is applied to the check valve and 'down' selected on the control valve.

5.5.3. Relief valves

Relief valves are used for protection and pressure regulation in every hydraulic system. As explained earlier in section 5.3, hydraulic pumps are positive displacement devices, and as such some external means is needed to prevent over-pressuring a hydraulic system.

The simplest relief valve is the spring and ball arrangement of fig.5.21a. The spring holds the valve shut until a preset pressure (set by the adjustable spring tension) is reached, when the valve cracks open. If pressure increases further, flow increases further. The valve is therefore an infinite position valve, and has the symbol of fig.5.21b.

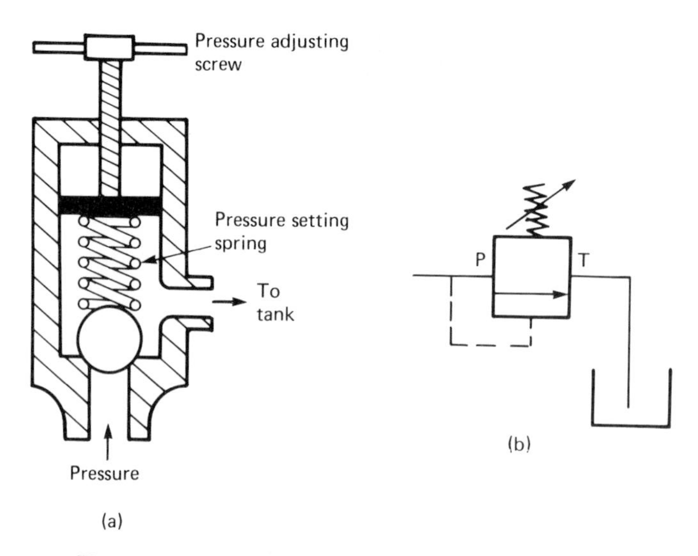

Fig.5.21 The relief valve. (a) Construction. (b) Symbol.

The pressure at which flow commences is called, not surprisingly, the 'cracking pressure'. There will also be, for a given valve, a pressure when full flow is obtained. Again not surprisingly, this is called the 'full flow pressure'. The difference is called the 'pressure override' and is an indication of the pressure regulation the valve will provide.

Pressure override is related to the spring tension in a simple relief valve. When a small, or precisely defined, override is required, a balanced piston relief valve (shown in fig.5.22) is used.

The piston in the valve is free moving, but is normally held in the lowered position by a light spring, blocking flow to the tank. Fluid is permitted to pass to the upper chamber through a small hole in the piston. The upper chamber is sealed by an adjustable spring loaded poppet. In the low pressure state, there is no flow past the poppet, so pressure on both sides of the piston is equal and spring pressure keeps the valve closed.

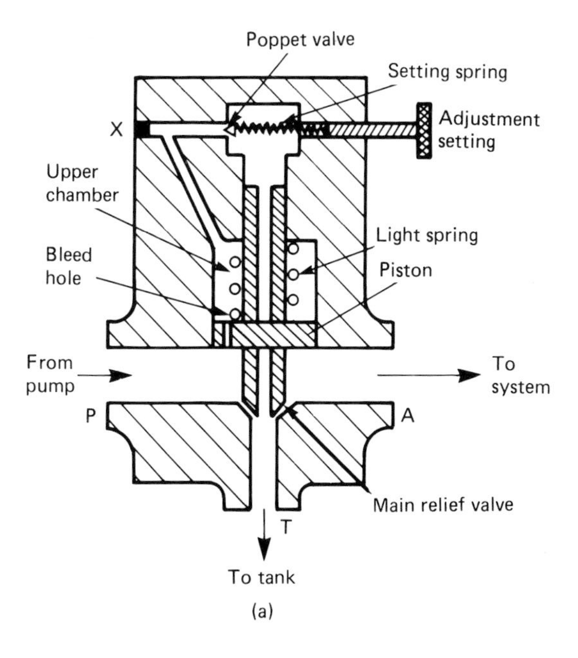

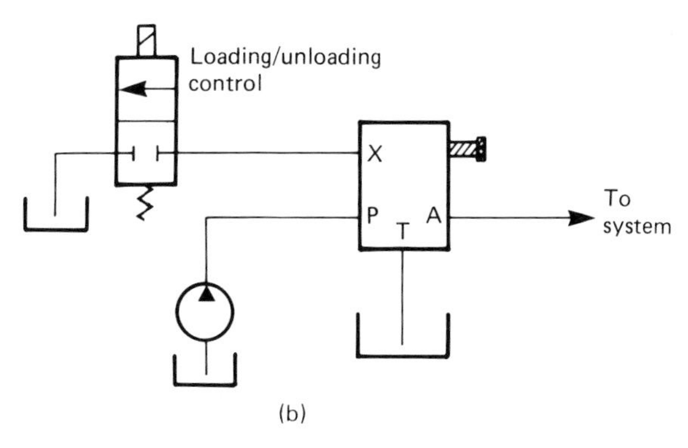

Fig.5.22 Balanced piston relief valve. (a) Construction. (b) Used as unloading/ regulation valve.

When the fluid pressure rises, the poppet cracks and a small flow of fluid passes from the upper chamber to the tank via the hole in the piston centre. This fluid is replenished by fluid flowing through the hole in the piston. With fluid flow, there is now a pressure differential across the piston, which is acting only against a light spring. The whole piston lifts, releasing fluid around the valve stem until a balance condition is reached. Because of the light restoring spring, a very small override is achieved.

The balanced piston valve can also be used as an unloading valve. The plug X is a vent connection, and if this is removed fluid will flow from the main line through the piston. As before, this will cause the piston to rise and flow to be dumped to the tank. Controlled loading/unloading can be achieved by the use of a finite position valve connected to the vent connection, as in fig.5.22b. When the solenoid is energised, pump output is directed direct to the tank.

5.5.4. Finite position (changeover) valves

The basic principle of finite position valves was described earlier in section 5.5.1 and fig.5.16. Almost all have a pressure P connection, a tank T connection, and plant connected ports denoted A, B. On/off valves are achieved by blocking one port.

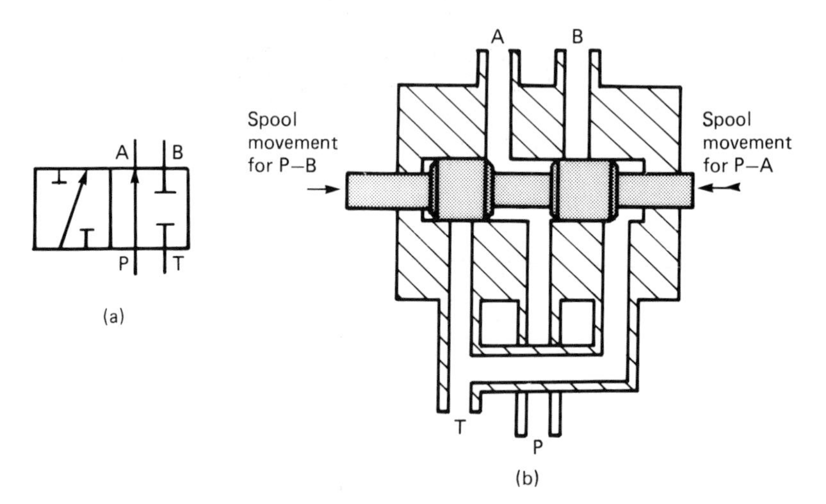

Fig.5.23 Two way spool valve. (a) Symbol. (b) Construction.

The commonest type is the spool valve shown in fig.5.23. The spool moves inside the valve body, and raised portions called 'lands' block ports to give the required operation. The valve illustrated changes over pressure between the A and B ports, the A port being selected with the spool to the left and the B port with the spool to the right. The tank connection is not used, and only serves to drain leakage from the valve.

Changes in valve operation are achieved by utilising spools with different land patterns, whilst maintaining the same valve

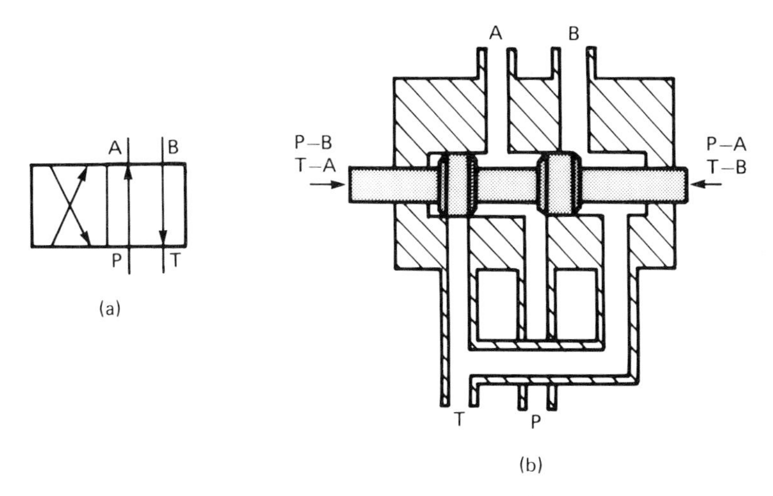

Fig.5.24 Four way spool valve. (a) Symbol. (b) Construction.

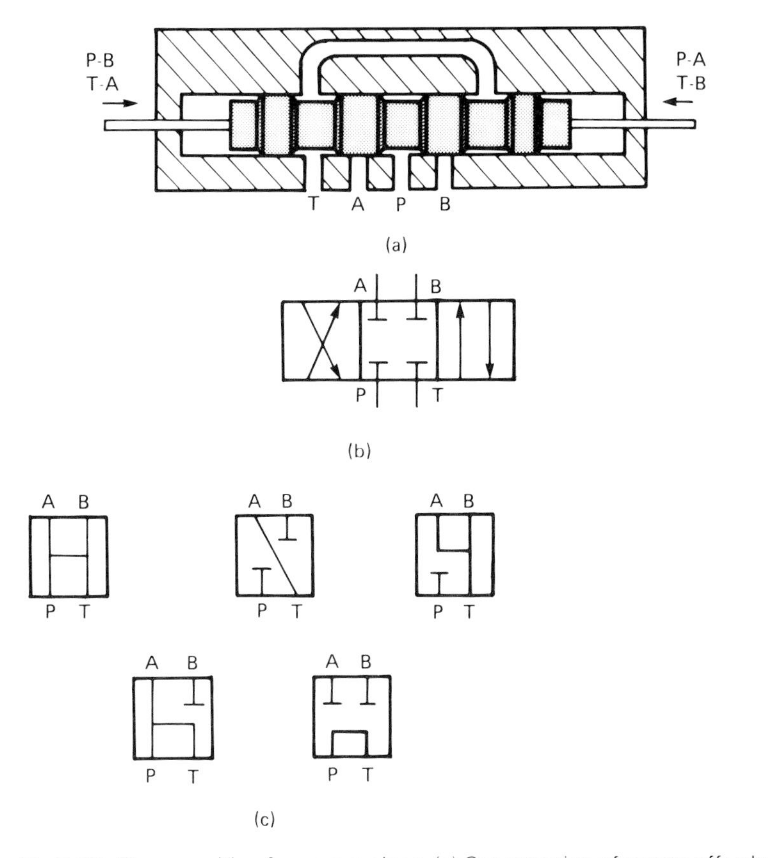

Fig.5.25 Three position four way valves. (a) Construction of centre off valve. (b) Symbol. (c) Common centre position connections.

body. Figure 5.24 shows a changeover valve using the same valve body as in fig.5.23. This approach obviously simplifies valve manufacture.

Three position valves are constructed in a similar manner. Figure 5.25a shows a three position changeover valve with centre off position. Three position valves can be obtained with many different centre positions, some of which are shown in fig.5.25c. These different valve configurations are obtained by utilising the same valve body with different land patterns on the spools.

Spool movement can be achieved manually by a lever or striker, electrically by solenoids at each end of the spool, or hydraulically by pilot pressure applied to the spool end.

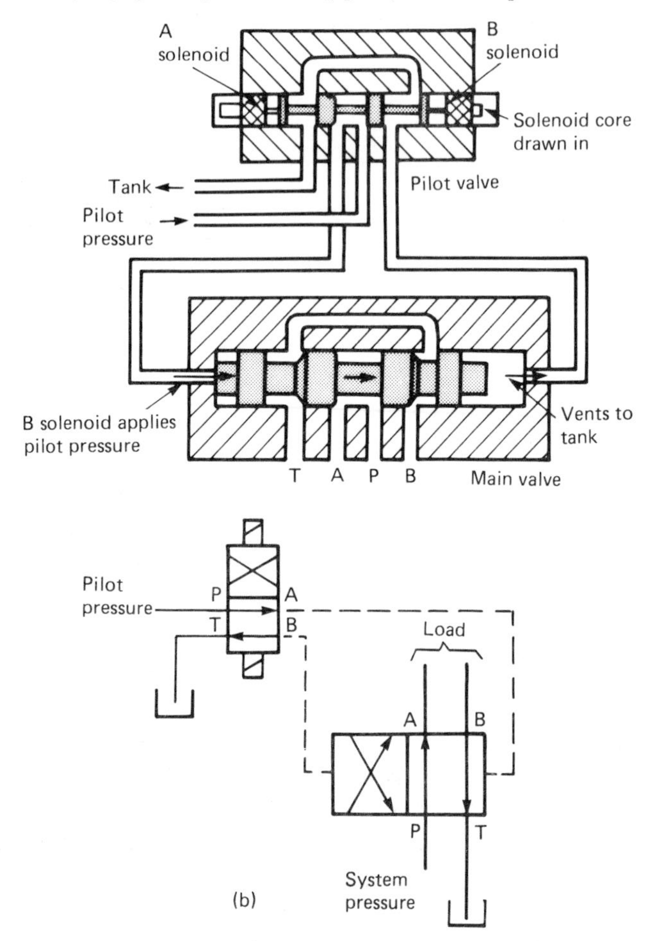

Fig.5.26 Pilot operated valve. (a) Construction. Power applied to B solenoid has moved pilot spool to left. This applies pilot pressure to left hand end of main spool, shifting spool to right. (b) Symbol.

Valves for use at high pressure or high flows require a higher spool force than can be reasonably obtained from a solenoid. In these cases a two stage valve is used, as in fig.5.26a. Solenoids operate a small valve which in turn applies pilot pressure to shift the large spool in the main valve. This arrangement is called pilot operation and is shown schematically in fig.5.26b.

5.6. Linear actuators

Actuators are the output components of a hydraulic system. Linear actuators, which produce a straight line motion, are described in this section. Rotary actuators (motors) are described in section 5.7.

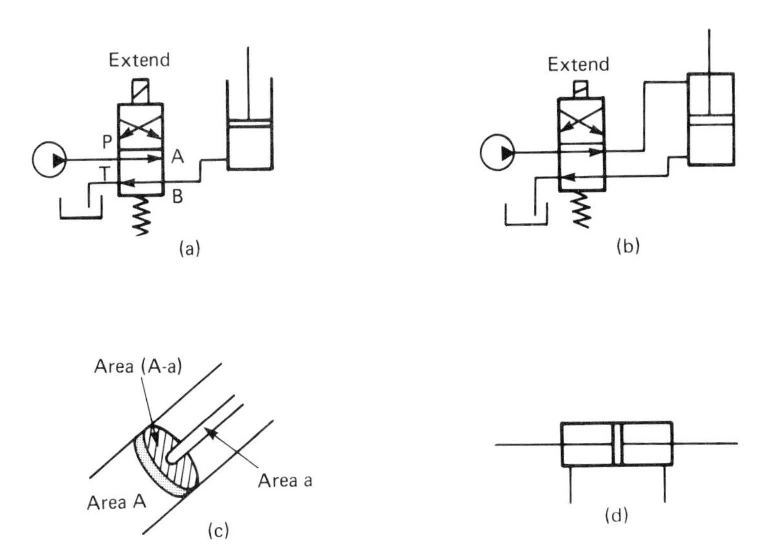

Fig.5.27 Linear actuators. (a) Single acting cylinder. (b) Double acting cylinder. (c) Effect of piston rod. (d) Non-differential cylinder.

All linear actuators are based on fig.5.27 and consist of a ram which is moved by the introduction of fluid to the enclosed space below the piston. Figure 5.27a is called a 'single acting' cylinder; it is extended by hydraulic pressure, but returns under gravity which pushes fluid back to the tank. Single acting cylinders must be mounted vertically unless spring return is employed.

Figure 5.27b uses hydraulic pressure to extend and retract the cylinder. This arrangement is known as a 'double acting' cylinder, and has a power stroke in each direction. Examination of fig.5.27c shows that the area of piston available for the extend stroke is greater than for the return stroke by the area of the connecting rod. The extend stroke is therefore lower than the return stroke (for constant flow rate) but is capable of exerting a greater force. Such an actuator is often called a 'differential' cylinder.

The double rod cylinder of fig.5.27d has equal areas for extend and retract strokes, and is consequently called a 'non-differential' cylinder.

Cylinders are as simple as fig.5.27 suggests, consisting of a barrel, piston, rod, caps, glands and piston seals. They are classified by available stroke, piston area and maximum operating pressure. Force, speed, pressure and flow rate are related by:

$$\text{speed} = K \times \frac{\text{flow rate}}{\text{piston area}} \tag{5.7}$$

$$\text{force} = L \times \text{pressure} * \text{piston area} \tag{5.8}$$

where K and L are scaling factors.

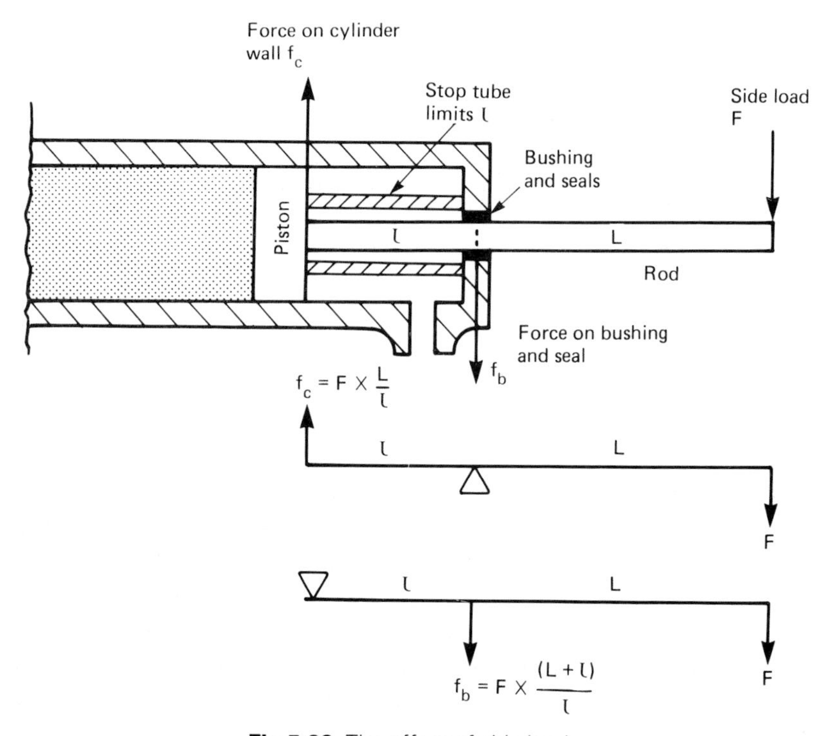

Fig.5.28 The effect of side loads.

Stroke is obviously less than the overall piston length, and is often further reduced by internal bushing if side loads are expected. The rod pivots around the end cap, and if the piston travels to the end of the bore considerable side load magnification results. The internal bush reduces this magnification as shown in fig.5.28, but also restricts the stroke.

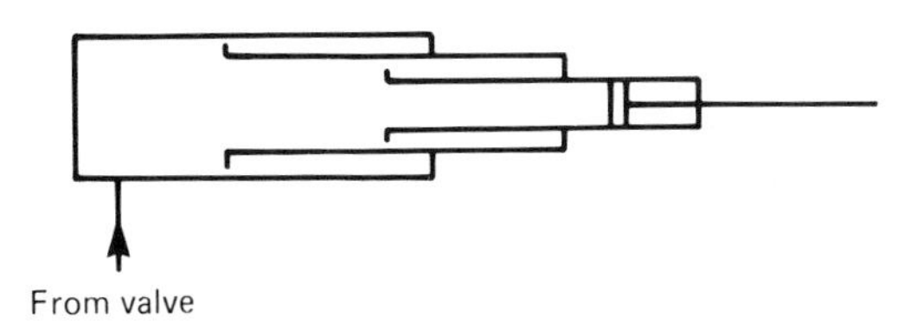

Fig.5.29 Telescopic cylinders.

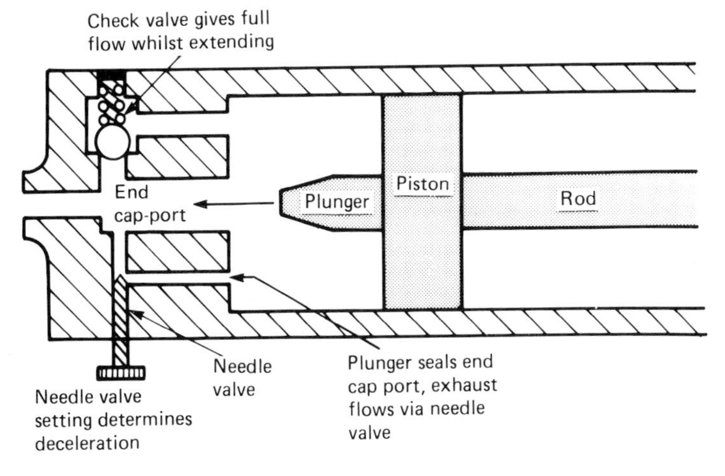

Fig.5.30 Cylinder cushioning.

Where long strokes and minimal cylinder size are needed, telescopic cylinders as in fig.5.29 can be used. Unfortunately these are single acting and have poor tolerance to side loads.

Cylinder cushions as in fig.5.30 are often employed to absorb the shock as the piston reaches the end of the stroke. Progressive deceleration occurs as the plug enters the end cap and reduces the outlet flow. The check valve allows free flow and full pressure to start reverse movement.

5.7. Rotary actuators (hydraulic motors)

5.7.1. Introduction

A rotary hydraulic actuator is usually called a hydraulic motor, and can be used in the same applications as an electric motor. It does, however, have several notable advantages. Hydraulic motors are physically smaller than equivalent electric motors (albeit with the need to have a hydraulic power pack located somewhere close) and give far better control of speed and torque at low speeds.

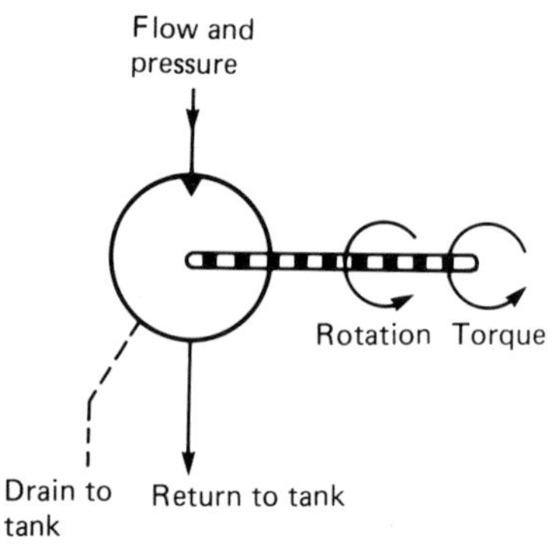

Fig.5.31 Graphical symbol for a rotary actuator.

The symbol for a hydraulic motor is shown in fig.5.31. The inverted arrow denotes a motor (compared with the pump symbol of fig.5.3). A motor accepts hydraulic fluid at a certain flow rate and pressure, and produces output shaft rotation and torque. Hydraulic motors are rated according to their displacement (i.e. the volume of fluid required for one rotation of the output shaft) torque rating and maximum working pressure. They are all related for any given motor:

$$\text{speed} = \text{flow rate} \times \text{displacement} \tag{5.9}$$

There is no relationship between speed and torque or pressure provided the mechanical load is constant.

$$\text{torque} = K \times \text{pressure} \tag{5.10}$$

where K is a constant for the motor called the 'torque rate' (e.g. a given motor could have a torque rating of 50 lb in/100 psi).

The 'torque rate' is related to the motor size, and motors with larger displacements require a lower operating pressure for a given torque.

It follows from equations 5.9 and 5.10 that speed is controlled by flow rate, and torque by pressure. These equations are, of course, theoretical, and in any practical system an allowance for losses (typically 30%) must be made. Flow control is discussed later in section 5.8.1.

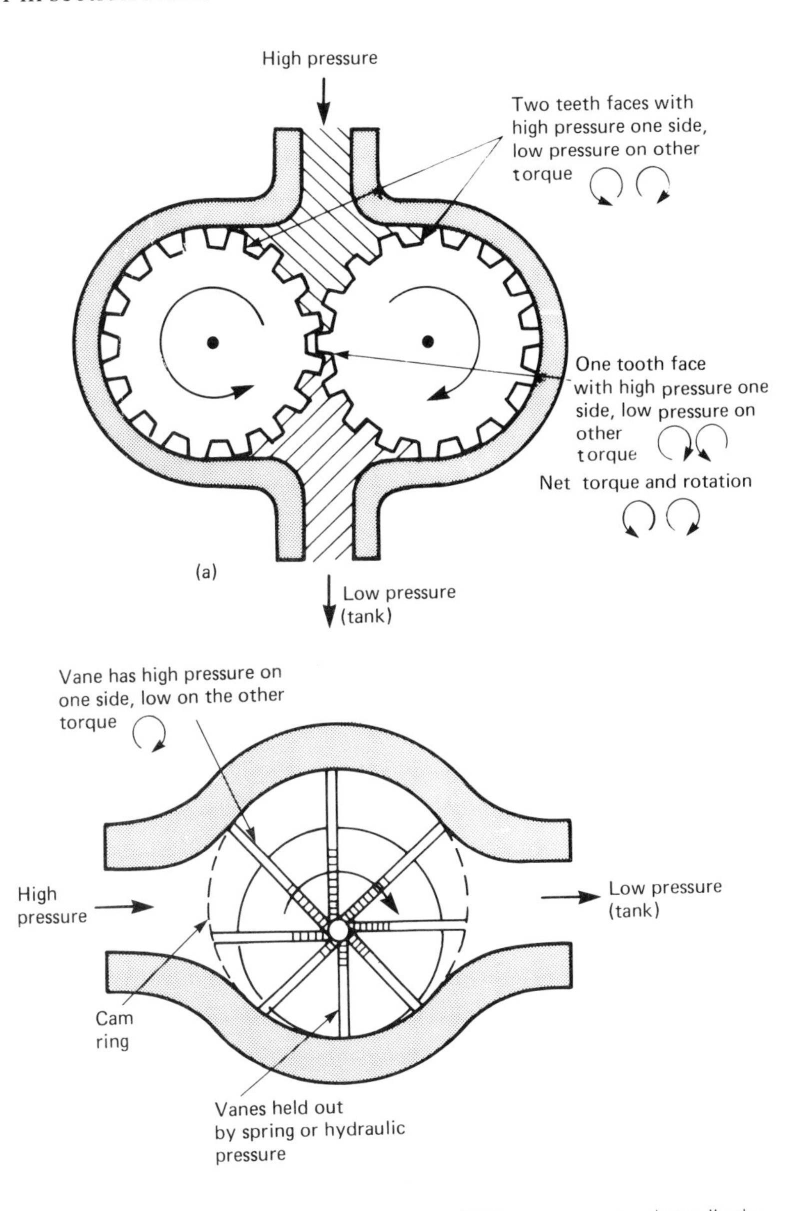

Fig.5.32 Hydraulic motors. (a) The gear motor. (b) The vane motor (usually the balanced construction of Fig.5.10 (b) is used).

5.7.2. Practical motors

Electrical motors and generators have, in many cases, almost identical construction, and a given machine could function as a generator (mechanical power applied to shaft, electrical power obtained from terminals) or as a motor (electrical power applied to terminals, mechanical power obtained from shaft). It is not, therefore, surprising to find that hydraulic pumps and motors have almost identical construction.

Figure 5.32, for example, shows the gear and vane pumps rearranged as motors. The piston pumps of fig.5.12 can also be used as the basis of mechanically speed variable motors where the swash plate of shaft angle varies the displacement and hence the speed (note that from equation 5.9 decreasing the displacement increases the speed for a given flow rate, but the decreased displacement reduces the available torque).

All hydraulic motors experience some internal leakage of oil. As this oil is essentially static, the pressure inside the casing would eventually build up to full line pressure with the possibility of internal damage to the motor. A drain line, shown dotted on fig.5.31, is therefore included to allow leakage oil to return to the tank, preventing internal pressure build-up.

5.8. Miscellaneous topics

5.8.1. Flow control

With linear and rotary actuators, speed is controlled by flow rate, and force (or torque) by pressure. Relief valves, described in section 5.3, can be used for pressure regulation. This subsection discusses methods of flow control.

There are essentially three types of flow control, illustrated in fig.5.33. 'Meter in' controls the flow of fluid to the actuator (and is possibly the most obvious method of speed control). Meter in can only be used, however, when the load opposes the actuator. It cannot be used where the load could run away (controlling the descending speed of the cylinders on a car inspection ramp, for example). In these circumstances the 'meter out' circuit of fig.5.33b is used.

Where directional valves are used, the flow control can be placed on the pump, or tank, side of the directional valve. Figure

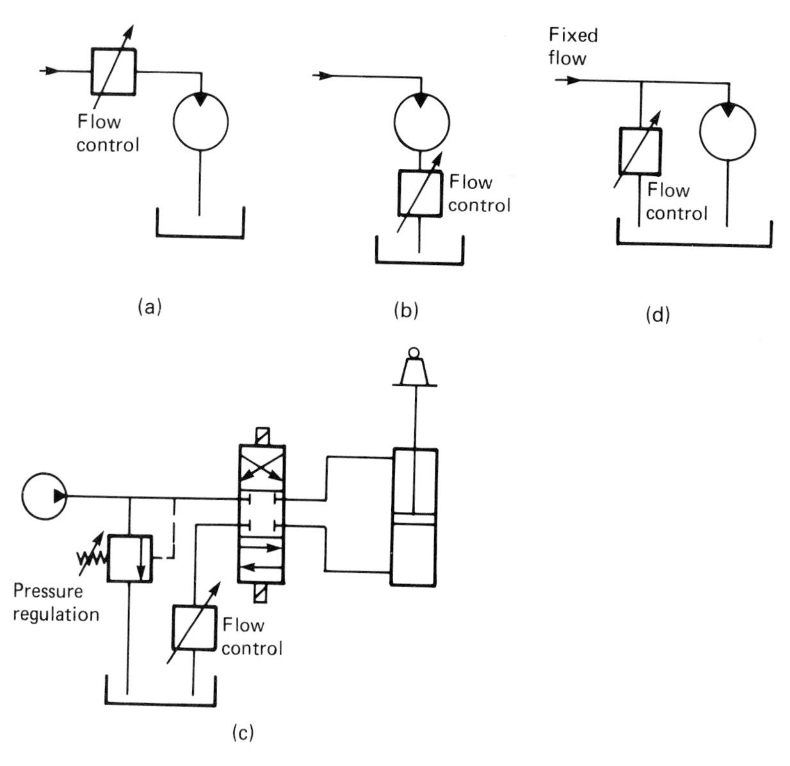

Fig.5.33 Flow control. (a) Meter in circuit. (b) Meter out circuit. (c) Piston speed control using meter out. (d) Variable bleed.

5.33c shows a circuit with meter out flow control connected to a lifting cylinder.

An alternative circuit sometimes encountered is the bleed off circuit of fig.5.33d where the flow to the actuator is controlled by bleeding off excess fluid. This method of speed control is not as accurate as meter in/out circuits. Where a single actuator and pump are linked, speed control can also be achieved by varying the pump displacement.

Flow control is achieved by applying a restriction in a pipe, and consequently has the symbol of fig.5.34a. Simple flow control devices are just an adjustable obstruction (e.g. a needle valve) and often incorporate a check valve to give unobstructed flow in the opposite direction, as in fig.5.34b. Such valves, however, exhibit a pressure drop which varies as the square of the flow, and consequently causes the actuator speed to vary with load.

An ideal flow controller would control flow whilst exhibiting a constant pressure drop. Such a device is shown in fig.5.34c. The

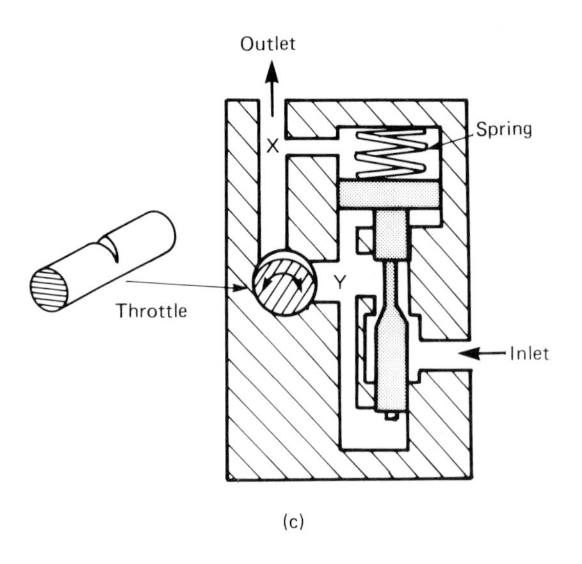

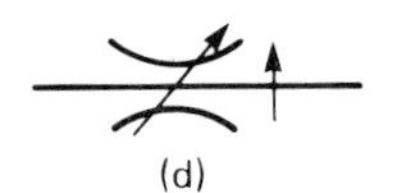

Fig.5.34 Flow control valves. (a) Flow control graphical symbol. (b) Uni-directional flow control. (c) Pressure compensated restrictor flow control valve. (d) Symbol for pressure compensated flow control valve.

throttle is a notch in a rotatable shaft. A movable piston controls the inlet of fluid to the valve by the raised land. The piston experiences a downward force due to the spring and the fluid pressure at point X, and an upward force from fluid pressure at point Y (the lower chamber makes the upper and lower piston areas equal).The piston will move up or down until the differential pressure between X and Y matches the spring compressive force.

The throttle therefore controls the flow through the valve, and the spring tension the pressure drop across the valve. This arrangement is known as a pressure compensated flow control valve, and is denoted by the symbol of fig.5.34d. Usually a check valve is incorporated to allow free flow in the reverse direction.

5.8.2. Servo (proportional) valves

Variable flow control can be achieved by the use of movable spool valves, similar to fig.5.25 except that the spool can take an infinite range of positions, not just two or three. There is little problem positioning the main spool with direct manual action, but remote electrical or pilot operation requires some form of feedback of main spool position to ensure a linear relationship between input signal and output flow. Such valves are generally called proportional, or servo, valves.

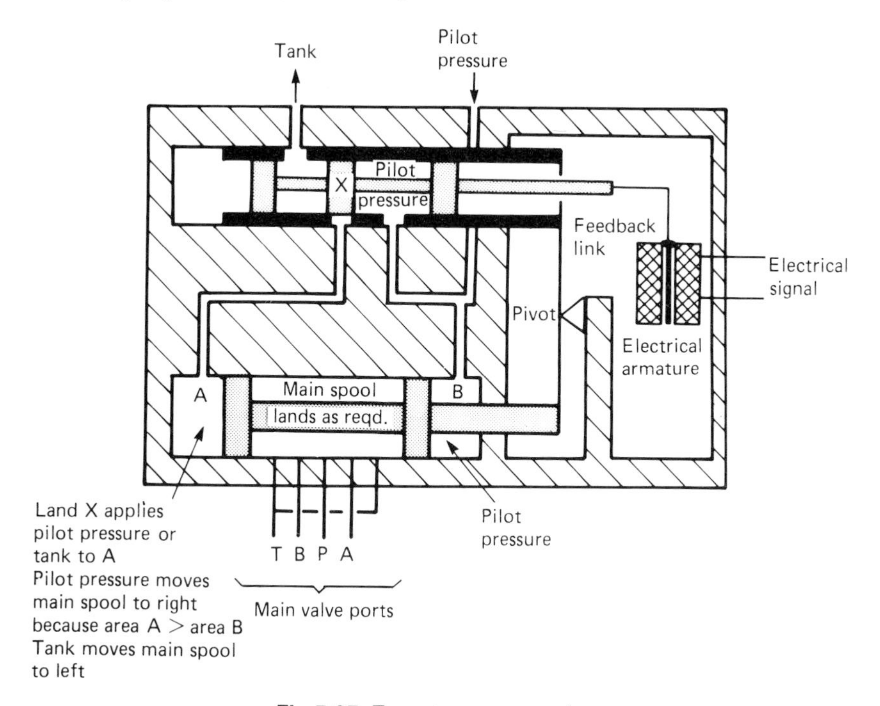

Fig.5.35 Two stage servo valve.

A common arrangement is shown in fig.5.35. The input electrical signal shifts the spool of a pilot valve a distance proportional to the input current (the restoring force being provided by a spring). The pilot valve sleeve is mechanically linked to the main spool, and pivots at a fulcrum.

Pilot pressure can be applied to either end of the main spool, but because of the difference in areas equal pressure will cause the spool to move to the right. Movement of the pilot spool either applies equal pressure to both ends (main spool moves to right, pilot sleeve moves to left) or pressure to end B, end A to tank (main spool moves to left, pilot sleeve moves to right). Only when the sleeve exactly matches the centre land on the pilot spool does the main spool stop.

This following action causes the main spool to follow precisely the movements of the pilot spool regardless of load and pressure variations. Note that the pilot and main spools move in opposite directions, with the ratio of relevant movements being set by the fulcrum position.

Figure 5.35 shows an electrical input signal (sometimes called a torque motor pilot). Obviously the movement of the pilot spool could equally well be controlled by an applied variable pilot pressure to the end of the pilot spool.

Figure 5.36a is a fully position controlled valve. The main spool position is measured (usually by an LVDT) and compared with

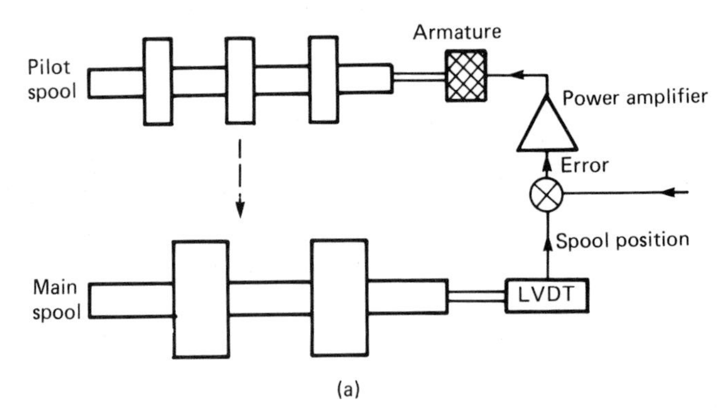

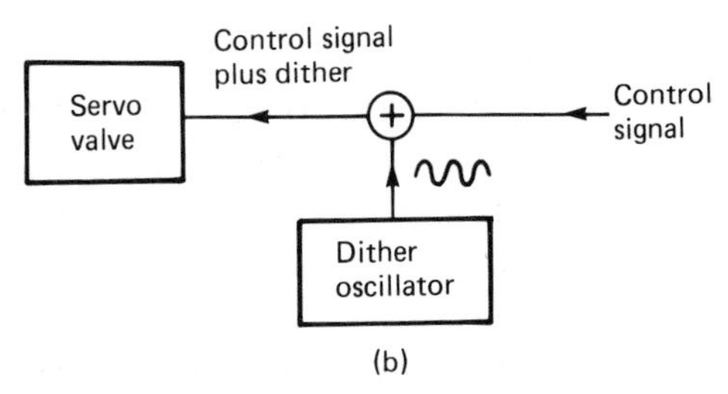

Fig.5.36 Electrical modifications to basic servo valve. (a) Position controlled servo valve. (b) Overcoming stiction with dither.

the input signal. Any error is amplified and used to shift the pilot spool until the positional error is zero.

Servo valves are precision devices, and operate with very small displacement of the pilot spool. They are particularly prone to sticking and erratic operation resulting from dirty or 'gummy' oil. A common practice with electrically controlled pilot valves is the addition of a 'dither' signal to the applied input, as in fig.5.36b. This is simply a small sinusoidal signal (at 50 or 60 Hz) which does not affect the mean valve position, but serves to keep the valve permanently in slight oscillation. This helps displace dirt and overcome stiction.

5.8.3. Accumulators

Accumulators are used to 'store' pressure in a hydraulic system, and are used in several circumstances. The first is where a large volume of fluid is required for very short periods. In this application the accumulator allows a pump to be used of lower capacity than the peak, thereby saving on initial pump and piping (as well as running) costs.

Accumulators also are used to reduce running costs where instant response is needed. If a pump/unloading valve arrangement is used (as in fig.5.13) there is a short delay whilst the pressure builds up after the pump has been put on load. An accumulator linked unloading circuit, described later, gives instant response combined with the power saving of a pump unloading valve.

Finally an accumulator can act as a 'buffer' to absorb shocks and transient pressure peaks which occur when fluid is stopped or reversed. In a non-accumulator system these shocks manifest themselves as loud bangs and hammering noises, which can often be severe enough to cause failures of piping or valves.

There are two types of accumulator, shown in fig.5.37. The piston/spring uses the compressive force of the spring as the pressure store, and the gas filled a pressurised inert gas. The gas is usually separated from the hydraulic fluid by a flexible bladder. In both cases the stored pressure and fluid are available, on demand, to the rest of the system.

Figure 5.38a shows a typical accumulator pressure control system. Two pressure switches are connected to the system, set for a slight differential of a few psi. The higher set switch opens

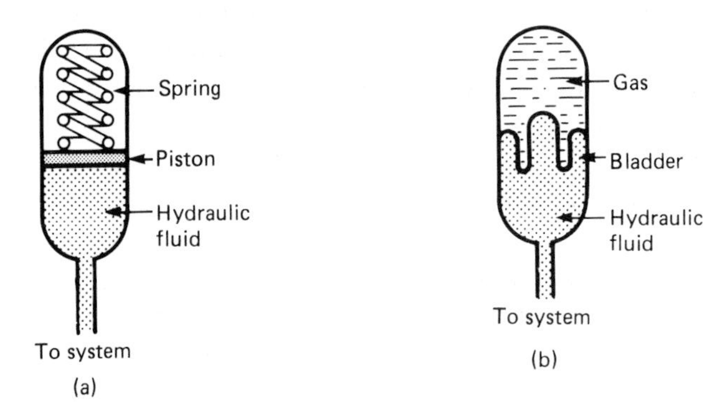

Fig.5.37 Accumulators. (a) Piston/spring. (b) Gas filled.

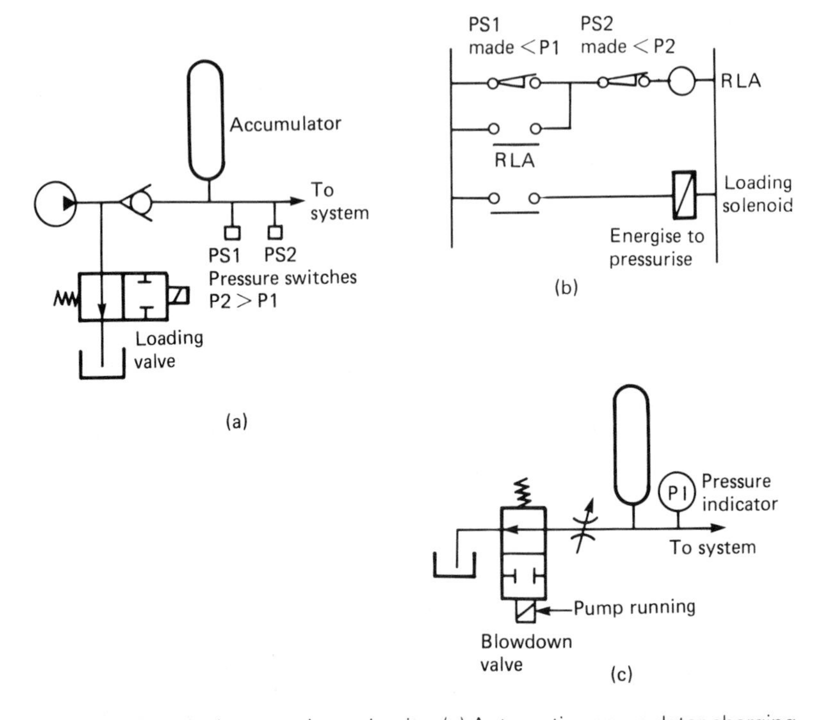

Fig.5.38 Practical accumulator circuits. (a) Automatic accumulator charging. (b) Electrical circuit. (c) Safety considerations.

the unloading valve, and the lower set switch closes it with the circuit of fig.5.38b. The system pressure thus cycles between the two pressure switch settings. The pump only comes on load when required to deliver fluid, but the accumulator gives instant pressure and flow on demand. Instant response with no wasted power is thereby achieved.

Accumulators can stay fully charged for days, or even months, and this brings a possible hazard to hydraulic systems. It is *very* dangerous to work on a pressurised hydraulic system; oil at high pressure can easily blind or maim, or even cause a fatality. An accumulator based system must include a mechanism by which the stored pressure can be released, and indicating devices which allow system pressure to be checked before any maintenance work is commenced. The author writes from personal experience of having been totally covered in oil during work on a system which was thought, incorrectly, to be depressurised.

A common arrangement is the automatic blow down circuit of fig.5.38c. The relief valve is held shut by the volts applied to the coil of the pump motor contactor. When the pump stops, or trips, the spring return on the relief valve causes accumulator stored fluid, and pressure, to be released to tank. The flow control allows the blow down rate to be adjusted. The pressure indicator should be checked to ensure blow down has been achieved before any work on the system is started.

5.8.4. Tanks and reservoirs

Hydraulic systems utilise a closed system of fluid, and as such require a tank or reservoir to hold the fluid between its return from the system and the pump intake. The reservoir is more than a simple tank, however, as examination of fig.5.39 will show. The tank must hold sufficient volume to allow for volume changes

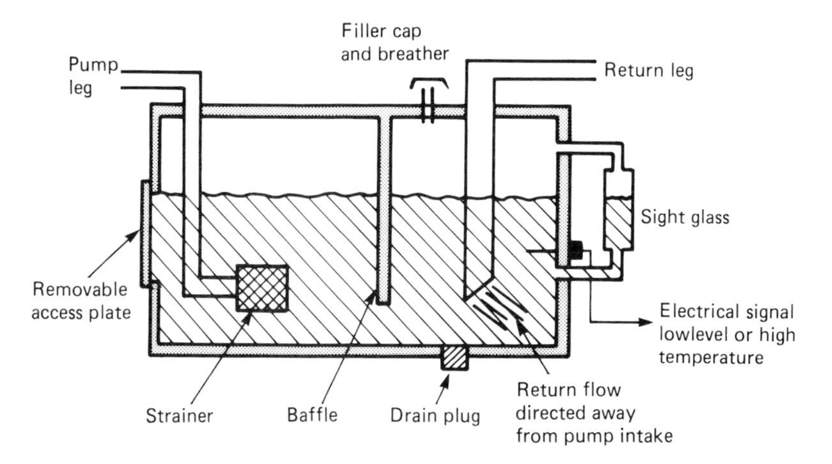

Fig.5.39 A hydraulic tank.

caused, for example, by single acting cylinders and temperature effects. The depth must be such that the pump inlet does not cause a whirlpool effect at the surface which would draw air into the system. A common rule of thumb is a tank volume of three times the volume delivered by the pump per minute.

The reservoir serves to cool the oil, and to assist this the inlet and outlet tank halves are separated by a baffle to make the oil take a circuitous route round the walls. If extra heat removal is necessary a water cooled heat exchanger can be included in the return line or the tank itself. The baffle also serves to reduce turbulence, and allows contaminants to settle out.

Inevitably a layer of sludge will form in the tank bottom. A drain plug and access plate should therefore be provided for regular cleaning. It is essential that this sludge is not drawn into the system, so a coarse strainer is provided on the pump inlet line. The return line is also angled so that the flow is directed away from the bottom and does not stir up the sludge.

If the tank level falls (as it will, because no system is leak free!) there is a danger that a whirlpool will form, drawing air into the pump. This will cause poor performance and even damage. A visual sight glass and low level float switch are provided to allow the level to be examined and a remote alarm given. Two level switches are sometimes provided. The first indicates a low level alarm, and the second, lower, switch stops the pump. A fluid overtemperature alarm is often incorporated with the level switch.

5.8.5. Filters

Cleanliness is next to godliness in hydraulics. It is generally dirt that causes sticking valves, failure of seals and premature wear. Particles of dirt as small as 20 microns can cause problems (a micron is one millionth of a metre; the naked eye can just resolve 40 microns). Filters are specified in microns or meshes per linear inch (sieve number).

The inlet line in the tank will be fitted with a strainer, but this will be a coarse wire mesh element switchable for removing relatively large metal particles and similar contaminants. A separate filter is need to remove the finer particles. This filter can be installed in three places as shown in figs.5.40a, b and c.

Inlet line filters protect the pump, but must be designed for low

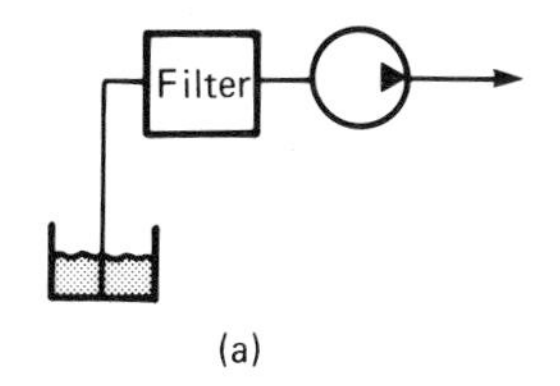

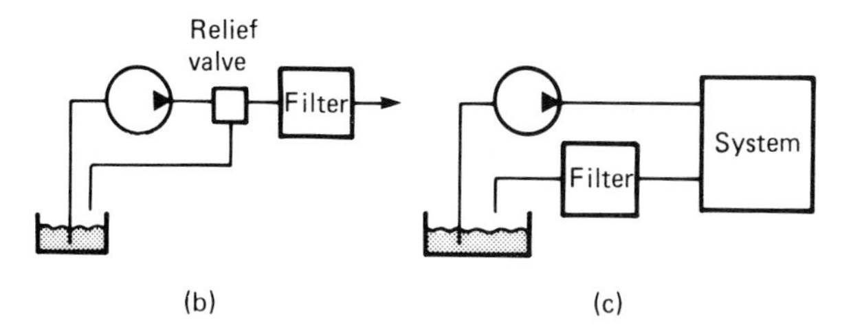

Fig.5.40 Filter positions. (a) Inlet line filter. (b) Pressure line filter. (c) Return line filter.

pressure drop or the pump will not be able to raise fluid from the tank. Low pressure drop implies a coarse filter or a large physical size. Pressure line filters which protect the valves and actuators can be finer and smaller, but must be able to withstand the full system operating pressure. Return line filters can be very fine and, paradoxically, serve to protect the pump by limiting the size of particle returned to the tank.

Filters can also be classified as full or proportional flow. In fig.5.41a, all the flow passes through the filter. This is obviously efficient in terms of filtration, but will incur a large pressure drop.

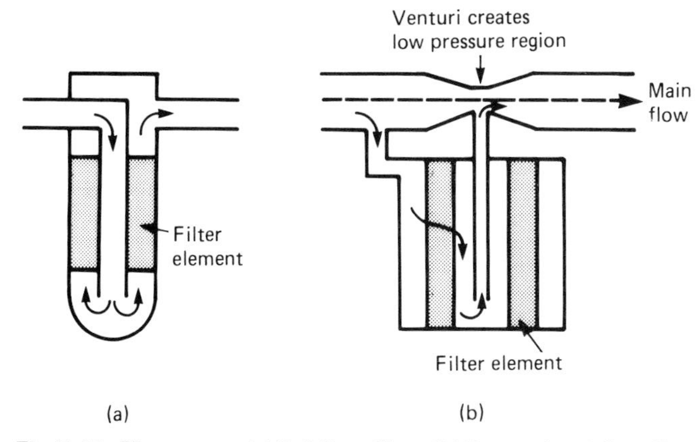

Fig.5.41 Filter types. (a) Full flow filter. (b) Proportional flow filter.

This pressure drop will increase as the filter becomes increasingly polluted, so a full flow filter usually incorporates a relief valve which cracks when the filter becomes unacceptably blocked. The filter should, of course, have been changed before this state was reached.

In fig.5.41b the main flow passes through a venturi, creating a localised low pressure area. The pressure differential across the filter element draws a proportion of the fluid through the filter. This design is accordingly known as a proportional flow filter, as only a proportion of the main flow is filtered. It is characterised by a low pressure drop, and does not need the protection of a pressure relief valve.

The pressure drop across the filter element is an accurate indication of its cleanliness, and many filters incorporate a

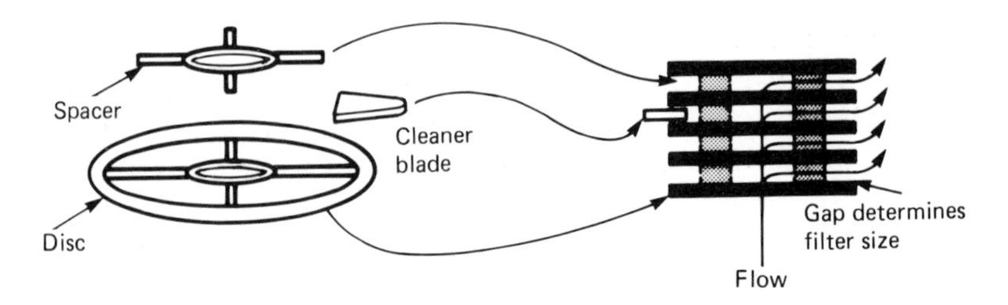

Fig.5.42 Edge type filter.

differential pressure meter calibrated with a green (clear)/amber (warning)/red (change overdue) indicator. Such types are called indicating filters.

The filtration material can be mechanical or absorbent. Mechanical filters are relatively coarse, and utilise fine wire mesh or a disk/screen arrangement as in fig.5.42. Absorbent filters are based on porous materials such as paper, cotton or cellulose. The filtration size can be made very small as the filtration is done by the pores in the absorbent material. Mechanical filters can usually be removed, cleaned and refilled, whereas absorbent filters are usually replaceable items.

5.8.6. Hydraulic fluids

Hydraulic fluid is used as the transmission medium, and its characteristics determine the performance and reliability of any hydraulic system. The fluid must meet several, often conflicting, requirements. To eliminate losses it must flow freely and it should be incompressible so that actuators respond instantly to valves.

The fluid also acts as a lubricant. In pumps, valves, actuators, etc., a thin film of fluid is employed where moving parts slide over each other. This thin film must also act as a seal around the lands of a valve spool, for example, stopping high pressure fluid leaking to the adjacent chamber.

In operation, hydraulic fluid experiences a wide range of temperature. At start-up, after a long weekend shutdown, temperature near freezing may be encountered. In operation, heat is generated from pipe friction and the action of pressure regulators, valves, actuators,etc., causing a significant temperature rise, possibly to the extent that cooling is necessary. The fluid must not notably change its characteristics with temperature changes. The nature of the fluid should also encourage heat loss to the pipe and reservoir walls.

A major cause of hydraulic problems is the sticking of valve spools. This is caused by dirt and a sticky, gummy deposit which can form when some oils stand or undergo temperature changes. Hydraulic fluid must have stable characteristics with age.

Water is present in all hydraulic systems, entering via the tank breather and mixing with the fluid as condensation forms with temperature changes. When water and oil are mixed, a white emulsion can form which, again, leads to sticking valves. Water can also lead to foaming, and cause cavitation and erosion damage. Hydraulic fluid should not change its characteristics in the presence of small amounts of water.

Hydraulic fluid, like all oils, is defined by its viscosity. This is a measure of its resistance to flow. Fluidity is the measure of how easily a fluid flows. Treacle has high viscosity and low fluidity. Petrol has low viscosity and high fluidity. Selection of the viscosity for hydraulic fluid is a compromise between sealing and lubrication (which requires high viscosity) and friction losses and speed of response (which require low viscosity).

Viscosity can be defined in absolute terms in poise or centistokes, or in relative terms in SUS (Saybolt universal seconds) or SAE (Society of Automobile Engineers) numbers which were developed to specify the viscosity of motor oil over a range of temperatures. SAE numbers are the most common.

Most hydraulic fluids are petroleum based oils. These are, however, inflammable and a hazard in industries with open flames or where welding is common. Fire resistant fluids are based on water/oil emulsions, or water/glycol mixes. These require regular testing as evaporation reduces the water content. Some synthetic fluids (e.g. phosphate esters) are also fire resistant but rather expensive. All fluids have additives to stabilise the characteristics with age and temperature and to reduce the formation of gummy deposits.

Chapter 6

Pneumatics and process control valves

6.1. Basic principles

6.1.1. Introduction

Chapter 5 described the use of hydraulics as an alternative to electrically powered actuators. Hydraulic systems utilise a liquid (usually oil or water) as the power transmission medium. It is also feasible to use a gas for transmitting force. Systems using gas are called pneumatic systems (derived from the Greek words for hidden gas). Industrial pneumatic systems are usually based on air.

Pneumatic applications can be loosely split into three groups. The first is linear and rotary actuators, where pneumatic devices are employed to operate devices such as control valves, rams, cylinders, etc. The second group is the use of pneumatics for process control signals. A 4 to 20 mA current is commonly used to represent process variables–a liquid flow from 0 to 3000 gallons per minute, say. In a similar way, a process variable can be represented as a pressure change. The commonest standard is 3 to 15 psi (or the equivalent 0.2 to 1 bar). Many pneumatic transducers are available which give a pneumatic 3 to 15 psi output signal as a representation of a process variable. Pneumatic three term controllers (with pneumatic pressure set point, pneumatic process variable and pneumatic output signal) are also available, allowing closed loop control with non-electrical signals. Such schemes are particularly attractive in applications with explosive atmospheres.

Pneumatics can also provide an alternative to digital logic and relay sequencing. Such systems, called fluidics, are not particularly common at the time of writing, but again are attractive in hazardous areas.

6.1.2. Fundamentals

In many respects there are similarities between hydraulics and pneumatics. In particular, both use the pressure of an enclosed fluid as a force transmission medium. Equation 5.1 relates the force produced by a fluid of pressure P acting on an area A, and can be rearranged as

$$\text{force} = P \times A \tag{6.1}$$

which is the fundamental equation for pneumatic systems.

In most pneumatic systems, gauge pressure measurements are used (i.e. with respect to atmosphere). This allows direct calculation of forces where cylinders are open to atmosphere on one side. In fig.6.1, for example, the piston has an area of 20

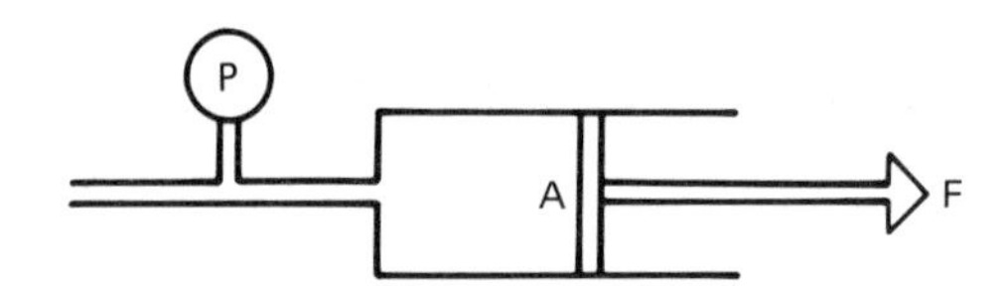

Fig.6.1 Relationship between force and pressure.

square inches, and is experiencing a force of 30 psig on one side and atmospheric pressure on the other. The net force is 30 psi × 20 square inches = 600 pounds.

The concepts of work, energy and power were defined in section 5.2. These apply equally to pneumatic and hydraulic systems.

6.1.3. Gas laws

The major differences between pneumatic and hydraulic systems arise from the compressibility of the fluid used as the transmission medium. Hydraulic fluid is, for all practical purposes, incompressible, whereas a gas can easily be compressed.

The behaviour of a gas subject to pressure, volume or temperature changes is described by three gas laws.

The first, known as Boyle's law, is illustrated in fig.6.2a, where a fixed enclosed mass of gas can be compressed by a movable piston. The temperature of the gas is kept constant by a

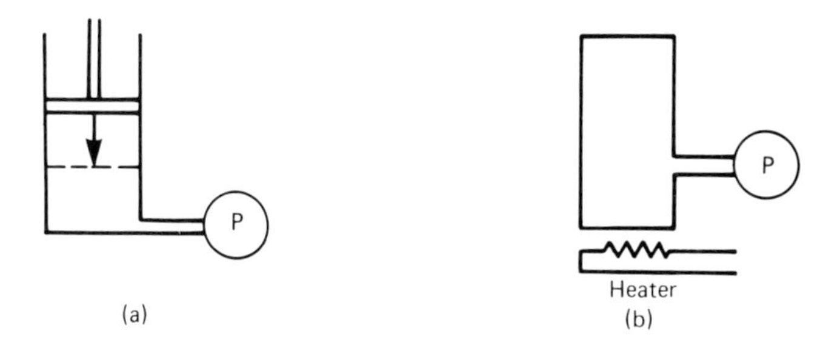

Fig.6.2 Gas laws. (a) Boyle's law. (b) Charles' law.

surrounding heat sink (this is known as adiabatic conditions). Boyle's law states that the pressure and volume are related by:

$$PV = K \tag{6.2}$$

where K is a constant. Equation 6.2 can be rearranged to describe conditions before (subscript b) and after (subscript a) compression:

$$\frac{P_a}{P_b} = \frac{V_b}{V_a} \tag{6.3}$$

Halving the volume doubles the pressure. It should be noted that equations 6.2 and 6.3 use absolute pressure. As most pneumatic systems use gauge pressure, conversions are necessary before and after the application of Boyle's law. For example, reducing the volume of gas at 30 psi to one third of its initial volume results in a pressure of:

$$3 \times (30 + 14.7) \text{ psia}$$

i.e. 134.1 psia or 109.4 psig.

Charles' law is concerned with the relationship between the pressure and the temperature of a fixed volume of gas, as in fig.6.2b. Increasing the temperature causes a pressure rise which is given by:

$$\frac{P}{T} = L \tag{6.4}$$

where L is a constant. As before, this can be more conveniently written to relate the pressure at temperature T_a, T_b as:

$$\frac{P_a}{T_a} = \frac{P_b}{T_b} \tag{6.5}$$

Both pressure and temperature must be given in absolute terms: i.e. temperature in degrees Kelvin (or Rankine). For example, a cylinder containing gas at 30 psig is heated from 20°C to 100°C. The resulting pressure is:

$$(30 + 14.7)*(100 + 273)/(20 + 233) \text{ psia}$$

which is 65.9 psia or 51.2 psig.

The compression of a gas is usually accompanied by a rise in temperature–commonly experienced when a bicycle pump gets warm in use. Figure 6.2a was a special case where the container was surrounded by a heat sink to maintain constant temperature. The generalised equation, called the ideal gas law, relates pressure, volume and temperature by:

$$\frac{P_a V_a}{T_a} = \frac{P_b V_b}{T_b} \tag{6.6}$$

Absolute units must, again, be used for pressure and temperature.

Equation 6.6 does not directly describe what happens to pressure and temperature, say, when the volume of a mass of gas changes. The work done will be converted partly to a change in pressure, partly to a change in temperature. The resulting conditions will be determined by factors such as rate of heat transfer, which are outside the scope of equation 6.6.

It should be noted that although equations 6.2 and 6.6 are usually exemplified by increases in pressure and temperature and decrease in volume, the equations are valid for all changes. A decrease in pressure or volume is accompanied by a fall in temperature, for example; a phenomenon used in refrigeration.

6.1.4. Differences between hydraulic and pneumatic systems

Figure 6.3 shows how a simple extend/retract system would be implemented in hydraulics (fig.6.3a) and pneumatics (fig.6.3b). The hydraulic system pump takes oil from a tank, with the pump outlet pressure being determined by a pressure relief valve which

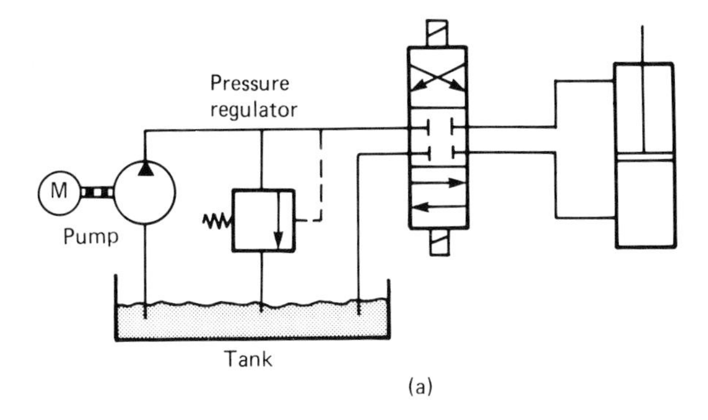

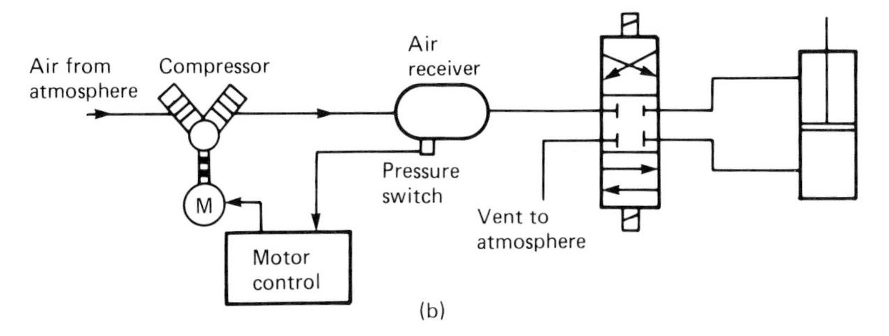

Fig.6.3 Comparison between hydraulics and pneumatics. (a) Hydraulic system. (b) Pneumatic system.

returns excess fluid to the tank. The direction control valve connects pressure to one side of the piston and drains oil back to the tank from the other side of the piston. A hydraulic system is essentially a closed system, the oil being recycled between the tank and the plant. The pump runs continuously, with pressure regulation being determined by the pressure relief valve.

A pneumatic system uses compressed air, and this is usually stored in a pressure vessel called an air receiver. Air, from the atmosphere, is delivered to the air receiver by a motor driven compressor. Unlike a hydraulic pump, the compressor is controlled by a pressure switch on the air receiver and either starts/stops on demand or vents to atmosphere when the receiver is charged. The cylinder movement is again controlled by a directional valve, but air returned from the cylinder is simply vented to atmosphere. A pneumatic system is an open system, the fluid being obtained from and returned to the atmosphere.

Pneumatic systems also require clean dry air (the Three Mile Island nuclear incident was initiated by water in pneumatic lines).

A practical pneumatic system has additional air treatment elements not present in a hydraulic system.

6.1.5. Elements of a pneumatic system

Figure 6.4 is a more detailed version of fig.6.3b, and contains all the elements normally found in a pneumatic system. Air is drawn to the inlet side of the compressor via a filter. This is necessary to

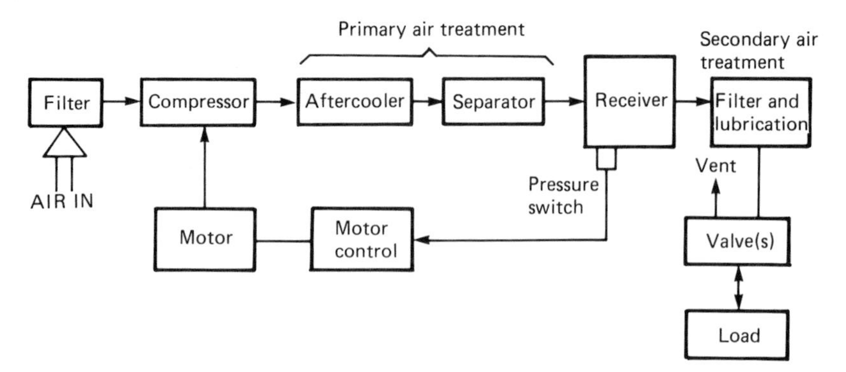

Fig.6.4 Components of a pneumatic system.

remove dust and insects which could damage the seals and valves of the compressor. As the air is compressed to decrease the volume and increase the pressure, its temperature rises according to equation 6.6. The compressor is therefore followed by a cooler to reduce the air temperature. Water vapour in the air tends to condense out at this point. Water can cause corrosion and line blockage, so the aftercooler is followed by a drier (sometimes called a separator or primary air treatment).

The air receiver stores the compressed air, and its pressure is controlled by a pressure switch acting directly on the starter of the electric motor for the compressor. The air receiver usually has a safety relief valve which will act if the pressure switch fails.

Ideally, the air should contain a slight oil mist to lubricate the system components. This is provided by secondary air treatment which incorporates further filtration, water removal and the introduction of the oil mist.

The items of fig.6.4 are discussed in detail in the sections which follow.

6.2. Compressors

6.2.1. Introduction

A compressor is used to compress atmospheric air to give the desired air pressure. Like hydraulic pumps, compressors can be classified into positive displacement devices which move a fixed volume of air per revolution of the input shaft, and dynamic devices which accelerate the air velocity. Large volume low pressure compressors used in pneumatic conveying, for example, are often dynamic devices and are called 'blowers'.

A plant pneumatic system is usually dealt with as a works-wide central service (in the same way as electricity, gas, steam, water, etc.) and is served by a compressed air distribution system fed from a central compressor station. Usually several compressors are installed in parallel to allow for servicing and give redundancy against failure.

6.2.2. Reciprocating piston compressors

The reciprocating piston compressor has a superficial resemblance to a motorcar engine (fig.6.5a). The crankshaft is rotated by an external electric motor, causing the piston to move cyclically up and down the cylinder bore. An inlet and outlet valve are mounted in the cylinder head. Unlike the valves in an internal combustion engine, these are not operated mechanically by cams but are operated directly by pressure variation above the piston.

As the piston falls, a partial vacuum is formed in the bore above the piston. Air pressure causes V1 to open, and air is drawn in from the atmosphere via the inlet filter and silencer. The partial vacuum keeps V2 firmly closed.

As the piston reaches its lowest position, the cylinder is filled with air at approximately atmospheric pressure. On the upward stroke, this air is compressed causing V1 to close. When the pressure inside the cylinder head exceeds the outlet receiver air pressure P_o, valve V2 will open and air is delivered to the receiver.

There is therefore an inlet and outlet stroke per cycle, and a mass of air whose volume is the swept cylinder volume at atmospheric pressure is delivered per cycle.

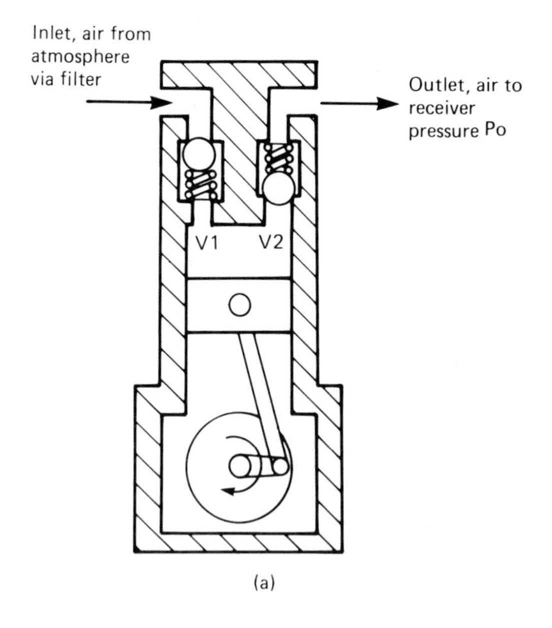

(a)

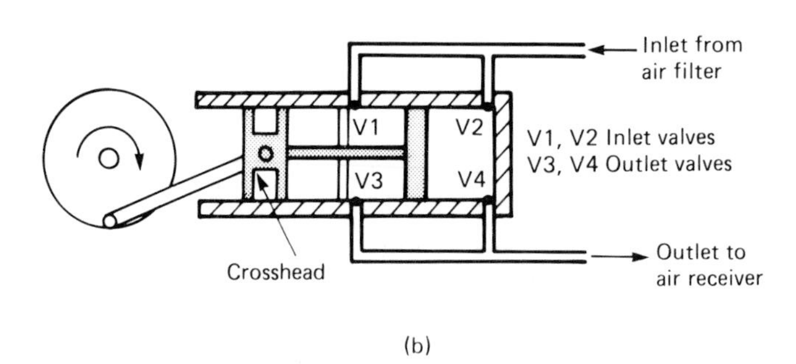

(b)

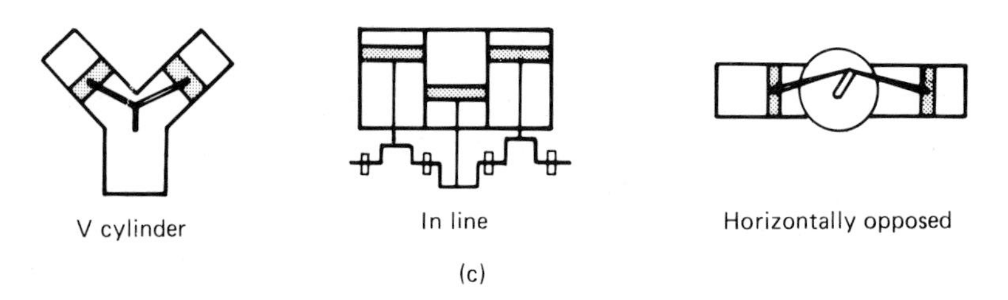

(c)

Fig.6.5 Reciprocating compressors. (a) Piston compressor. (b) Double acting compressor. (c) Cylinder arrangements.

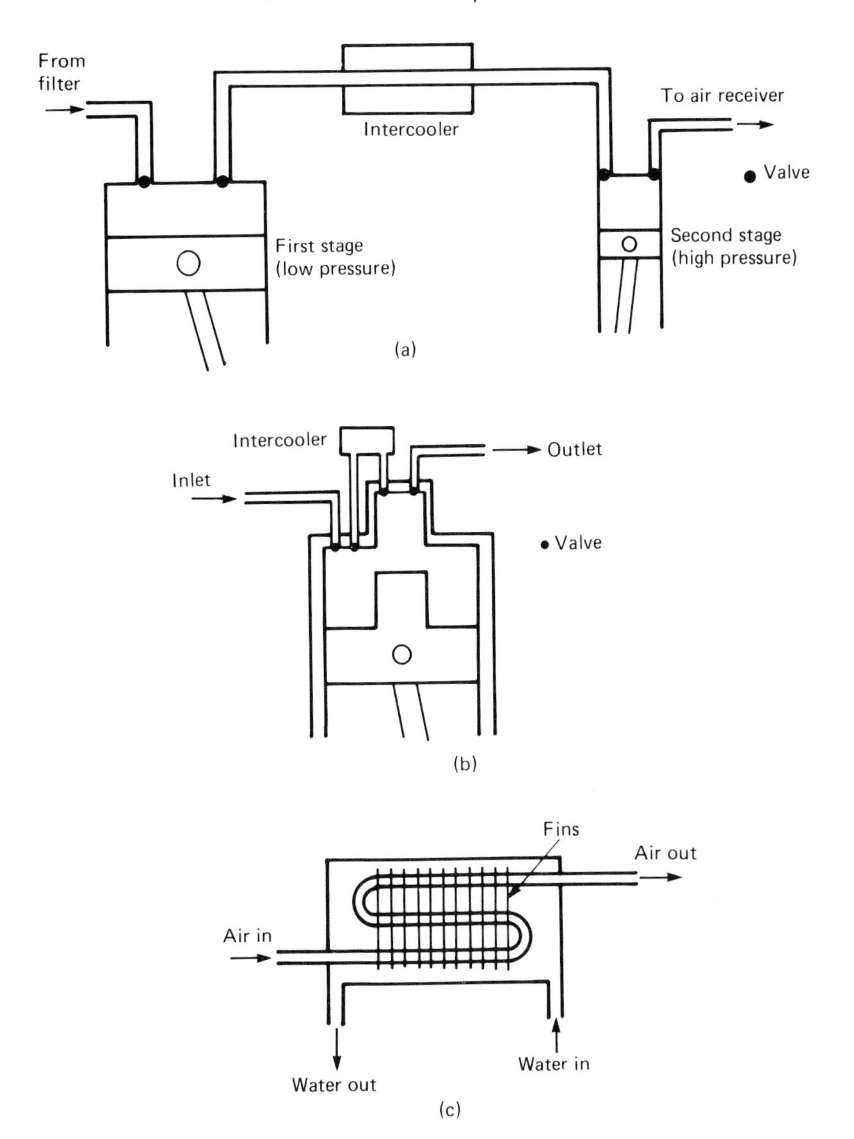

Fig.6.6 Multistage compressors. (a) Two stage compressor. (b) Combined two stage compressor. (c) Intercooler.

Ball valves are shown in fig.6.5a for simplicity, but in practice valves are constructed of 'feathers' of spring steel or disks seating on to annular inlet ports.

A modification of the basic reciprocating compressor is the double acting arrangement of fig.6.5b. This has two inlet and two exhaust strokes per revolution. The crosshead and guide serve to keep the piston rod parallel to the cylinder bore.

Two or more cylinders can be connected in parallel in a variety of ways–V, in line, horizontally opposed, etc., as shown in fig.6.5c. Such arrangements are called single stage compressors, and can be used up to about 7 bar.

Where higher pressures are required, a multistage compressor is used, as in fig.6.6a or b. When air is compressed, its temperature rises in accordance with equation 6.6. This temerature rise increases the power required to drive the compressor. In a multistage compressor, therefore, it is usual to cool the air between stages by a device called an intercooler. This can be a tubed water cooled heat exchanger, as in fig.6.6c, or an air cooled finned pipe or radiator. The bodies of large compressors are also water cooled or finned to assist heat removal. The effects of cooling on power consumption are dramatic. To provide 5 m^3 of air at 7 bar per minute requires a shaft input power of about 20 kW for a single stage compressor and will result in a temperature rise of 200°C. If a multistage compressor with cooling is used, the shaft power is only about 10 kW. Cooling must not, however, be overdone or water vapour will condense out and cause considerable damage to the second stage of the compressor.

6.2.3. Rotary compressors

Rotary compressors do not use reciprocating pistons, and as such are smaller, quieter and easier to maintain than reciprocating compressors. The air supply from a rotary compressor is also smooth and non-pulsating. These benefits are gained, however, at the expense of slightly reduced efficiency and lower operating pressures. This section describes positive displacement rotary compressors.

The vane compressor of fig.6.7a operates in a similar manner to the hydraulic vane pump of section 5.4.3. The rotary screw compressor of fig.6.7b uses two intermeshing counter-rotating screws. These mesh with a clearance of a few thousandths of an inch, and are driven by timing gears. As the screws rotate, pockets of air are carried from the inlet port to the outlet port.

The liquid ring compressor of fig.6.8a is a variation on the vane compressor. This device uses many vanes which rotate inside an eccentric casing. The casing is filled with a liquid, usually water, which is flung out by centrifugal force to form a ring which

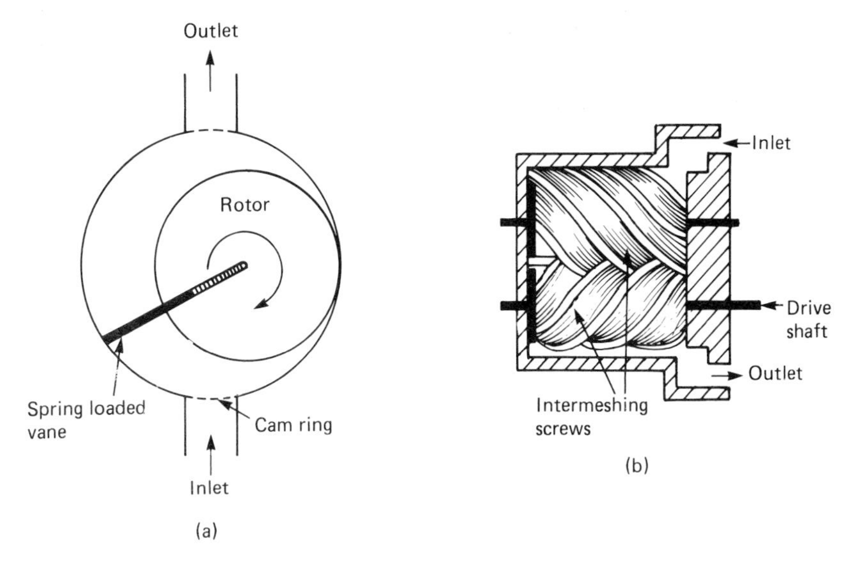

Fig.6.7 Various compressors. (a) The vane compressor. (b) The screw compressor.

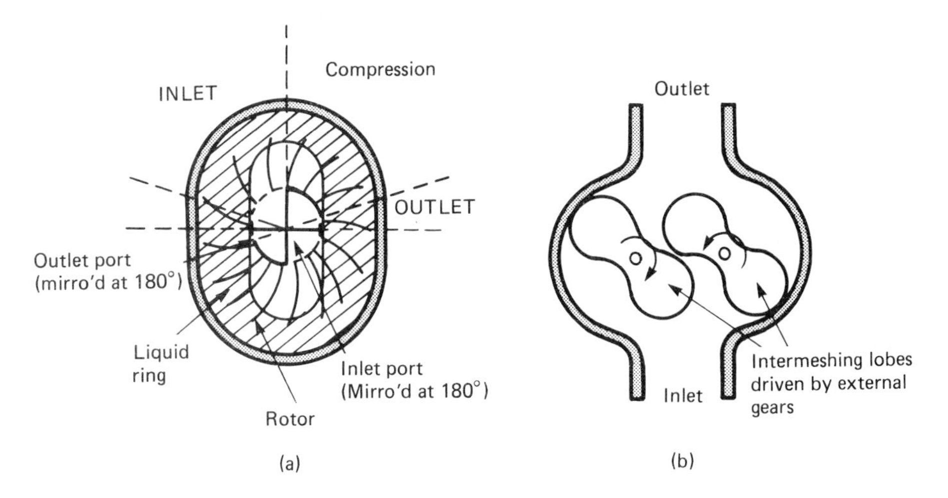

Fig.6.8 Further compressors. (a) Liquid ring compressor. (b) Lobe compressor.

follows the contour of the casing. The volume entrapped by the liquid between vanes therefore increases and decreases as the shaft rotates, delivering air from the inlet ports to the outlet ports.

The final type of rotary compressor is the lobe (impeller) compressor of fig.6.8b. This operates in a similar manner to the hydraulic gear pump of section 5.4.2, sweeping pockets of air between the impeller blades and the casing. The lobe compressor is a high volume low pressure device, operating typically up to 2 bar (30 psig). The operating pressure is mainly limited by leakage between blades and between blades and the casing.

Rotary compressors can, like reciprocating compressors, have their output pressure range increased by the use of multistages with intercoolers to increase efficiency.

6.2.4. Dynamic compressors

Many applications, such as gas/air burners and process air, require a large volume low pressure air supply. The reciprocating and rotary compressors described so far are essentially low volume high pressure devices. Dynamic compressors are non-positive displacement devices, and as such there is a direct shaft route from the load to the supply side; if the drive shaft stops, the pressure can decay back from the load through the compressor.

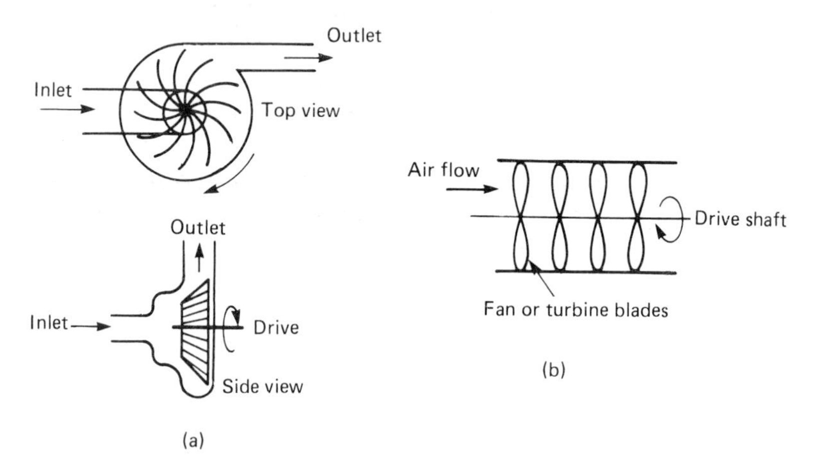

Fig.6.9 Non-positive displacement compressors (blowers). (a) Centrifugal type. (b) Axial type.

Dynamic compressors can be classified into two types, shown in fig.6.9. The centrifugal types use centrifugal force to transfer air from an axial inlet port to a peripheral outlet port. An axial compressor is essentially a series of in line fans or turbine blades.

Dynamic compressors, often called blowers, operate at high speed (up to 10,000 rpm) and can deliver well in excess of 100,000 m^3 of air per minute.

6.2.5. Practical considerations

Because a compressor can be used over a range of pressures, manufacturers usually specify a compressor by its 'free air capacity'. This is the volume of air drawn into the compressor at atmospheric pressure per unit time (e.g. cubic feet or cubic metres per minute). A maximum outlet pressure will also be specified.

The air consumption rate on the plant can be calculated from the volume of actuators and their rate of use. Knowing the working pressure, the volume of air used (with a healthy allowance for losses) equation 6.3 gives the required free air capacity for the compressor. Sizes should always be chosen conservatively.

Small compressors are usually stopped and started on demand from a pressure switch as shown in fig.6.4. Large compressors are often run continuously with an outlet vent valve being controlled by the air receiver pressure switches. Compressors are usually started off load to limit the starting current in the electric drive motor.

Input air filtering is essential to long compressor life. Filter cartridges can use replaceable paper or gauze elements, or cleanable mesh filters. The elements should never be cleaned in petrol; this can turn compressors into diesel internal combustion engines, with dire results! Filter elements should be examined as part of regular maintenance schedules.

6.3. Air treatment

Air always contains a certain amount of moisture in the form of water vapour. The amount of water present in the air depends on the temperature, pressure and atmospheric conditions. For a given volume of air at a certain temperature and pressure there is a maximum amount of water that can be held in the form of vapour. Air in this state is said to be saturated. The amount of water vapour in a volume of air expressed as a percentage of the maximum water vapour that could be held (i.e. saturated air) is called the relative humidity. Obviously the relative humidity of saturated air is 100%.

We intuitively feel the relative humidity of air by describing days with high relative humidity as 'heavy', 'humid', or 'sticky', the latter description arising because sweat does not readily evaporate from the skin when the relative humidity is high. Low relative humidity is experienced as dry, crisp weather and (in housewives' terms) a good drying day as moisture easily evaporates from clothes.

The amount of water vapour that can be held in a given mass of air is dependent on temperature and pressure, increasing with increasing temperature and decreasing with increasing pressure. Effectively, the mass of water vapour that can be held in a given volume at a certain temperature is independent of pressure. Figure 6.10 shows what happens when the mass of water vapour held in 10 m^3 of saturated air at atmospheric pressure is taken to various temperatures and gauge pressures. At 1 bar and 40°C, for example, the air will hold about 0.26 kg of water vapour. If this volume of saturated air is cooled to 20°C, still at 1 bar, it can only hold just over 0.1 kg of water vapour, the excess of 0.16 kg appearing as moisture in the form of condensation.

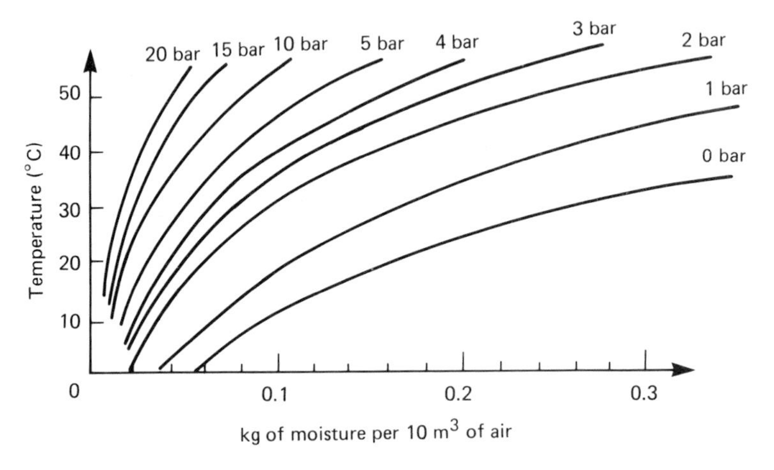

Fig.6.10 Moisture curves.

Figure 6.10 also allows us to predict what happens with non-saturated air undergoing temperature and/or pressure changes. At 50% relative humidity at atmospheric pressure and 20°C, 10 m^3 of air will contain about 0.08 kg of water vapour. If this air is compressed to 3 bar and allowed to cool back to 20°C, it can only hold just over 0.04 kg of water vapour; the remaining 0.04 kg will condense out.

If the pressure of a volume of non-saturated air is held constant whilst cooling takes place, the relative humidity of the air will increase until at some temperature the air becomes saturated and condensation occurs. The temperature at which condensation occurs is called the dew point. From fig.6.10, for example, 10 m^3 of air containing 0.1 kg of water vapour at 0 bar will have a dew point of about 10°C.

When air is compressed in a pneumatic system, the increased pressure results in a large fall in the amount of water vapour that can be held in the air. In the immediate region of the compressor the air temperature is sufficiently raised to prevent condensation, but as the air cools in the receiver and the rest of the piping, moisture will appear. In extreme cases, the moisture will appear as the air pressure drops suddenly which, by equation 6.6, is accompanied by a temperature fall that can be sufficient to form ice particles.

Water particles cause severe problems in pneumatic systems, leading to rust, rapid wear and pitting, and the formation of a sticky oil/water emulsion that can jam valve spools and block fine orifices in process control devices. There is little that can be done to prevent the condensation, and air treatment consists of condensing and removing the water after the compressor but before it can cause problems, as in fig.6.11.

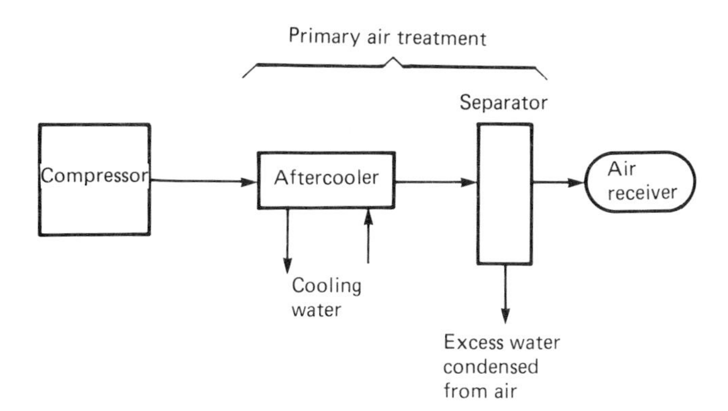

Fig.6.11 Air dryer.

In many systems, all that is required is a simple aftercooler following the compressor which cools the air sufficiently for the water to condense out. The water vapour, which now exists as a mist or droplets, is removed in a strainer or separator. The aftercooler is a simple shell and tube heat exchanger, with water, chilled brine or ethylene glycol coolant. In small systems, finned air coolers may suffice.

Pneumatic systems can be roughly split into bulk air systems, for pneumatic driven tools, actuators etc., and instrument air supplies for controllers. An aftercooler and moisture separator will suffice for bulk air supplies, but instrument air often requires very dry air.

The moisture content can be reduced further by the use of a refrigerated dryer, shown in fig.6.12. This chills the air to just above 0°C in the refrigerator heat exchanger which condenses

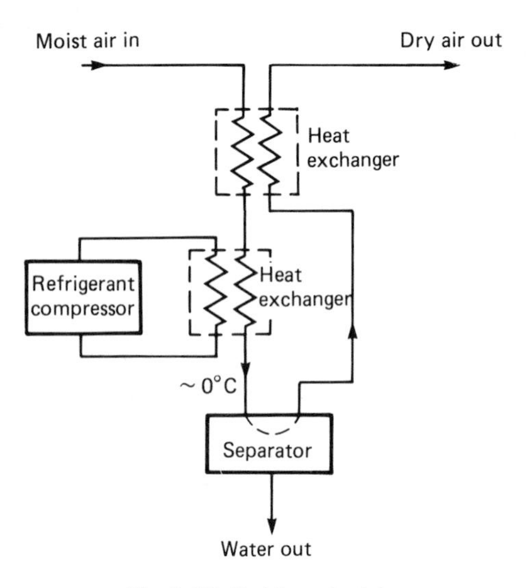

Fig.6.12 Refrigerated dryer.

almost all the water vapour. The air leaving the refrigerator pre-chills the incoming air. The water droplets are removed in the moisture separator.

Where absolutely dry air is required, chemical dryers are used. Various chemicals absorb water from the air by deliquescence or absorption. A deliquescent dryer is shown in fig.6.13a. As the dryer chemical removes the water vapour it turns to a liquid which collects at the bottom of the vessel for periodical draining. Fresh desiccant chemical must be added from time to time.

An absorption dryer uses chemicals which exist in a hydrated and dehydrated state (e.g. copper sulphate). These absorb water vapour, which can be released again by heating. Absorption dryers, an example of which is shown in fig.6.13b, use two columns which are sequenced between drying and regeneration. As shown, column A is drying the air and column B is being regenerated by the heater.

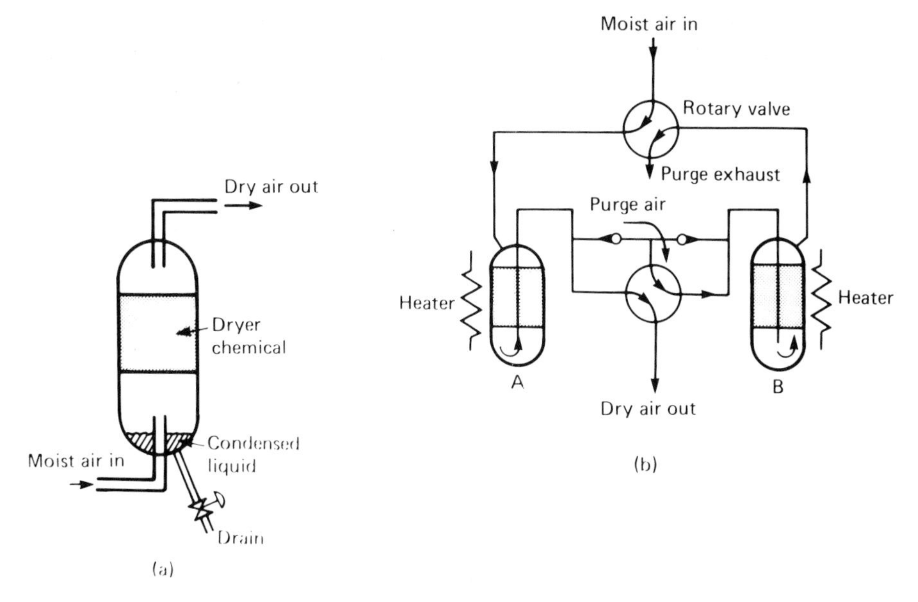

Fig.6.13 Chemical dryers. (a) Deliquescent dryer. (b) Absorbtion dryer.

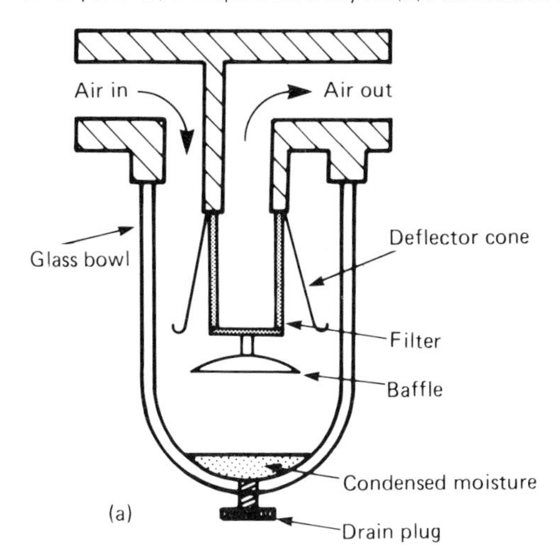

Fig.6.14 Air filter and water trap. (a) Construction. (b) Swirl introduced by deflector cone. (c) Symbol.

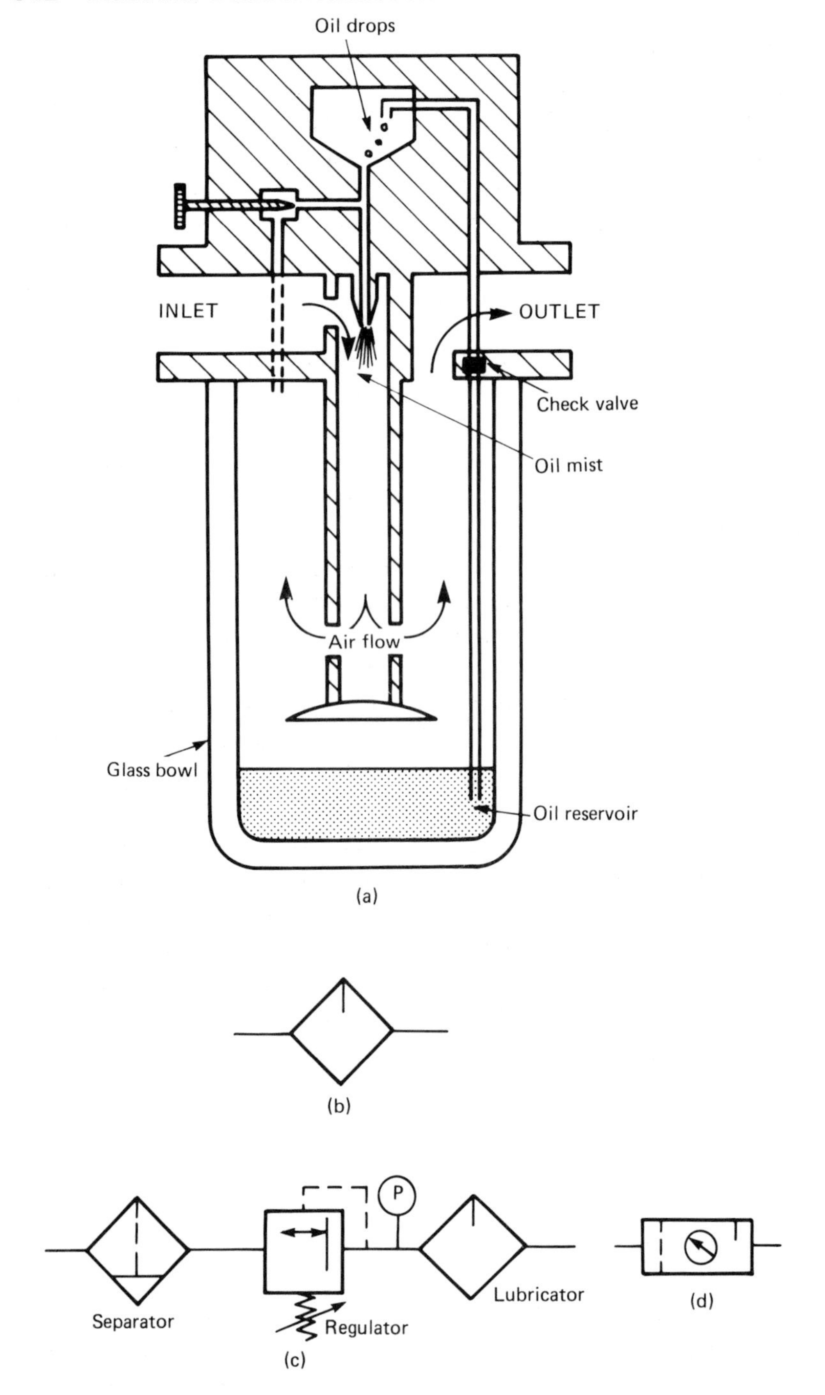

Fig.6.15 Lubricators and service unit. (a) Construction. (b) Symbol for lubricator. (c) Service unit. (d) Composite symbol.

Aftercoolers and refrigerated dryers condense the water vapour to a mist or droplets which must be removed by a moisture separator. These are simply vessels in which the air is caused to suddenly reverse direction or swirl, as in fig.6.14. The heavier water particles are flung out and down, to collect in the bottom of the trap where they are drained manually periodically or automatically by a float operated drain valve. Similar separators can also be used to remove oil and other contaminants.

Air treatment may also include the introduction of a carefully controlled amount of oil mist into the air, although process control pneumatics usually require oil free air. This oil lubricates and protects main parts, but can only be added after the air has been thoroughly dried and cleaned or the oil and water will form a trouble-causing sticky emulsion.

Figure 6.15 shows a typical lubricator. This operates on a similar principle to the petrol/air mixing in the carburettor of an internal combustion engine. As the air passes through the lubricator, its velocity is increased by a venturi ring. This causes a fall in pressure and a partial vacuum in the upper chamber, drawing oil up the riser tube. The oil emerges from a jet to mix with the air. The pressure drop, and hence the oil flow, depends on the air flow rate giving a consistent air/oil ratio over a wide flow range. The needle valve adjusts the pressure difference across the oil jet, and hence the air/oil ratio. On leaving the mixing region the air follows a circuitous route to remove any large oil particles, as described for the moisture separator.

Moisture separators, filters and lubricators are frequently combined in one unit along with a pressure indicator and pressure regulator, as shown with the symbolic representation of fig.6.15b. Such assemblies are frequently called service units.

6.4. Pressure regulation

6.4.1. Introduction

Pneumatic systems, like hydraulic systems, require pressure regulation. In hydraulic systems, pressure regulation is mainly achieved by relief valves which bypass excess fluid back to the tank. Pressure regulation in pneumatic systems takes many forms.

The pressure in the air receiver is determined by a pressure switch which controls the compressor output, either by starting and stopping the compressor drive motor or by operating a loading valve. The air receiver is protected against failure of the main pressure switch by a relief valve. This is set higher than the main pressure switch, and is sized to be capable of handling the full compressor capacity.

Flow velocities in pneumatics can be quite high, leading to substantial pressure drops. Pneumatic distribution systems usually operate at a higher than required pressure, with local pressure regulation as shown. This pressure regulation can take three forms, shown in fig.6.16.

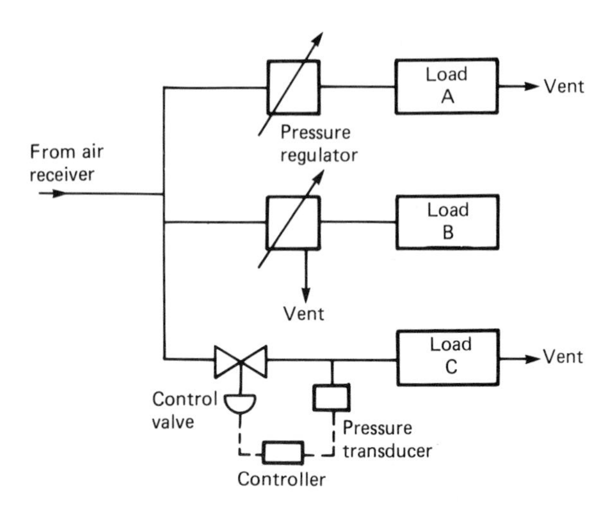

Fig.6.16 Types of pressure regulator.

Load A uses a regulator which controls pressure by a variable restriction, essentially controlling the flow to maintain constant pressure. This arrangement requires some minimum flow. If the load takes no air the pressure downstream of the regulator will rise to the supply pressure. Pressure can only be reduced by air being passed through the load. Such devices are called 'non-relieving regulators'.

Load B uses a three port regulator which can vent air from the load to reduce pressure when required. This type of regulator can handle a 'dead-end' load, and is called a 'relieving regulator'.

The final load uses a large air volume beyond the capacity of a simple in line regulator. This requires a pressure regulator valve controlled by a separate regulator.

6.4.2. Relief valves

At first sight a relief valve, illustrated in fig.6.17, appears similar to a hydraulic (or pneumatic) check valve. The difference is each case is the strength of the spring. In a check valve the spring is relatively weak and serves only to seat the valve. In a relief valve the spring is strong and determines the pressure at which the valve cracks. The spring tension is adjustable to set the relief pressure.

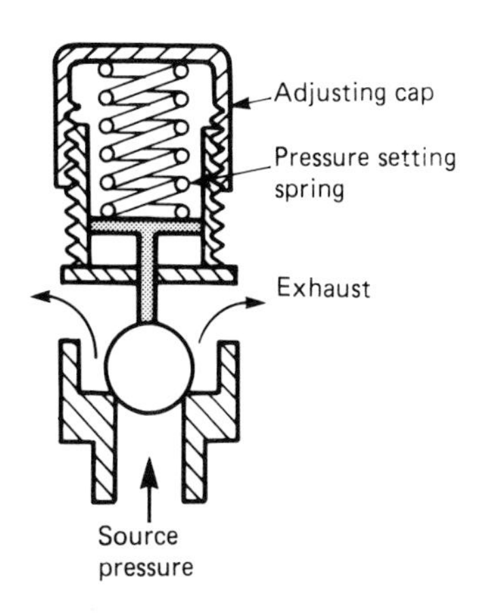

Fig.6.17 Relief valve.

Once cracked, the flow is a function of the excess pressure, an increase in pressure leading to an increase in flow. As stated earlier, a relief valve must be capable of handling the full line flow.

Safety valves fulfil a similar purpose to relief valves, but behave slightly differently. A safety valve goes to fully open once the set pressure is reached (unlike the flow/pressure relationship of the relief valve), and it is usual for a safety valve to remain open until reset manually.

6.4.3. Pressure regulating valves

Figure 6.18 shows the construction of a non-relieving pressure regulator, as connected to load A in fig.6.16. The flow through the valve is controlled by a poppet connected to a spring tensioned diaphragm. The outlet pressure is applied to the lower face of the diaphragm.

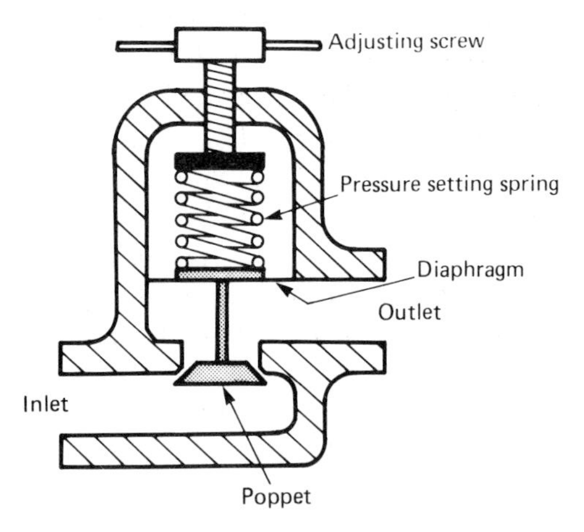

Fig.6.18 Non-relieving pressure regulator.

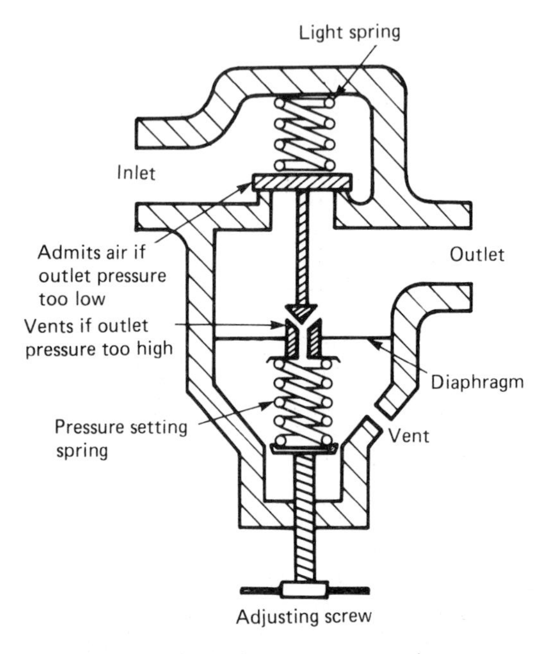

Fig.6.19 Relieving pressure regulator.

If the outlet pressure falls the spring forces the poppet down, increasing the air flow and hence the load pressure. Similarly a rise in outlet pressure results in the diaphragm moving up and reducing the flow. The valve balances when the outlet pressure acting on the diaphragm balances the spring force.

A relieving pressure regulator is shown in fig.6.19. This can deal with dead-end loads similar to load B in fig.6.16. If the outlet pressure falls, the poppet valve is pushed up by the adjustable spring, admitting air to the load. The valve at the base of the valve stem is closed, blocking off the vent.

If the outlet pressure rises, the diaphragm is forced down by the increased pressure. This causes the poppet valve to close and the valve between the stem and diaphragm to open, allowing air to pass from the load to the vent port and thereby reducing the outlet pressure.

The valve will balance with the diaphragm just admitting sufficient air to keep the load at the pressure set on the tensioning spring.

Improved performance can be obtained by using a pilot operated regulator. Figure 6.20 shows a relieving regulator for simplicity, but the technique can also be used for a non-relieving regulator.

The outlet pressure is compared with the preset spring force at the pilot diaphragm. Inlet air is bled through a restriction, and

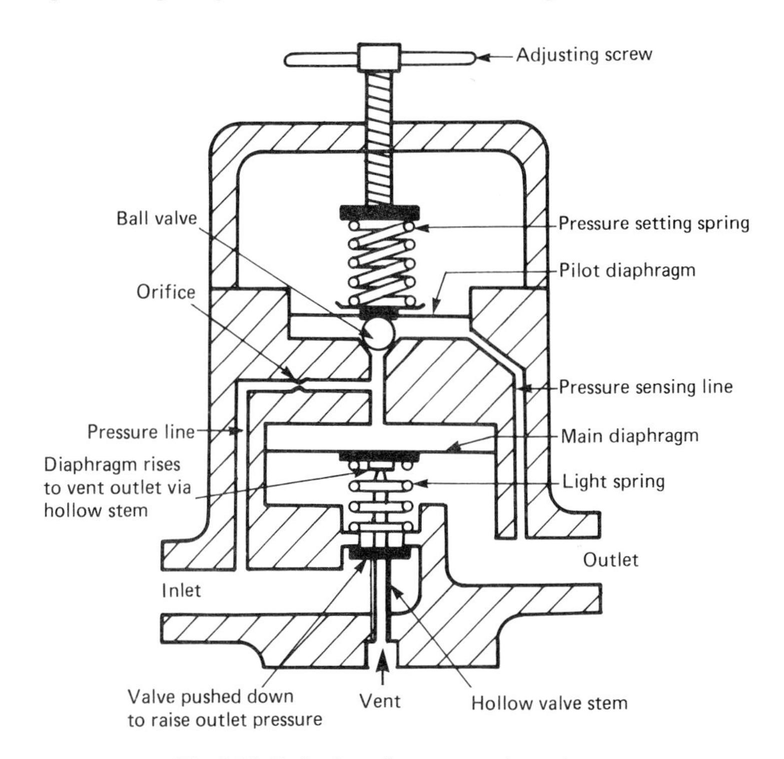

Fig.6.20 Relieving pilot operated regulator.

either applied to the main diaphragm or vented according to the movement of the pilot diaphragm. If the outlet pressure falls the pilot diaphragm will move down, sealing the ball vent valve. The supply pressure now is applied to the main diaphragm, allowing more air to flow to the load.

If the outlet pressure rises, the pilot diaphragm is lifted and the ball valve vents the air bleed. The main diaphragm now rises, closing the poppet valve and opening the vent through the centre of the poppet spool.

Figures 6.18, 6.19 and 6.20 use spring compression to set the required pressure. If remote pressure setting is required, a pilot pressure can simply be applied to the sensing diaphragm. The pilot pressure then determines the regulation pressure.

6.5. Control valves

6.5.1. Introduction and symbols

Pneumatic control valves are used to control the flow of air to and from actuators and other devices driven by compressed air. These valves are similar to the hydraulic valves described in chapter 5, but obviously differ in detail of construction, seal, material, etc.

A valve is described by its number of connections (called ports), and the number of control positions. Symbols for pneumatic valves are similar to those described for hydraulic valves in section 5.5.1. Figure 6.21a is therefore a two port, two

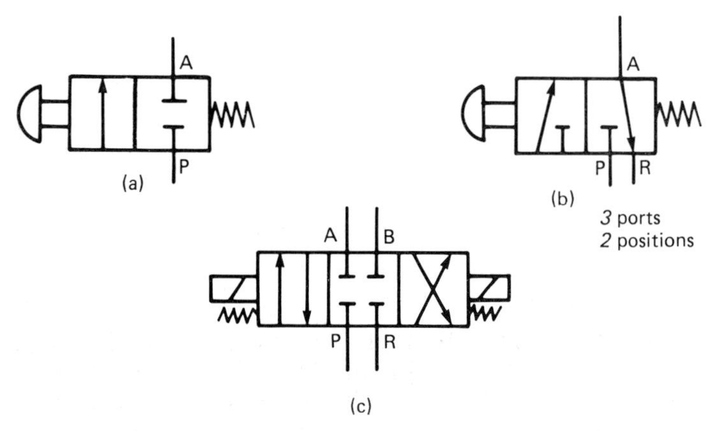

Fig.6.21 Valve descriptions. (a) 2/2 way valve. (b) 3/2 way valve. (c) 4/3 way valve.

position valve (written 2/2 way valve), fig.6.21b a three port, two position valve (3/2 way) and fig.6.21c a 4/3 way valve.

The supply port is usually labelled P (for pressure), and vent ports R, S, etc. Working ports are designated A, B, C, etc., and control ports (e.g. pilot lines) Z, Y, X, etc. The control positions are sometimes labelled a, b, c, etc., with the normal de-energised position denoted 0.

Actuator symbols (solenoid, spring, button, etc.) are identical to those outlined in section 5.5.1.

6.5.2. Valve types

There are essentially three types of control valve: the poppet valve, the spool valve and the rotary valve. Figure 6.22 shows a

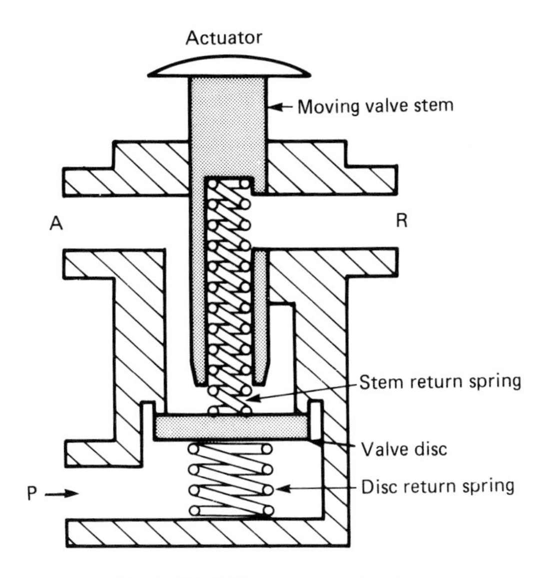

Fig.6.22 3/2 way poppet valve.

3/2 way poppet valve. Button actuation is shown, but it could equally well operate by solenoid or pilot pressure. In the de-energised state, the A and R ports are connected by the hole through the centre of the plunger, and the P port is blocked. When the button is pressed, the plunger descends and contacts the valve disk, sealing off the A and R ports. Further movement pushes the valve disk off its seat, connecting the P and A ports. The valve thus acts as in fig.6.21b.

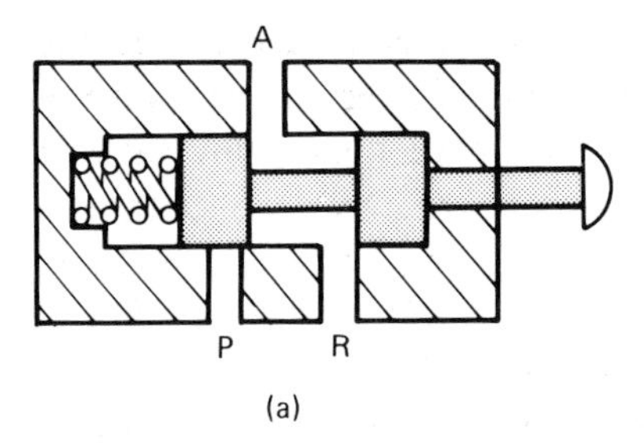

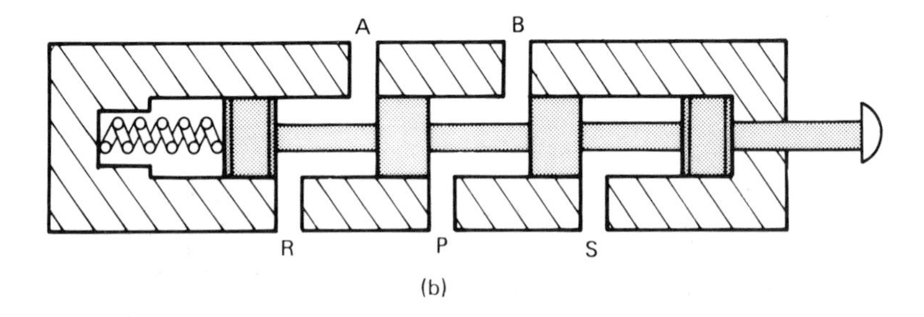

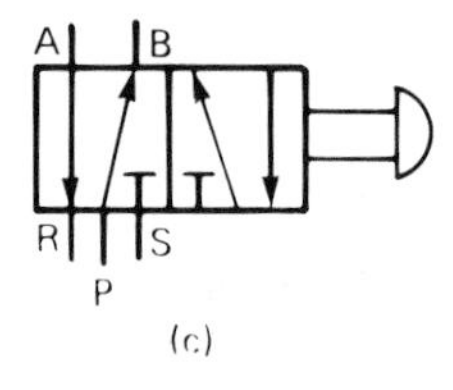

Fig.6.23 Spool valves. (a) 3/2 way valve. (b) 5/2 way valve. (c) symbol for 5/2 way valve.

Spool valves utilise a moving spool with raised lands which block or uncover the required ports. Figure 6.23a shows the spool valve equivalent of the poppet valve of fig.6.22. In the de-energised state, as shown, ports A and R are connected. When the button is pressed, the spool moves over to connect ports A and P and block port R.

Spool valves require less operating force than poppet valves, as pressure forces are equal and opposing on the land faces. The only force the actuator has to overcome is the restoring spring force. On the poppet valve, the actuator has to overcome the spring force plus the air pressure acting on the valve disk.

The action of a spool valve can also be interchanged easily. The valve of fig.6.23a can be converted to normally energised by swapping the P and R connections. If this were tried on the poppet valve of fig.6.22, the air pressure would open the valve disk, connecting all three ports simultaneously.

Poppet valves are limited to relatively simple operations, whereas spool valves can be constructed to almost any desired complexity. Figure 6.23b shows a 5/2 way valve with the operations denoted by the symbol of fig.6.23c.

The final valve type, the rotary valve, is shown in fig.6.24. These utilise a rotating spool with drilled passages which align with ports in the valve casing to give the required action. The valve shown is a 4/3 way valve.

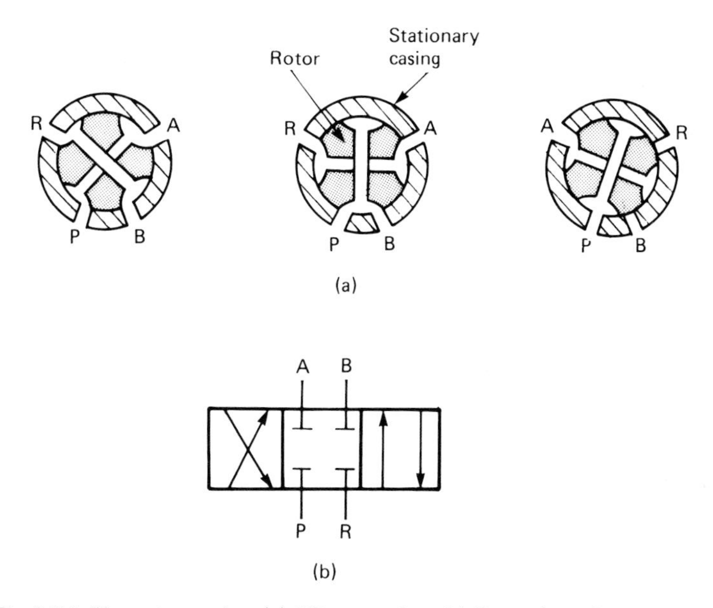

Fig.6.24 The rotary valve. (a) 4/3 way valve. (b) Operation of rotary valve.

Where the valve construction results in uneven forces on the valve disk or spool (as is inherent in poppet valves) a pilot valve can be used. A simple application is shown in fig.6.25. The actuator causes the small pilot valve to open. This puts full pressure on to the control spool. As this has larger area than the main valve disk, the valve opens. Releasing the actuator vents the space above the pilot spool, causing the valve to close. Operating force is the product of the line pressure and the (small) pilot valve disk.

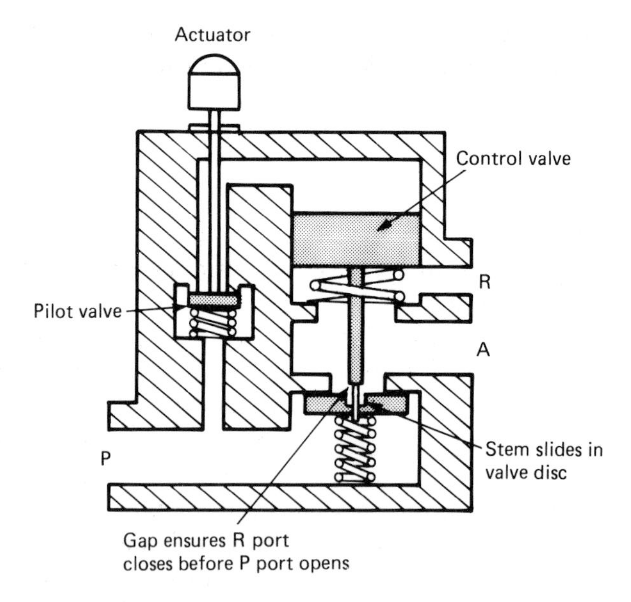

Fig.6.25 3/2 way pilot operated poppet valve.

6.6. Actuators

6.6.1. Linear actuators

A pneumatic cylinder, or ram, is used where a pneumatically controlled linear motion or force is required. Figure 6.26a shows a simple cylinder. Air is introduced to the right of the piston. This produces a force on the piston given by:

$$F = P \times A \tag{6.7}$$

where P is the gauge pressure and A the piston area. The force available at the shaft is slightly less than the force at the piston because of the opposing force from the return spring.

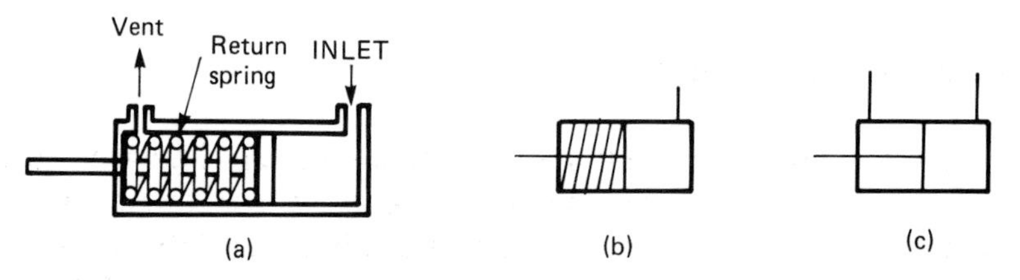

Fig.6.26 Linear actuators. (a) Single acting cylinder. (b) Symbol. (c) Double acting cylinder.

If the opposing force from the load is less than the shaft force the piston will move to the left with a velocity:

$$V = \frac{Q}{A} \tag{6.8}$$

where Q is the volume of air delivered to the piston per unit time. This flow rate is determined by the flow capacity of the valve controlling the cylinder.

If the space to the right of the piston is vented to atmosphere, the restoring spring will move the piston to the right. Normally the force available from the spring is small (to avoid reducing the force from equation 6.7 significantly) so a cylinder similar to fig.6.26a (called a single acting cylinder) can only deliver force in one direction. Typical applications are cylinders for clamping work under a machine tool or a lifting cylinder which returns under gravity. Single acting cylinders with spring return are, of necessity, longer than the stroke to give space for the spring to compress.

Double acting cylinders, as in fig.6.26c, are used where force is required in both directions of motion. It should be noted that the available force is not equal in both directions because the area on one side of the piston is reduced by the output shaft. Control of a double acting cylinder requires a 4/2 way or 4/3 way valve (see fig.6.21).

The construction of pneumatic cylinders is generally similar to that of hydraulic cylinders, described in section 5.6. Details such as stroke restrictors to reduce side loads and cylinder cushions are frequently found in pneumatic cylinders.

Valve actuators are a specialised form of actuator where a linear displacement proportional to input pneumatic pressure is required. These are described further in section 6.6.4.

6.6.2. Rotary actuators

Rotary actuators, or pneumatic or air motors as they are more commonly called, convert the motion and pressure of air flow to mechanical torque and rotational motion. In general the available torque is determined by the supply pressure, typically 100 psi, and the rotational speed by the air flow rate.

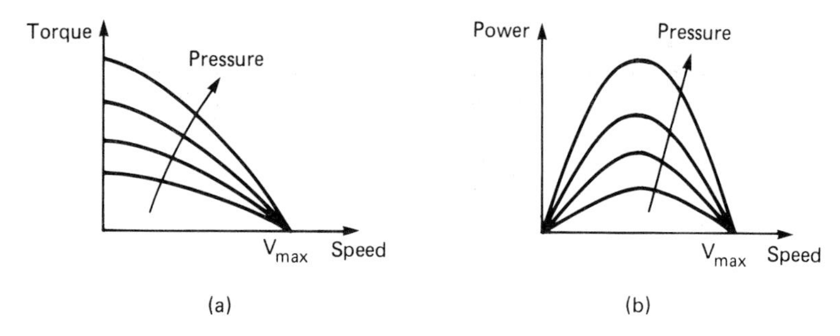

Fig.6.27 Torque/power speed curves for rotary actuator. (a) Torque. (b) Power.

Figure 6.27 shows the torque and power curves for a typical motor. It can be seen that the torque is maximum in the zero speed (stalled) state. Like a hydraulic motor (but unlike an electric motor) a pneumatic motor can be stalled indefinitely without damage. The power curve, fig.6.27a, is a maximum at approximately half the no-load maximum speed.

Power, torque and rpm are related by:

$$\text{power} = \text{K} \times \text{torque} \times \text{rpm} \tag{6.9}$$

where K is a scaling constant dependent on the units used. Imperial units are still commonly used in pneumatics, for which:

$$\text{HP} = \frac{\text{torque (inch-pounds)} \times \text{rpm}}{63025} \tag{6.10}$$

Rotary actuators use a considerable volume of air, typically 1 to 2 cubic metres per minute per kW at an operating pressure of 5 to 6 bar. Care must be taken to ensure that the compressor can supply an adequate volume of air and the piping does not introduce pressure losses.

The construction of pneumatic motors is generally similar to that of hydraulic motors, described in section 5.7. The commonest types are vane motors and axial or radial piston motors. Most are positive displacement motors, the exception being high speed turbines used in some hand tools.

6.6.3. Flow control valves

In a large number of pneumatic systems, the final actuator controls the flow of some fluid–liquid, gas or steam, for example.

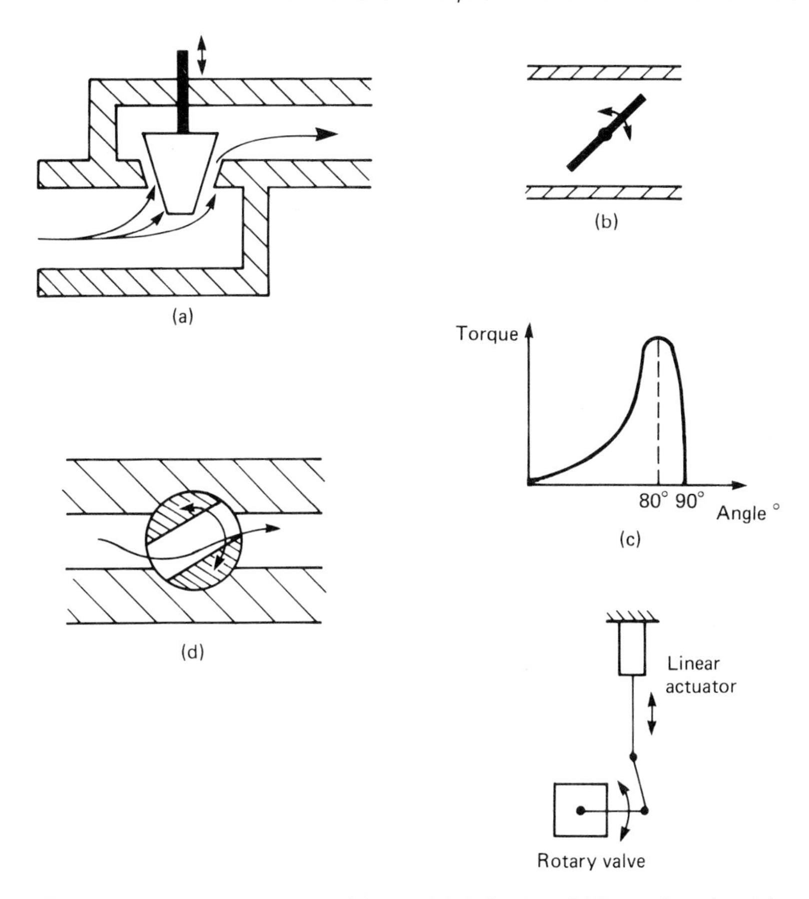

Fig.6.28 Flow control valves. (a) Plug (globe) valve. (b) Butterfly valve. (c) Torque on butterfly valve. (d) Ball valve. (e) Use of linear actuator for valve positioning.

The following section describes valve positioning actuators, but it is first useful to describe the operation of common flow control valves.

All control valves operate by inserting a variable restriction in the flow path. Figure 6.28 shows the three commonest arrangements. The plug, or globe, valve operates by moving a tapered plug, thereby varying the gap between the plug and the valve seat. The flow is controlled by linear movement of the valve stem. Normally the plug is guided by a cage (not shown for simplicity) to prevent sideways movement.

Butterfly valves, such as in fig.6.28b, utilise a circular disk which is rotated to vary the restriction. The leakage of a butterfly valve in the shut-off position is not as good as that which can be obtained with globe valves. Butterfly valves can, however, be constructed to almost any required size. Dynamic torque effects on the disk limit the travel to about 60° from the closed position. Figure 6.28c shows the torque acting on a butterfly valve related to valve position.

The ball valve of fig.6.28d uses a variable restriction obtained by rotating a ball with a through hole which moves within an accurately machined seat. Often a V notch hole is used. Ball valves have excellent shut-off characteristics. Both the butterfly and ball valve require a rotary shaft motion. This is usually obtained from a linear actuator acting on a lever, as in fig.6.28e.

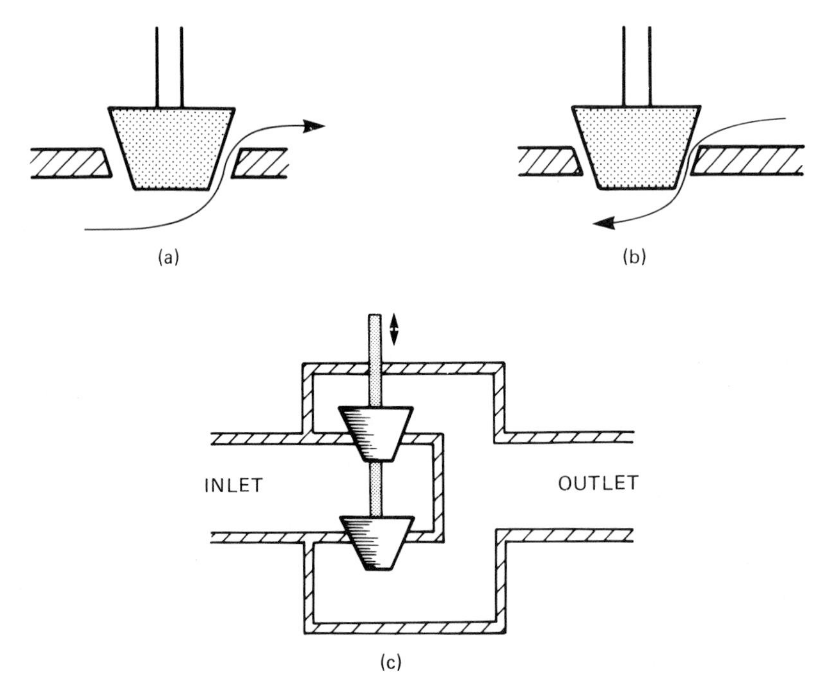

Fig.6.29 The bathtub plughold effect. (a) Flow assist opening. (b) Flow assist closing. (c) Balanced valve.

The dynamic forces on the restriction act on the actuator shaft. In fig.6.29a the flow opposes the opening of the valve, whereas in fig.6.29b the flow assists the closing. The latter case is particularly difficult to control at low flows as the plug tends to slam into the seat. This effect can easily be observed by trying to control the flow of water from a bath or basin with the plug. A balanced valve, with little or no reaction on to the actuator shaft, can be constructed with two plugs and seats, as in fig.6.29c. With careful design the opposing dynamic forces can be made to cancel. Such a valve, however, has a rather high shut-off leakage as manufacturing tolerances will cause one plug to seat before the other.

The valve characteristic relates the flow through the valve to the valve opening. Figure 6.30 shows the three commonly used characteristics. These are specified at constant pressure drop across the valve, a condition which is rare in real-life systems. The choice of valve characteristic is chosen to give a linear

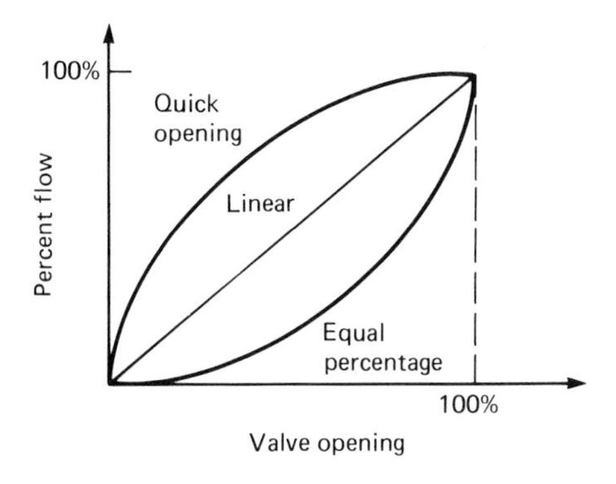

Fig.6.30 Valve characteristics.

flow/position characteristic in the specific application. The differential pressure across a valve in an installation can increase or decrease with increasing flow, dependent on the behaviour of the rest of the system. Quick opening and equal percentage valves compensate, to some extent, for change in differential pressure with flow to give a linear flow/position relationship.

Valve selection can be a complex procedure. Obvious considerations are the pressure, temperature, flow range/turndown and chemical composition of the controlled fluid. The differential pressure across the valve must be calculated from a knowledge of the characteristics of the valve and the rest of the system. With liquids, care must also be taken that cavitation or flashing does not occur in the low pressure region just downstream of the valve. Cavitation and flashing are accompanied by valve damage and excessive noise.

6.6.4. Valve actuators

A pneumatically controlled valve regulates flow by the movement or rotation of the valve shaft, as described in the previous subsection. The controlling signal is a pneumatic pressure, the shaft movement being proportional to the applied pressure. An input signal range of 0.2 to 1 bar, for example, could cause a shaft movement of zero to 50 mm and a flow change from zero to 1000 litres per minute. Such an arrangement could be represented by fig.6.31, with an actuator having a gain of 62.5 mm/bar and the valve a gain of 20 litres/min/mm.

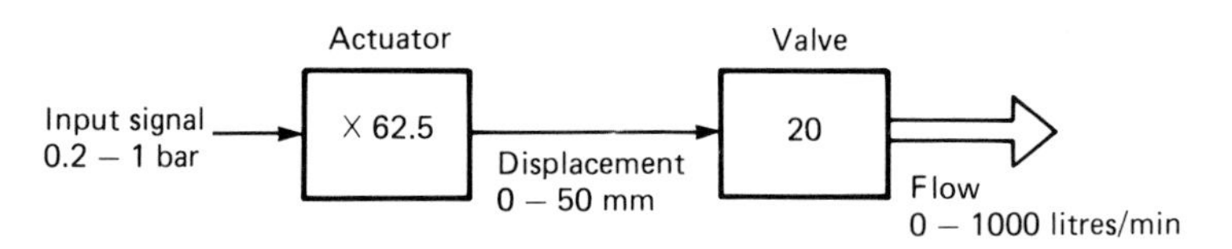

Fig.6.31 The gain of pneumatic components.

A valve actuator operates in a different manner to the linear actuators of section 6.6.1. A linear actuator produces a *force* which is proportional to the applied pressure. A valve actuator produces a *displacement* which is proportional to applied pressure.

A typical actuator is shown in fig.6.32a. The controlling signal is applied to the top of a piston which is sealed by a flexible diaphragm. The pressure produces a downward force, given by equation 6.7. This is opposed by a force from the restoring spring. As the control signal increases, the increased pressure causes the piston to move down until the force from the now more compressed spring again balances the force on the piston. Note that there is a fundamental difference between the spring action in fig.6.32a and the spring in fig.6.26a. The latter is a relatively weak spring to return the piston when the air space is vented.

The actuator gain (movement/applied pressure) is determined by the spring stiffness. The pressure at which the actuator starts to move (0.2 bar in the example above) is set by the pre-tension adjustment.

The action of the rubber diaphragm is shown in fig.6.32b. The diaphragm ensures that the effective piston area remains constant over the full range of actuator travel.

The shaft on the actuator of fig.6.32a extends for increasing pressure (and fails with the shaft fully in). In fig.6.32c the pressure is applied to the underside of the piston and the spring force is reversed. This arrangement gives a shaft motion which extends for decreasing pressure (and fails with the shaft fully

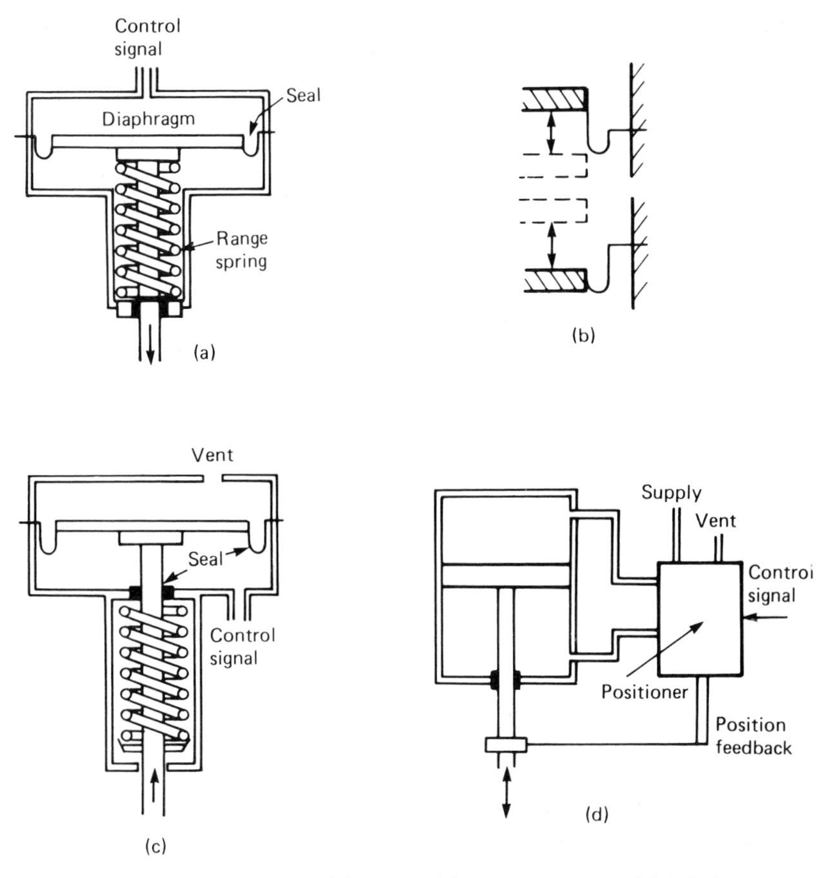

Fig.6.32 Valve actuators. (a) Fail up. (b) Diaphragm seal. (c) Fail down. (d) Piston actuator.

out). This is known as a reverse acting actuator. Note that an O ring seal is necessary on the valve shaft.

The net force acting on the actuator shaft is the algebraic sum of the PA force from the control signal, the restoring spring force and dynamic forces from the valve being controlled. The effect of valve forces will therefore be the creation of an offset error. This can be reduced by increasing the control signal pressure range or the diaphragm area. There are limits, however: the diaphragm can only withstand relatively low pressures, and the physical size of the actuator must be reasonable. Piston actuators such as in fig.6.32d are used, operating at high pressure, where large operating forces are needed. These are similar to the linear actuators of section 6.6.1, but incorporate a closed loop positioning measuring device and closed loop position control to make the shaft displacement correspond to the input low pressure control signal. Positioners are described in section 6.7.5.

6.7. Process control pneumatics

6.7.1. Signals and standards

Process variables (pressure, flow, temperature, level, etc.) are commonly represented by electrical voltages or current. A temperature measurement in the range 0 to 100°C, for example, could be represented by a current from 4 to 20 mA. In this scaling 50°C would be represented by a current of 12 mA.

Process variables can also be represented by pneumatic pressure. Liquid level in a tank varying between 0 and 4 metres could, say, be represented by a pressure from 0.2 to 1.0 bar (20 to 100 kPa). With this representation, a pressure of 0.8 bar would correspond to a level of 3 metres.

Pneumatic representation has advantages over electrical methods in certain circumstances. Explosive atmospheres are common in the chemical and petrochemical industries. The use of electrical transducers and actuators is potentially dangerous, and requires the complexity of zener barriers or intrinsically safe equipment which can be installed in dangerous areas without special precautions.

Early process controllers were employed before semiconductors were available, and the only available technology was pneumatics. A great deal of design and application experience has evolved around pneumatic control. Companies with existing pneumatic control understandably will stay with a level of technology that is readily comprehended by their staff.

The usual signal pressure range is 3 to 15 psig or the metric equivalent, 0.2 to 1 bar (20 to 100 kPa). Other signal ranges tend to be a simple multiple (e.g. 6 to 30 psig). Almost all use an offset zero (e.g. 0.2 bar in a 0.2 to 1 bar system) for speed of response and protection from damage to piping.

Figure 6.33a shows a simple level measurement system where a pneumatic level transducer drives a remote level indicator (which is, of course, a suitably scaled pressure gauge). The mechanism of the transducer need not concern us, but essentially it connects the signal pipe to the supply pressure to increase the reading, or to the vent to reduce the reading. In both cases, the indicator will

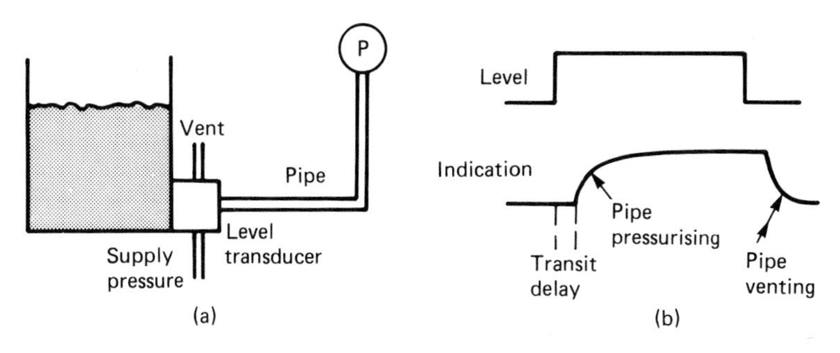

Fig.6.33 Response of pneumatic systems. (a) Pneumatic instrumentation system. (b) Response.

follow an exponential curve, as in fig.6.33b. The speed of response for increasing signal can be increased by raising the supply pressure (a supply of about 2 bar would be used for a 0.2 to 1 bar signal). If a zero pressure corresponded to zero signal, the response of the system to a decreasing signal would be very slow. The use of the offset zero gives a similar response for both increasing and decreasing signals.

If the signal line is fractured it will vent to atmosphere, giving a pressure of 0 bar. This is below instrument zero and will cause the indicator to read off scale negative. An offset zero thus gives protection against damage to the transmission path.

The main disadvantage of pneumatic systems is a slow speed of response. Suppose the transducer of fig.6.33a has to follow a step increase of level. This will require a step change of pressure. There are two components to the response of the indicator. There is an inherent transit delay as the pressure step can only travel at the speed of sound (330 ms^{-1}). Piping runs of several hundred metres are common in even medium size plants, so delays of the order of a few seconds are inevitable.

The speed of response can also be adversely affected if the pressure change is associated with a volume change at the indicator or actuator (moving a piston in a cylinder, for example). The change of volume must be supplied by the transducer, and will result in an exponential rise of pressure. A similar effect occurs because the increase in pressure is achieved by a transfer of air into the piping/indicator. The volume of air required is a function of the total system volume. Fast response from a pneumatic system therefore requires short runs, small volume systems and 'loads' which do not change their volume with pressure changes.

6.7.2. The flapper/nozzle

The heart of most pneumatic process control devices is an assembly which converts a small physical displacement to a pressure change. This is invariably a variation on the flapper nozzle shown in fig.6.34a.

An air supply, typically at 2 bar, is applied to a nozzle via a restriction. Air flowing from the nozzle will cause a pressure drop across the restriction and the output pressure will be lower than the supply pressure by an amount related to the flow from the nozzle.

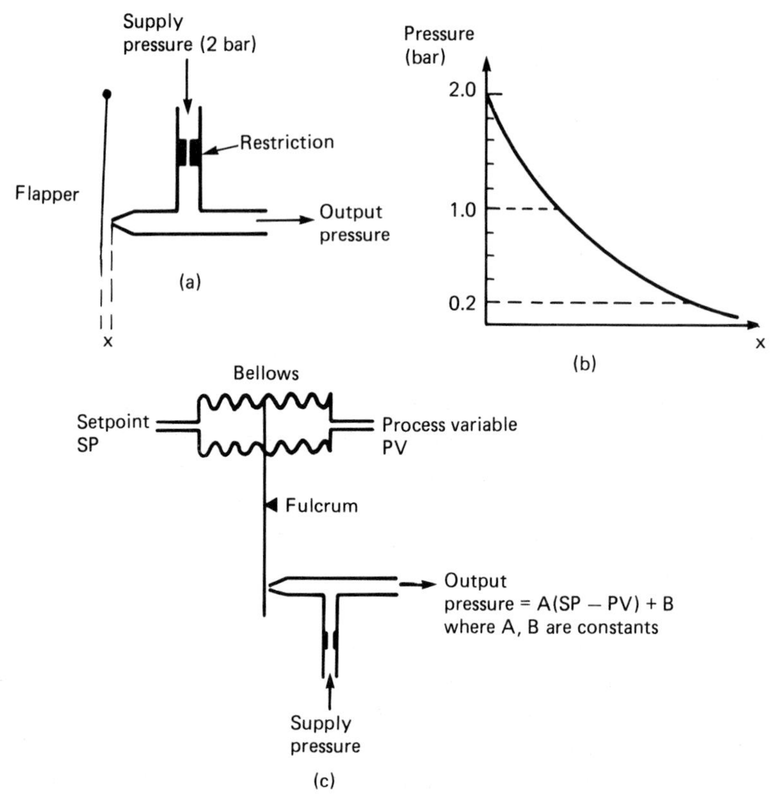

Fig.6.34 The flapper effect. (a) Arrangement. (b) Gap/pressure relationship. (c) Error amplifier.

The input displacement is applied to a flapper and varies the gap between the flapper and the nozzle. This varies the flow from the nozzle and consequently the output pressure. A typical relationship is shown in fig.6.34b; note the small range of the displacement. At the extremes of the curve the relationship is

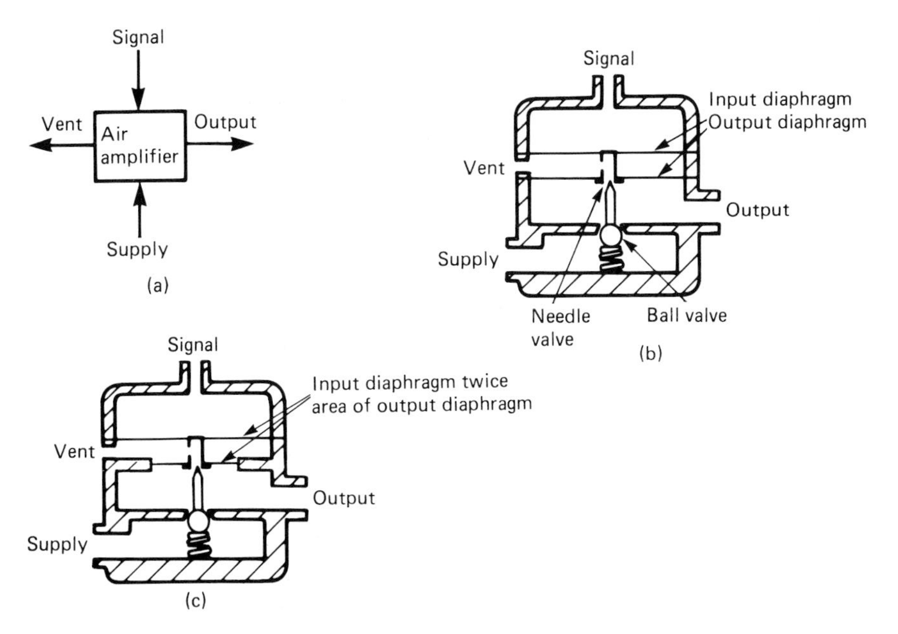

Fig.6.35 Pneumatic amplifiers. (a) Block diagram. (b) Unity gain (volume booster). (c) Air amplifier (gain × 2).

very non-linear, but can be considered to follow a straight line over the normal signal range of 0.2 to 1 bar.

Figure 6.34c shows an application which generates an error signal between pneumatic signals representing a set point and the process variable. These are applied to opposing bellows, and any difference will cause a deflection of the flapper and a corresponding change in output pressure. When the SP and PV signals are equal, the output pressure will be mid range (0.6 bar for a 0.2 to 1 bar system). Sections 6.7.4, 6.7.5 and 6.7.6 discuss further applications of flapper/nozzles.

The arrangement of fig.6.34a supplies an output pressure, but is incapable of supplying any significant volume of air. For this reason a flapper/nozzle is invariably followed by an air amplifier or volume booster, described in the following section. Force balance techniques, described below, are also used and effectively operate with a fixed gap to overcome the non-linearities of the flapper/nozzle.

6.7.3. Air amplifiers, boosters and relays

Air amplifiers, represented by fig.6.35a, are commonly used to increase or decrease pneumatic pressure by a fixed multiple or to supply increased volume of air. A 2× amplifier, for example, could be used to convert a 0.2 to 1 bar linear signal to a 0.4 to 2 bar signal.

The input signal controls the air flow from the supply to the load in such a way that the correct relationship is maintained. When the input signal falls, the output pressure is reduced by releasing air via the vent port.

Figure 6.35b shows a typical arrangement for a unity gain device. The signal is applied to the input diaphragm, and the supply to the output diaphragm. If the force from the input pressure is larger than the force from the output pressure, both diaphragms will move down, forcing the ball valve off its seat and admitting air to the output until the input and output pressures equalise.

If the input pressure falls both diaphragms move up, closing the ball valve and opening the exhaust valve. Air now vents until the output pressure falls to the current level. The output pressure thus follows the input pressure.

The input port requires negligible air volume changes, and as such is capable of being driven from a flapper nozzle. The output port can supply a large volume of air. Note, however, that the device will maintain pressure at the outlet port and cannot compensate for flow related pressure drops downstream of the device.

Figure 6.35b has equal area input and output diaphragms, and hence has unity gain. As such it is often called an air booster. If the input and output diaphragm have unequal areas, an air amplifier of non-unity gain is obtained. Figure 6.35c has a larger input diaphragm and acts as an amplifier. The gain is the ratio of the area. A reducing device has a larger output diaphragm.

Figure 6.36 utilises the force balance principle to boost the available air volume from a flapper/nozzle. The air relay of fig.6.36a applies the signal to a diaphragm, the extension of which is proportional to the applied pressure. The diaphragm motion is used to admit air to, or bleed air from, the outlet port. In this arrangement there is no feedback, and the inlet pressure controls the flow to, or from, the output.

The air relay of fig.6.36a is combined with a flapper/nozzle, as in fig.6.37b. The pressure variation from the nozzle is applied to the input of the air relay. The output is applied to the input of the air relay. The output is applied to a force balance diaphragm which opposes the force from the error signal (PV − SP). Any

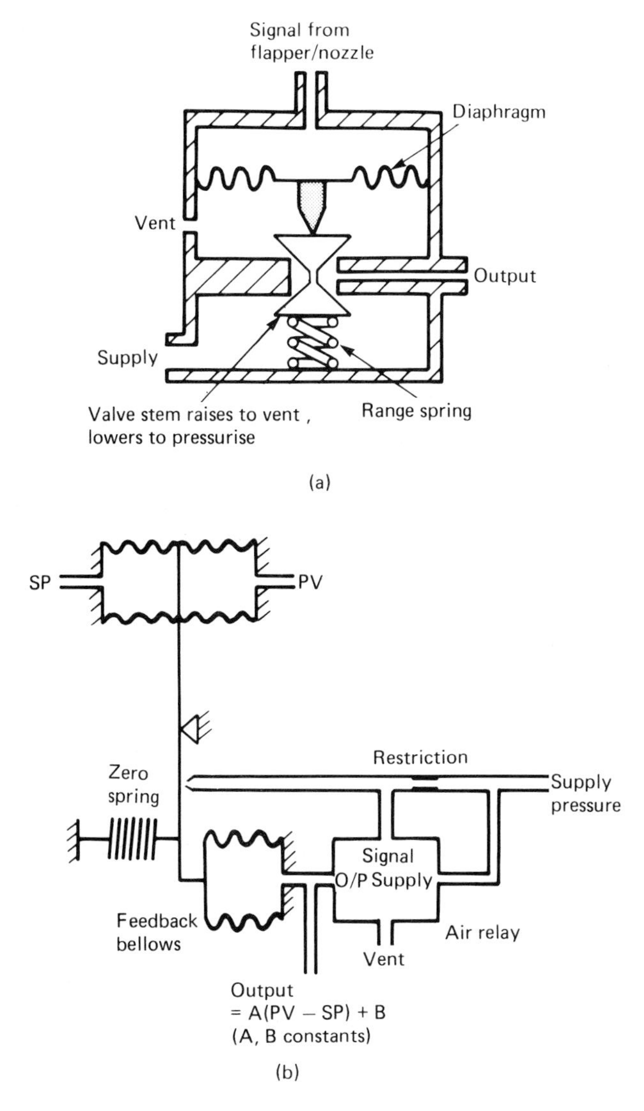

Fig.6.36 The force balance principle. (a) Air relay. (b) Proportional controller.

change in this signal will cause air to flow between the air relay and the force balance to the diaphragm until the flapper/nozzle gap is restored to its operating position.

Because fig.6.36 operates at a fixed flapper/nozzle gap, any non-linearities from fig.6.34b are overcome, and an output pressure linearly related to (PV − SP) is obtained. The output signal is taken from the air relay, and can hence deliver a reasonable air volume without introducing any error.

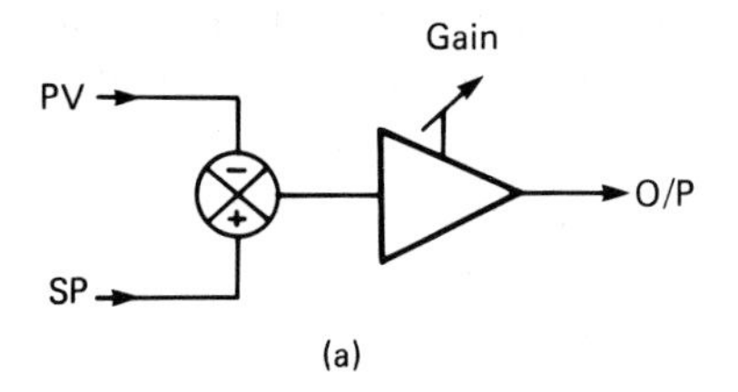

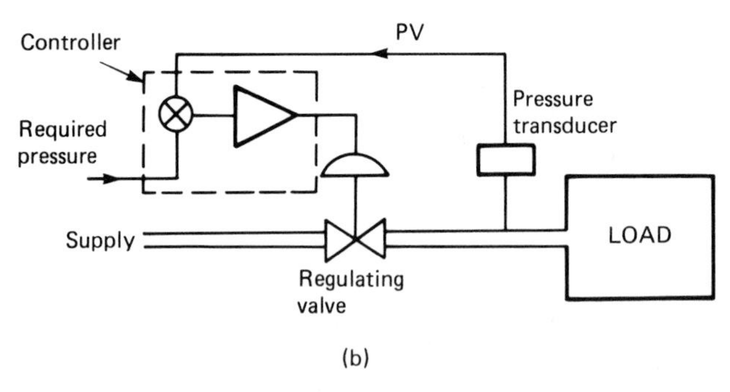

Fig.6.37 Use of controller to regulate pressure. (a) Representation of proportional controller. (b) Pressure regulating system.

6.7.4. Pneumatic controllers

Figure 6.36b can be represented by fig.6.37a. This is identical to a proportional controller (the principles of controllers are discussed in chapters 4 and 6 of Volume 3). The gain can be adjusted by moving the position of the pivot. Figure 6.37b shows the applications of a pneumatic proportional only controller to control a pressure regulating valve.

Figures 6.36 and 6.37 operate in the steady state with a fixed flapper/nozzle gap. To achieve this, a closed loop system will stabilise when the forces from the setpoint, PV and feedback bellows balance. This results in a small difference between the PV and setpoint pressure called an 'offset error', which is inherent in all proportional only controllers.

Electronic controllers are available to perform three term control, so called because the output signal is the sum of proportional, integral and derivative terms related to the error. These controllers perform the function:

$$V_o = K\left(E + \frac{1}{T_i}\int Edt + T_d\frac{dE}{dt}\right) \tag{6.11}$$

where V_o is the output signal, E the error, K the gain, T_i the integral time (sometimes referred to as reset or the inverse repeats per minute) and T_d the derivative time. K, T_i and T_d are adjusted by the process control engineer to give the fastest system response consistent with stability. The integral action of equation 6.11 overcomes the offset error of a proportional controller as the actuator signal will integrate the error. The derivative term adds stability and gives a large correcting signal when the error is changing rapidly.

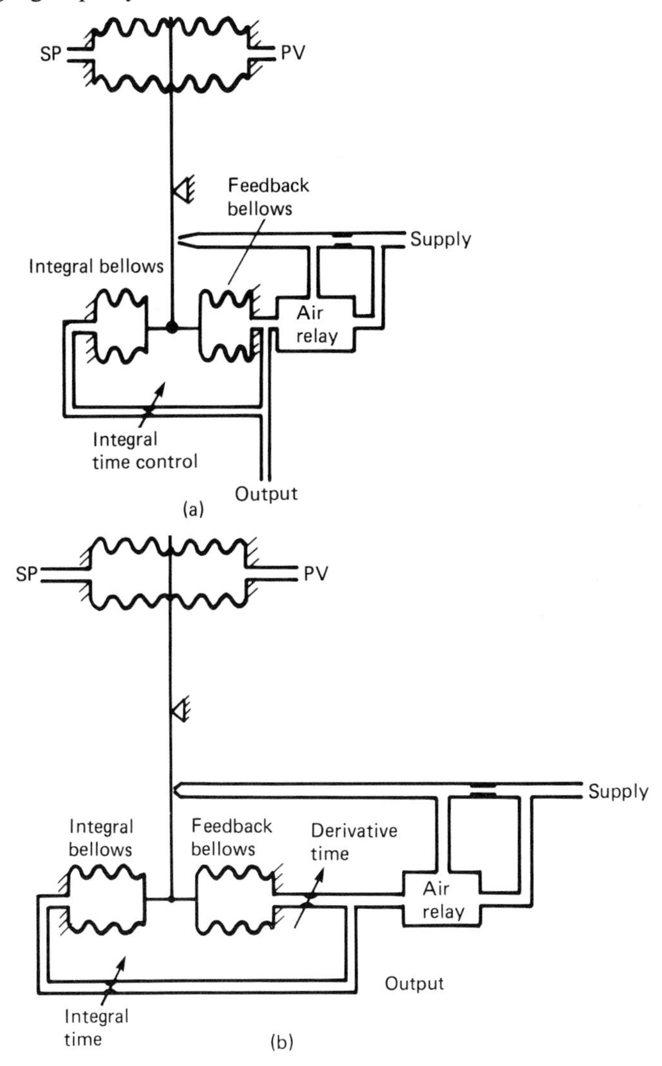

Fig.6.38 Pneumatic controllers. (a) Proportional plus integral (P + I) control. (b) Proportional plus integral plus derivative (PID, 3 term) controller.

Figure 6.38a shows how integral action can be achieved in a pneumatic controller. With the integral adjustment valve closed, the controller will act in proportional (P) mode and control with an offset error. Assume the controller is established in P mode, and the integral valve is opened. The integral bellows will oppose the action of the proportional bellows, changing the flapper/nozzle gap and the output signal. The system will stabilise with the flapper/nozzle at the steady state gap, and the set point and PV equal, and the contributions from the feedback and integral bellows equal. Under these conditions, the output pressure is just correct to make the PV and the setpoint equal with no offset.

Derivative action is obtained by restricting the flow to the feedback bellows, giving an output pressure related to the rate of change of error.

Adjustment of K, T_i, T_d in equation 6.11. is achieved by adjusting the beam pivot point (gain K) and the integral and derivative bleed valves. Unlike electronic controllers, pneumatic controllers often exhibit interaction between the three terms and are consequently more difficult to adjust.

6.7.5. Valve positioners

A pneumatic actuator such as those described in section 6.6.4. gives a displacement which is linearly related to a pneumatic signal. The performance of actuators can sometimes be improved by the incorporation of a separate actuator position control system, called a valve positioner.

Positioners are mandatory for double acting actuators, such as in fig.6.32d, and are useful in the following circumstances:

(a) When accurate valve position is required.
(b) To speed up the response of a valve (particularly in cascade systems where it is essential to have the inner loop faster than the outer loop).
(c) Where volume boosting is needed, e.g. where the device providing the control signal is incapable of driving the valve directly.
(d) Where a pressure boost is required to give the necessary actuator force.

Figure 6.39 shows a positioner using the force balance principle. The position of the actuator is converted to a force by a

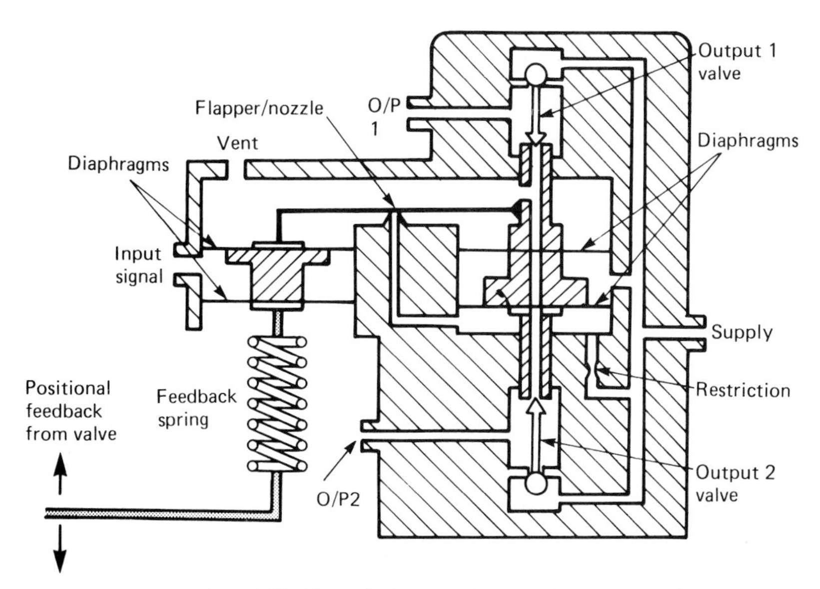

Fig.6.39 Force balance valve positioner.

range spring. This is coupled to an input diaphragm assembly which moves one side of the beam on the flapper/nozzle according to the difference between the spring force and the diaphragm force from the input signal pressure.

The pressure from the flapper/nozzle is fed to the pilot valve diaphragm. If the flapper/nozzle gap is too small, the pilot pressure will increase, moving the spool up. This admits air to the lower connection on the actuator and vents the upper connection, causing the actuator to rise, and reducing the spring force.

Conversely, if the flapper/nozzle gap is too large the pilot pressure will decrease, causing the spool to move down. Air will be admitted to the top actuator connection and vented from the lower, causing the actuator to fall and the spring force to increase.

In both cases the system will balance when the spring force matches the force on the input diaphragm from the input signal, i.e. the actuator position matches the required position.

The positioner zero is adjusted by altering the relative position of the shaft and spring, and the span by altering the effective spring constant.

An alternative positioner, shown in fig.6.40, uses the motion balance principle. The valve shaft position is converted to a small displacement by a cam, and applied to one end of a flapper/nozzle

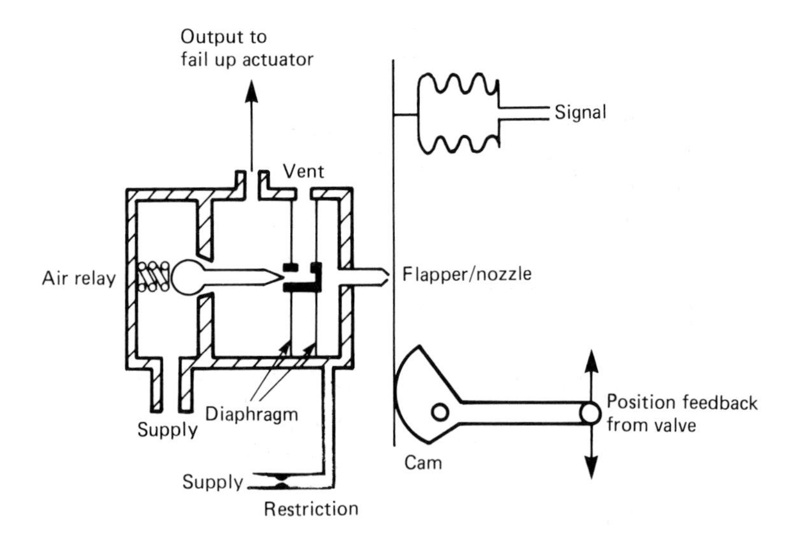

Fig.6.40 Motion balance positioner.

beam. The input signal pressure is converted to a displacement by bellows at the other end of the beam.

The nozzle pressure is applied to an air relay which admits air to, or vents air from, the actuator until the nozzle gap is correct. At this point the actuator position matches the input signal.

Positioners are usually equipped with gauges indicating supply pressure, signal pressure and actuator pressure(s). Often bypass valves are fitted to allow the signal pressure to be passed direct to the actuator as a temporary measure in the event of a failure of the positioner.

6.7.6. I/P and P/I converters

Pneumatic signals are used where large forces are required or explosive atmospheres are used. Electronic controllers and displays are widely used in control rooms. I/P (current-to-pressure) and P/I (pressure-to-current) converters provide the interface between these differing technologies.

Figure 6.41 shows a common form of I/P converter, again built around the force balance principle and the ubiquitous flapper/nozzle. The signal current is passed through four coils, wired as shown, and causes a rotary torque in the flapper arm. The torque is proportional to the signal current and causes a change in the flapper/nozzle gap.

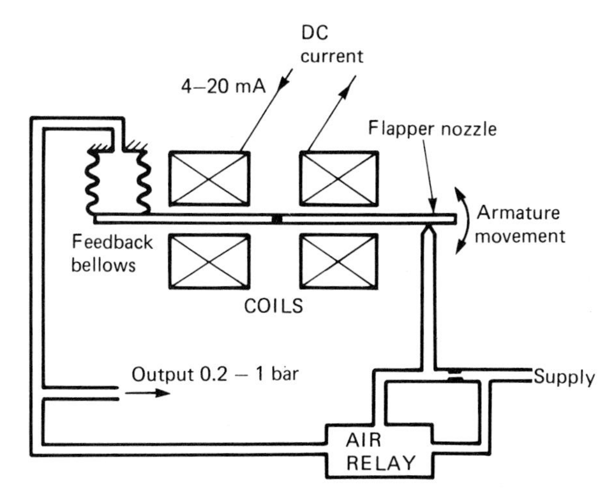

Fig.6.41 Current to pressure (I/P) converter.

The resulting pressure controls an air relay, causing the output pressure to rise, or fall until the feedback bellows returns the flapper beam to its balanced position. The output pressure thus follows the input current.

Pressure to current converters are simply pressure transducers (measuring gauge pressure). Pressure transducers are described in chapter 3 of Volume 1.

6.7.7. Fluidics

Sequencing applications are normally implemented with relays, digital ICs (such as TTL or CMOS) or programmable controllers. It is also possible to devise sequencing systems which are totally pneumatic. Such systems, usually called fluidics, are not particularly cost effective when compared with more conventional devices, but are useful where electronic circuits cannot be used (e.g. in explosive atmospheres or some medical applications).

Figure 6.42 shows a pneumatic equivalent of an electrical limit switch. With the pneumatic sensing outlet exposed, air bleeds to atmosphere and the output port is at atmospheric pressure. With the sensing outlet obstructed, diverted air causes a rise in pressure at the output port.

Logic gates use the 'wall attachment' or 'Coanda' effect shown in fig.6.43a. Fluid stream exiting from a jet with a Reynolds

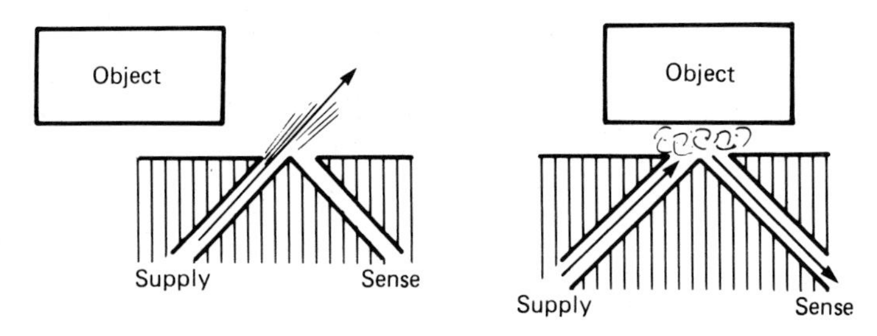

Fig.6.42 Pneumatic limit switch.

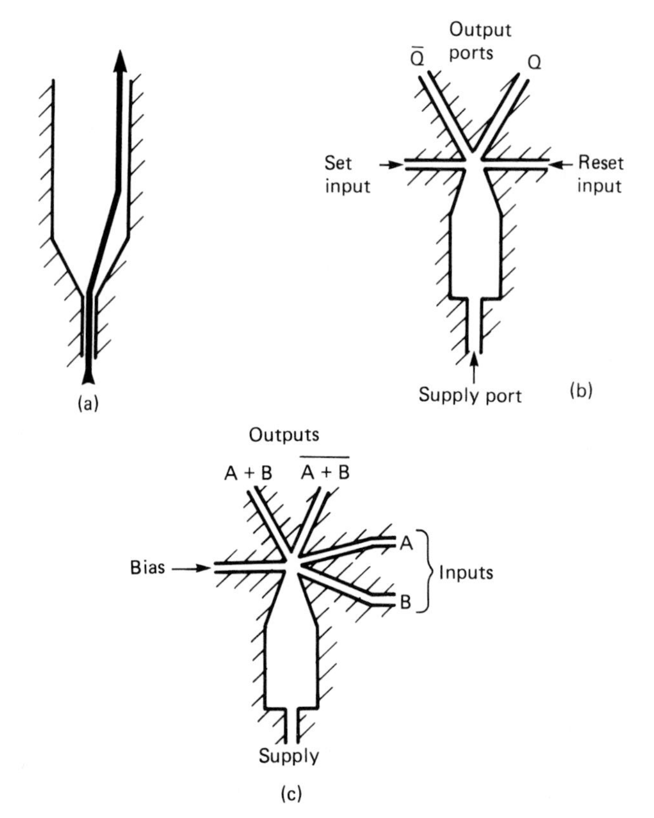

Fig.6.43 Fluidic logic. (a) The Coanda effect. (b) Set/reset flip flop. (c) OR/NOR gate.

number in excess of 1500 (see section 5.2.2, Volume 1) tends to attach itself to one wall. If disturbed from the wall for any reason, it will attach itself to the opposite wall.

Figure 6.43b shows a pneumatic SR flip flop. Control ports are used to shift the jet stream between output ports. Once switched, the stream will be stable until a control pulse is applied to the other input.

An OR/NOR gate is shown in fig.6.43c. The auxiliary inputs bias the jet to the right-hand wall. A signal applied to either input will cause the jet to switch sides. All the common logic elements (AND, NAND, OR, NOR, timers, shift registers, etc.) can be constructed using similar ideas.

Chapter 7
Recording and display devices

7.1. Introduction

All processes require some form of system to allow the operators to observe the plant operation. For a simple motor this can just be lights saying 'Running' and 'Stopped'. At the other extreme, a complete petrochemical plant may have a large mosaic mimic display or a series of computer driven VDUs. Whatever the

Fig.7.1 A typical layout of operator controls, with a mixture of Mimics, VDUs, keyboard, analog and digital instruments and miscellaneous switches and lights. Control layout is always a compromise between control requirements, visibility, available space and ease of use. (Photo courtesy of GEC Electrical Projects.)

complexity of the display system, the aim is to give the operator full knowledge of what is going on and to allow fault conditions to be quickly identified.

There is also often a need to provide permanent records of plant performance, sometimes called trend recording. These records can be used for management analysis, maintenance planning and production control. They are also essential in plants with very long time constants, where instantaneous values of plant variables are of far less importance than indication of direction and rate of change.

This chapter is concerned with devices that are used for the display and recording of plant data. These are often described by the rather grandiose expression 'man machine interface'.

7.2. Analog instruments

7.2.1. Pointer scale instruments

In many applications the simple moving pointer fixed scale instrument is the cheapest and most effective indicator. Most are constructed around the moving coil fixed magnet principle of fig.7.2. The current through the coil is determined by the external voltage and source resistance, and the resistance of the coil itself. This current produces a torque on the armature as described in section 2.2. The torque is opposed by a fine hair spring, so the armature will balance at an angle where the torque from the

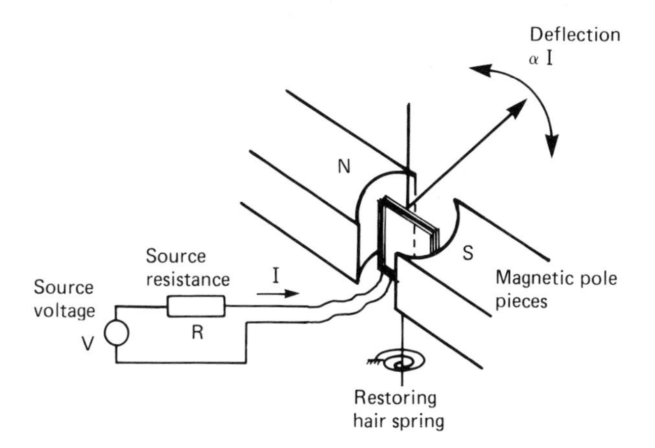

Fig.7.2 The moving coil meter.

spring matches the torque from the coil current. With careful design the angle of deflection can be made proportional to the current.

Moving coil instruments used to measure voltage have high coil resistance and low full scale current. This ensures the coil current is largely independent of the source resistance, and the meter does not affect (load) the voltage being measured. Current measuring instruments (milliameters and microameters) have low coil resistance to give a low voltage across the coil at full scale current. In both cases, the pointer deflection is proportional to coil current.

Any process variable can be represented as an electrical voltage or current by a suitable transducer (Volume 1 is concerned solely with transducers). Common signal ranges are 4 to 20 mA, 1 to 5 V and 0 to 10 V. The elevated zero signals have protection against open circuit or short circuit cable as the indicator will drive offscale negative.

The manufacture of, and standards for, electrical pointer scale instruments is defined in the UK by the British Standard BS 89. This describes nine classes of accuracy from ± 0.05% to ± 5% of full scale deflection (FSD). For industrial applications, accuracies of better than ± 1% are generally of little more than academic interest as they are beyond the resolution of the eye of the human observer.

The speed of response of a pointer scale instrument is inherently slow, with time constants of a few tenths of a second. BS 89 specifies that for a step change of 2/3 FSD input, the pointer should settle within 1.5% of FSD in 4 seconds. The construction of the meter makes it behave as a second order system characterised by a natural frequency and damping factor. Normally a damping factor of 0.7 is used, which gives the fastest response for a given natural frequency.

A meter scale should be chosen to be easy to read at the normal viewing distance. A useful rule of thumb is a scale length of 1/15 of the viewing distance (e.g. 20 centimeter scale length for 3 metre viewing distance). Figure 7.3 shows various types of scale; all have the same scale length, but take up different panel areas.

Experiment has shown that observers can quite accurately interpolate to 1/5 of a scale division provided the scale length is chosen as above. The use of 20 scale divisions over the full scale will therefore give a 1% resolution, which is sufficiently accurate

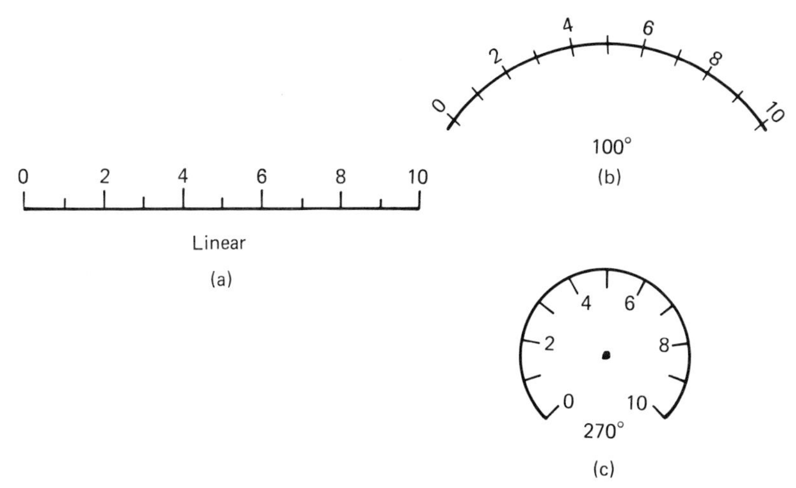

Fig.7.3 Different meter scales, all have equal scale length. (a) Linear. (b) 100°. (c) 270°.

for most applications. A greater number of scale divisions does not give increased resolution; the greater number of divisions merely clutters the scale and makes it difficult to read. The design and labelling of instrument scales is covered by BS 3693. Typical scales are shown in fig.7.4. Most have twenty minor divisions and four or five major divisions.

Meters should be scaled so that the normal indication is between 40% and 60% of FSD. If this causes an abnormal indication to go offscale, a non-linear scale can be used (see below). In many applications, the operator is required to detect abnormal conditions rather than accurately read individual indicators. This is simplified if the indicators are arranged in such a way that their pointers are in the same orientation in the normal operating conditions. Ideally the 9 o'clock position should be chosen for horizontal groups of meters and the 12 o'clock position for vertical grouping, as in fig.7.4c.

The moving iron meter of fig.7.5a is simpler and more robust than the moving coil meter. Two pieces of soft iron are under the magnetic influence of the coil. As current flows, the iron pieces become magnetised and repel each other. As before, a restoring force is provided by a hairspring giving a deflection which is a function of the current. It will be seen that the deflection does not depend on the direction of the current as the repulsion occurs because the two pieces of iron have the same sense of magnetisation. Moving iron meters can therefore be used on DC of either polarity or AC.

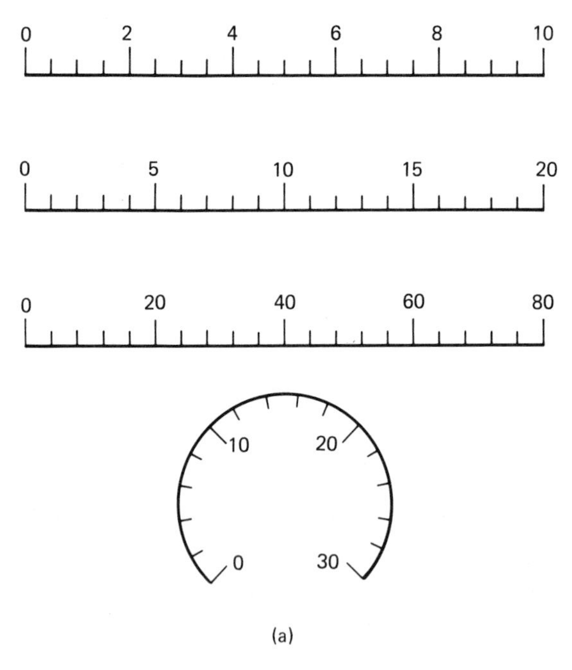
0
2
4
6
8
10
0
5
10
15
20
0
20
40
60
80
10
20
0
30

(a)

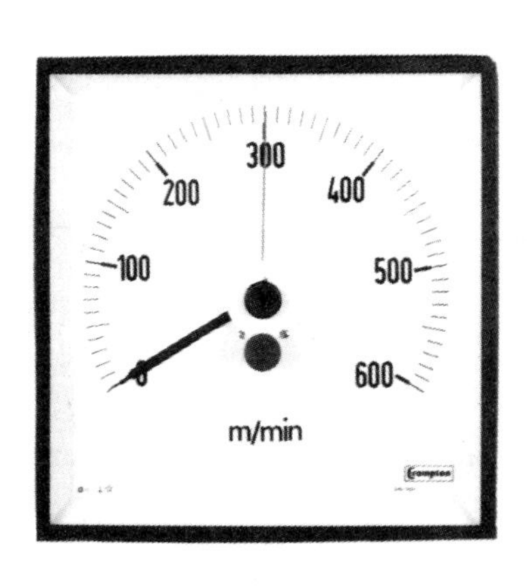
300
200
400
100
500
0
600
m/min

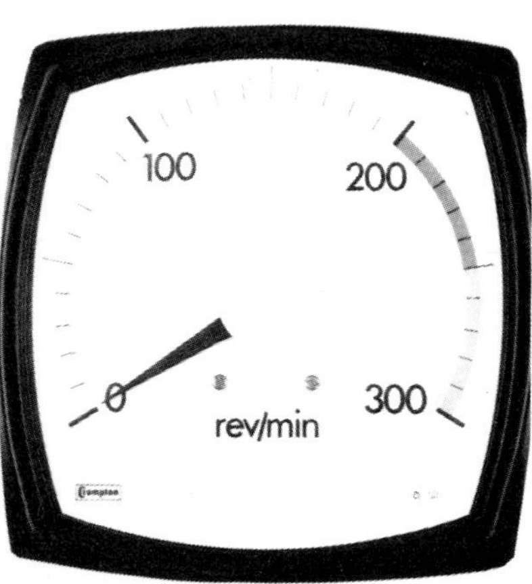
100
200
0
300
rev/min

100
50
150
0
200
50.2
kW
Crompton
244-DA

(b)

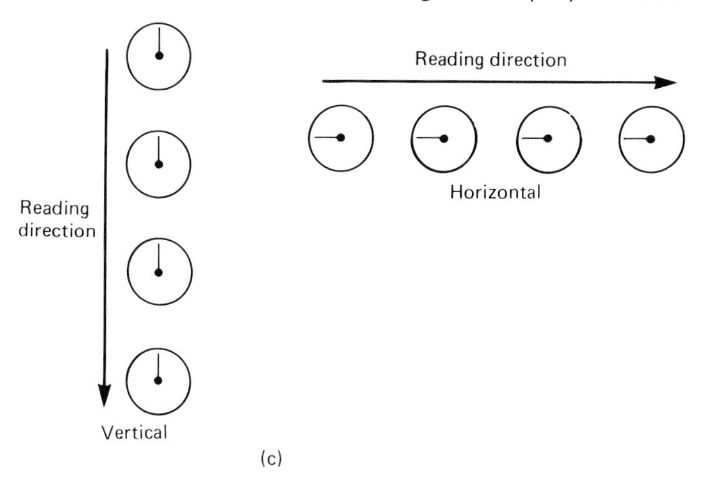

Fig.7.4 Moving pointer indicators. (a) Scale markings readable to about 1% resolution. (b) Examples of easy to read meter scales. LCD meters are more robust as well as combining the resolution of a digital display. (Reproduced from the catalogue of Crompton Instruments.) (c) Grouping of meters which are scanned for deviations rather than read precisely. Normal indication arranged as shown.

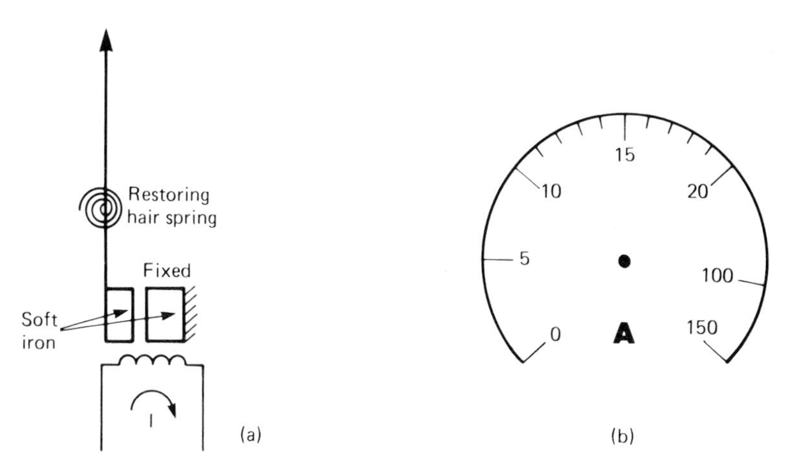

Fig.7.5 Moving iron meters. (a) Construction of meter. (b) Non-linear meter scale used with AC motors.

A moving iron meter is inherently non-linear. This characteristic can be used to give sensible scale deflections for meters where an abnormal condition causes a reading many times higher than a normal condition. Figure 7.5b shows a moving iron ammeter for a fan motor. This gives a normal reading around 50% of pointer deflection but can indicate about ten times overload under starting and fault conditions.

7.2.2. Bar graphs

A bar graph such as fig.7.6 is a useful alternative to a pointer scale instrument where a block of similar data is to be displayed. They are particularly useful where precise reading is not required, but a fault condition needs to be visually obvious.

Early bar graphs used a rotating tape scale, as in fig.7.7a, or a vertical drum with a spiral red/white division, as in fig.7.7b. In both cases the display was moved directly by a meter movement

Fig.7.6 Bar graph indicators showing cooling water temperature. The pattern and a rogue value can be clearly seen.

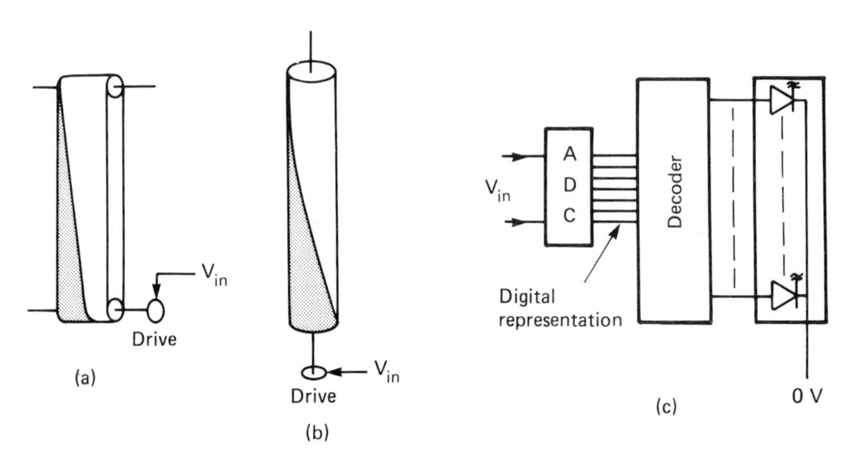

Fig.7.7 Bar graph indicators. (a) Belt drive. (b) Drum drive. (c) Electronic.

or a position control servo. Most modern devices, however, are electronic and based around the block diagram of fig.7.7c, where the input signal is digitised by an ADC and displayed by an array of LEDs or gas discharge tubes. A resolution of 0.5% is common, requiring 200 individual indicator segments. Scaling is achieved by changing the input sensitivity and the plastic indicator overlay.

Electronic bar graphs frequently incorporate alarm detection, giving an indication if the process variable goes outside preset limits. Commonly the bar display is flashed or changes colour for an alarm condition.

Bar graphs are generally more expensive than moving pointer instruments, and require a separate power supply. They are generally easier to read where accurate indication is not required.

7.3. Chart recorders

7.3.1. Introduction

Most plants require records of plant performance to be kept for subsequent analysis. Records of operating conditions may, for example, be needed to show that a plant is being operated within its designed parameters. The operation of unmanned plant such as pump houses must be recorded and checked at regular intervals. Recording can overcome problems with the speed of plant operation. High speed action can be captured and viewed at leisure, and slow speed changes over hours, or days, can be analysed for trends. Multichannel recording can show up relationships, expected and unexpected, between plant variables.

The commonest instrument is probably the chart recorder, which produces a continuous graph of a process variable plotted against time. This can exist in a variety of forms according to the speed and nature of the variables to be recorded.

The first consideration is chart speed which must be chosen to match the likely rate of change of the variable. Too low a speed will blur detail, the minimum resolvable time being limited by the pen width. Too high a speed makes trends difficult to see and is wasteful of paper. Chart speeds are available from about 1 mm per day to tens of millimetres per second for pens and several metres per second for UV recorders, described later. High speeds can obviously only be used intermittently. Chart drives are based around stepper motors with electronic speed control (see section

2.11) or synchronous motors with speed changes being effected by replaceable gear trains.

The response speed of the pen is also critical. Most pen positioning systems behave as a second order system, characterised by a natural frequency and a damping factor. Figure 7.8 shows the response and phase shift for a recorder driven by a sine wave input at various frequencies and damping factors. Figure 7.8 is normalised, i.e. the frequency axis is expressed in terms of the natural frequency fn.

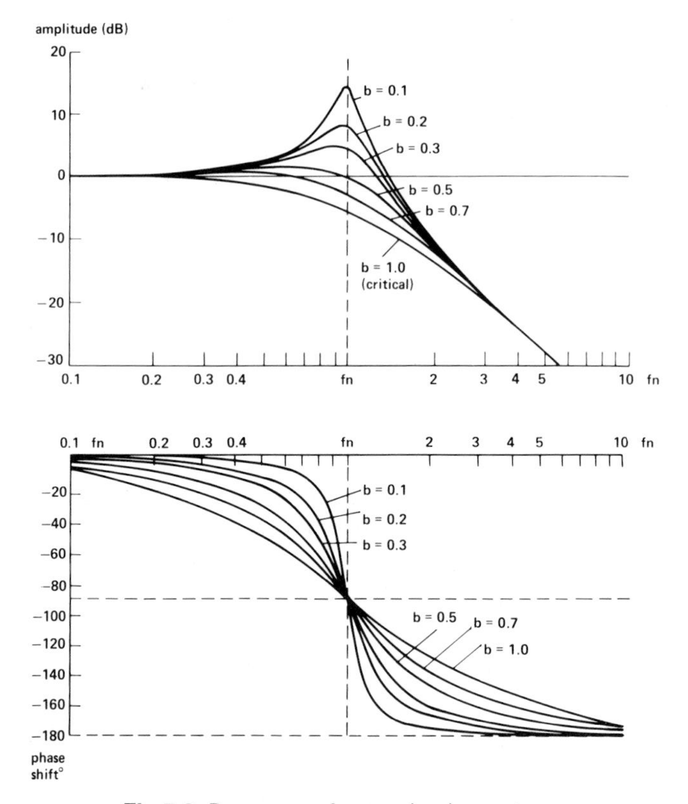

Fig.7.8 Response of second order systems.

Figure 7.8 shows that the widest frequency response is obtained with a damping factor of 0.7, and a recorder can be used up to about 0.5 fn without significant error.

Measured variables do not follow sine waves, but Fourier analysis shows that any regular waveform can be expressed as a sum of sine and cosine waveforms. With a knowledge of these frequency components, the natural frequency of the recorder can be specified to give the required accuracy. In practice, the specification of natural frequency for a chart recorder is usually more empirical than analytical and a figure is chosen which is related to the expected rate of change of the plant variable. Commonly the time taken for a full scale movement across the chart is specified. Too low a natural frequency will result in an inability to follow fast changes and a loss of detail.

Pen friction can limit the natural frequency and stiction results in a minimum pen movement, i.e. a minimum resolvable error. Early chart recorders used capillary tubes and ink reservoirs. These needed regular maintenance and cleaning (turning white-collar workers into blue-spotted-collar workers in the process). The modern tendency is to combine ink/pen cartridges, usually in the form of felt tip pens.

At high chart and pen speeds the viscous friction of pens becomes excessive, and other recording methods are used. Heat sensitive paper changes colour when heated, and is used in conjunction with a heated stylus. Electrosensitive paper changes colour under the action of an electric current (which burns away the top paper layer), and is used with a stylus connected to a current source. Both methods allow very high chart speeds, but the specialist paper is quite expensive.

Chart recorders usually operate on the standard signal ranges of 4 to 20 mA, 1 to 5 V and 0 to 10 V, the signal standards with offset zeros giving protection against cable faults and transducer failure. Temperature recorders for use with thermocouples and resistance thermometers usually connect direct to the sensor and incorporate the drive circuits (for PTRs), cold junction compensation (for thermocouples) and linearisation. The recorder can usually be arranged to drive offscale high or low in the event of a cable or sensor fault.

The accuracy of a chart recorder is specified in terms of FSD, with a deadband that is the smallest signal which will move the pen. Typical figures are, respectively, 1% and 0.2% of FSD. Speed of response can be specified in terms of natural frequency (for galvanometric instruments) or time to traverse the chart (for servo recorders).

7.3.2. Galvanometric and open loop recorders

In the simplest (and cheapest) chart recorders the process variable signal is used directly to move the pen. In the circular

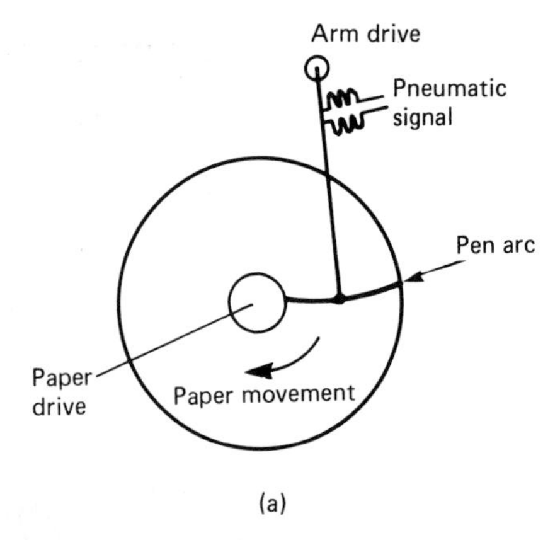

(a)

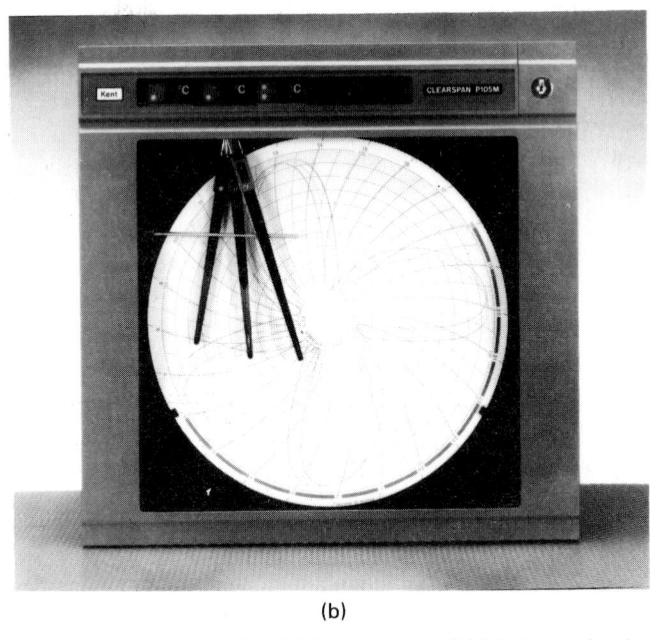

(b)

Fig.7.9 Circular chart recorder. (a) Construction. (b) Multipen circular chart recorder. (Photo courtesy of Kent Industrial Measurements.)

chart recorder of fig.7.9a, for example, a pressure signal moves the pen via a set of bellows. The records on a circular chart are difficult to read and interpret because of non-linearities from the circular chart and the arc of the pen. The main advantage is the relationship of chart to a set period (e.g. a shift, a day, a week) and an ease of filing (a simple dowel).

Strip chart recorders are easier to read, but direct driven recorders have inherently non-linear traces. Most are based on

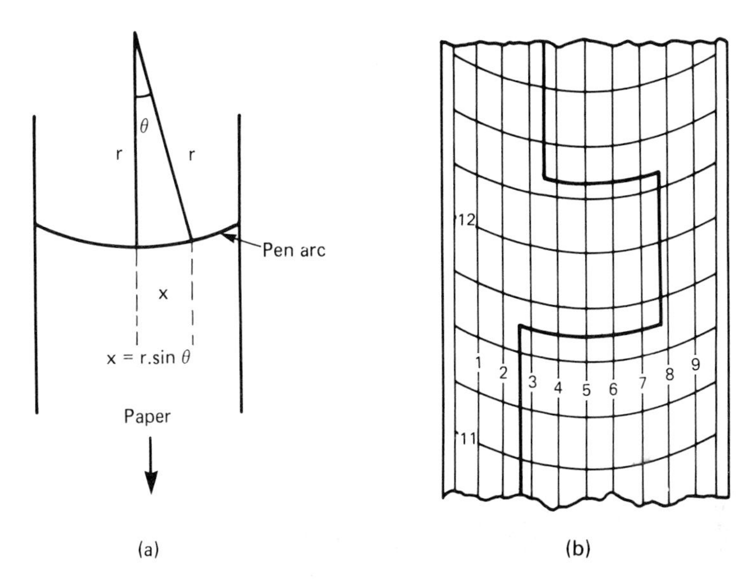

Fig.7.10 Direct drive chart recorder. (a) Construction. (b) Sample chart.

the moving coil mechanism described in section 7.2.1, with the pen being connected to the end of the pointer. The pen tip traverses an arc, as shown in fig.7.10a, with radius equal to the pen arm. This requires chart paper pre-printed with arcs, as in fig.7.10b. Although these are easier to read than circular charts, it can still be difficult to infer relationships. The distances along the arcs are directly proportional to the input signal, but the distance from the centre line is given by:

$$x = R \sin \theta \qquad (7.1)$$

To minimise error θ is kept small, usually by employing a long arm.

A curved grid can be avoided by passing the recording paper over a straight knife edge, as in fig.7.11. Obviously a pen cannot be used, heat sensitive or electrosensitive paper (see section 7.3.1) being required. Although this arrangement gives a rectilinear chart, it is non-linear with grid spacings getting larger away from the centre line. As shown in fig.7.11b:

$$x = R \tan \theta \qquad (7.2)$$

As before, errors can be reduced by using a small θ, usually obtained by making the pen arm as long as possible.

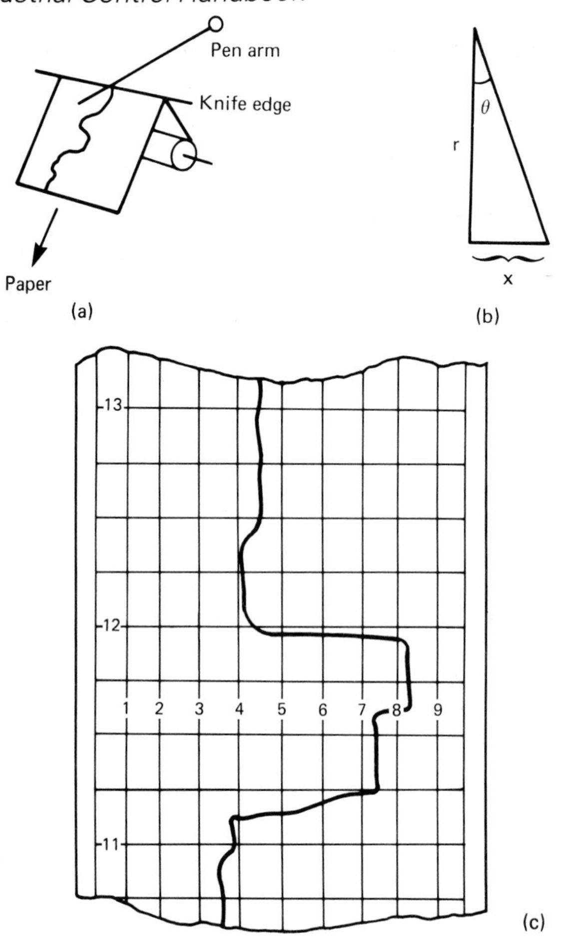

Fig.7.11 Knife edge recorder. (a) Construction. (b) Deflection. (c) Sample chart.

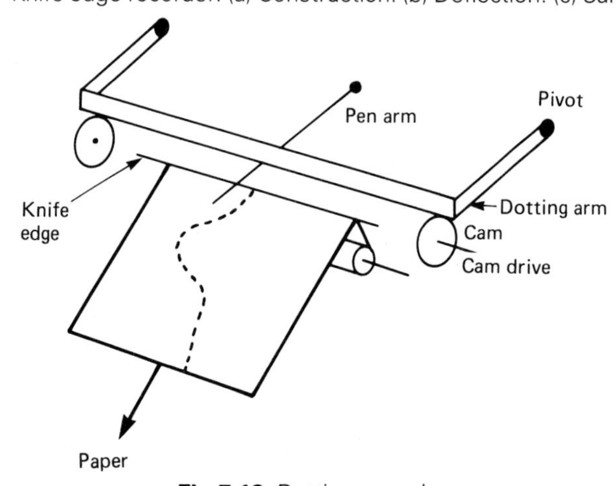

Fig.7.12 Dotting recorder.

The torque available from galvanometric recorders is small, and pen friction is a possible source of error as well as a limiting factor on speed of response. A common technique is to use a 'dotting' pen technique, illustrated in fig.7.12. The pen is normally allowed to move freely just above, but not contacting, the paper. A clamping bar above the pen is supported by motor driven cams. At regular intervals the bar is released on to the pen arm, simultaneously clamping it and leaving a dot on the paper.

An alternative approach is the use of a power amplifier to boost the input signal, allowing a more powerful galvanometer to be used.

7.3.3. UV recorders

The natural frequency of a galvanometric device is given by:

$$f_n = \frac{1}{2\pi} \sqrt{\frac{K}{J}} \text{ Hz} \tag{7.3}$$

where K is the restoring string constant, and J the moment of inertia of the armature assembly. A high natural frequency is required for high speed measurements, and this can be obtained by increasing K or reducing J.

Ultraviolet, or UV, recorders (sometimes called oscillographs) use mirror galvanometers which have very low inertias. A typical arrangement is shown in fig.7.13. A mirror galvanometer is encased in a permanent magnet block. Collimated light from a UV lamp is reflected from the mirror on the galvo to produce a spot on the UV sensitive paper. Current passing through the galvo deflects the mirror, and hence the trace on the paper. Figure 7.13 has the same inherent non-linearity as fig.7.11, but this is reduced to acceptable levels by the use of a long path length.

Galvanometers can be purchased with natural frequencies well in excess of 50 kHz. Equation 7.3 shows that once a practical lower limit of inertia has been reached, the natural frequency can only be raised by increasing the spring constant K, which in turn implies a decrease in sensitivity. High frequency galvos tend to be less sensitive than their lower frequency cousins.

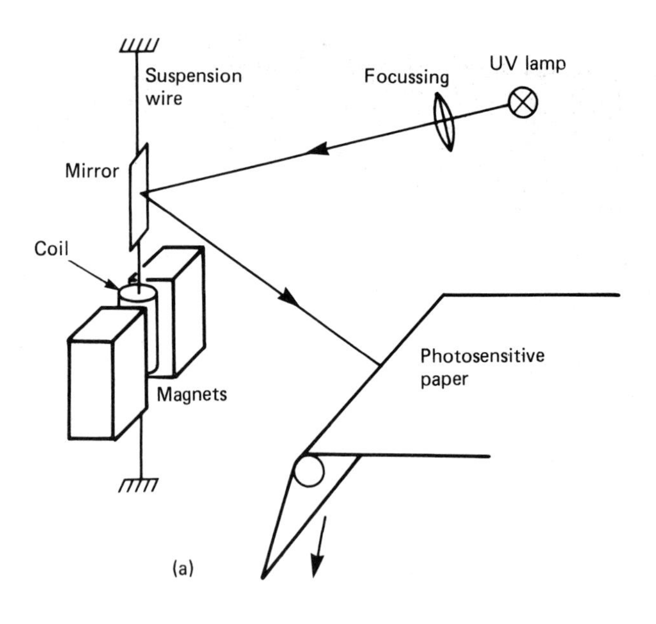
Suspension wire
Focussing
UV lamp
Mirror
Coil
Magnets
Photosensitive paper

(a)

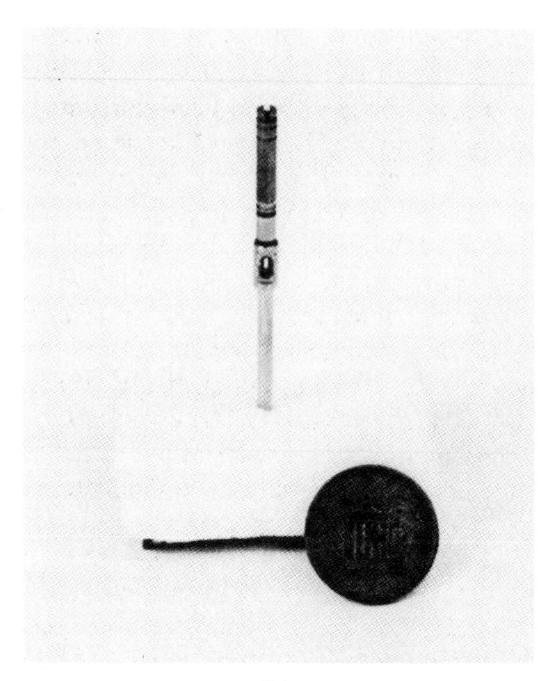

(b)

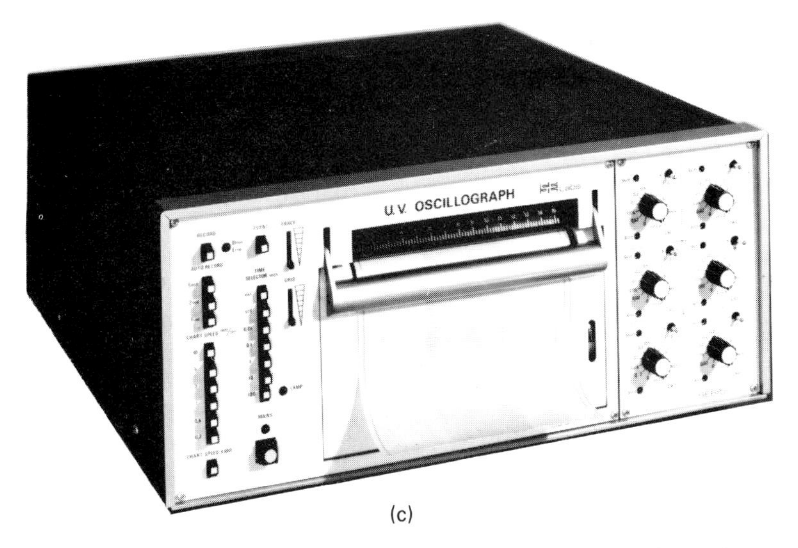

(c)

Fig.7.13 The ultraviolet recorder. (a) Construction. (b) A UV recorder galvanometer. The small size gives a good frequency response. (c) UV recorder SE6151 manufactured by SE Labs. (Photo courtesy of Thorn EMI Datatech.)

Damping is determined by viscous damping within the body of the galvo, and the impedance of the signal source. As the galvo armature moves in the magnetic field it acts as a generator, and the induced voltage opposes the applied current, consequently providing a damping force. The lower the source impedance, the higher the damping factor. The damping factor can be increased by connecting a damping resistor across the galvanometer leads. A damping factor of 0.7 gives the best response.

Galvanometers are current operated devices, and if connected directly to a voltage source must be used in series with external current determining resistance boxes. More expensive UV recorders incorporate voltage-to-current amplifiers to separate the voltage source and the galvos. The use of integral amplifiers also makes the damping factor independent of the signal source.

Commercial recorders generally have magnet blocks that can take several galvos (six being typical) to give multitrace recording. The user generally purchases the galvanometers separately to cover a range of sensitivities and natural frequencies. Chart speeds of several metres per second are possible, but at these speeds a roll of paper will be consumed in less than a minute.

The photographic paper used is self-developing in a few seconds in natural and incandescent light (but develops rather slowly in fluorescent light). Exposure to sunlight will fog the paper, and the trace slowly disappears in natural light. Fixing sprays are available to make the trace permanent.

7.3.4. Servo recorders

Servo recorders use a closed loop position control system to drive the pen. This gives an accurate, fast and linear response, and overcomes errors from pen friction. The principle is shown in fig.7.14a. The pen friction is measured by a linear potentiometer, which produces a wiper voltage proportional to the pen position. This is compared with the input signal by an error amplifier, which in turn drives the servo motor to move the pen until the error voltage is zero. The error amplifier drives the servo motor via a power amplifier to overcome the effect of pen friction.

Figure 7.14a uses a rotating servo motor. Some recorders use the linear motor of fig.7.14b. The pen carriage is connected to a coil which can move linearly between the poles of a permanent magnet. If a current is passed through the coil the carriage will move, being attracted to one pole and repelled from the other. The carriage movement can therefore be controlled by the direction and magnitude of the coil current. The arrangement of fig.7.14b has the advantage of only one moving part and is consequently more reliable and robust.

Stepper motors, described in section 2.11, are also used as positioning devices in microprocessor based recorders.

7.3.5. Multichannel recorders

The need often arises for several variables to be recorded on one chart to show interactions between variables or purely to reduce costs and panel space. Multipen recorders giving continuous records can be overlapping as in fig.7.15a, or side by side as in fig.7.15b. Both methods have disadvantages; the overlapping pens need to be offset to allow them to cross giving a time shift between traces, and the side by side arrangement reduces the chart width per record and hence the resolution.

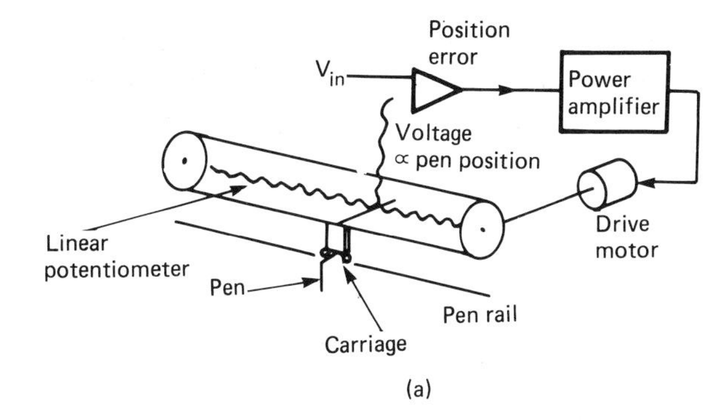

(a)

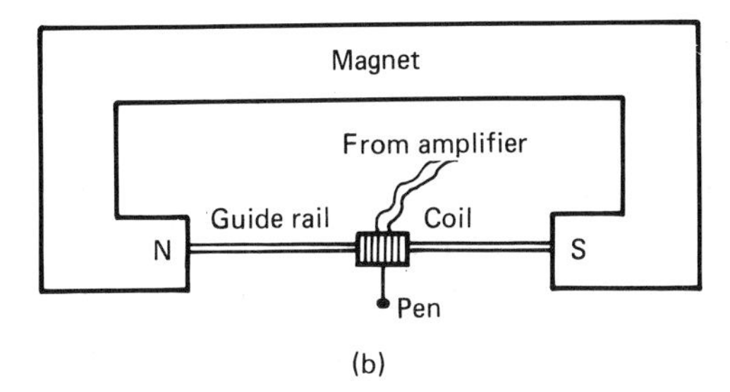

(b)

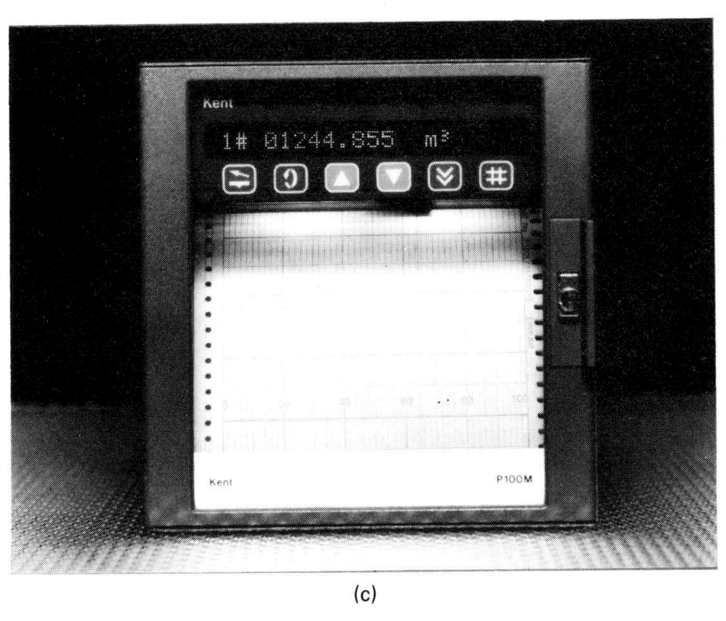

(c)

Fig.7.14 Servo chart recorders. (a) Block diagram. (b) Pen motor. (c) Modern servo chart recorder with readout in engineering units. (Photo courtesy of Kent Industrial Measurements.)

(a)

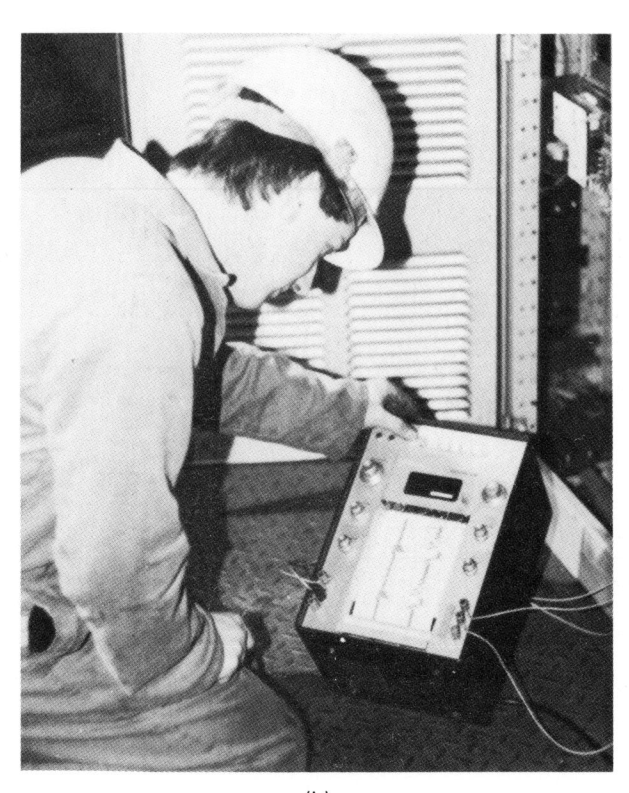

(b)

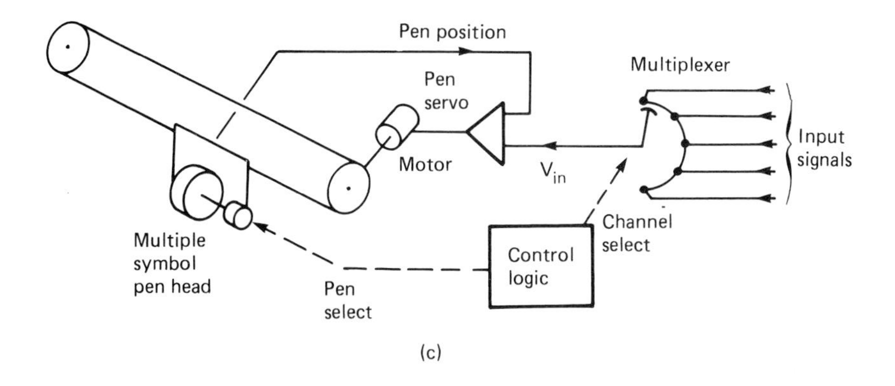
Pen position
Multiplexer
Pen
servo
Motor
V_{in}
Input
signals
Channel
select
Control
logic
Multiple
symbol
pen head
Pen
select

(c)

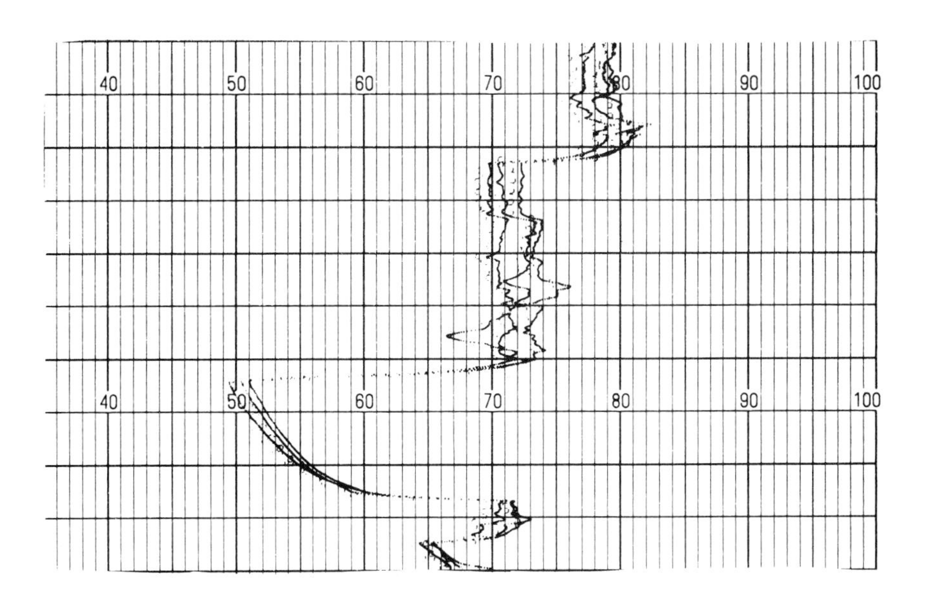
40
50
60
70
80
90
100
40
50
60
70
80
90
100

(d)

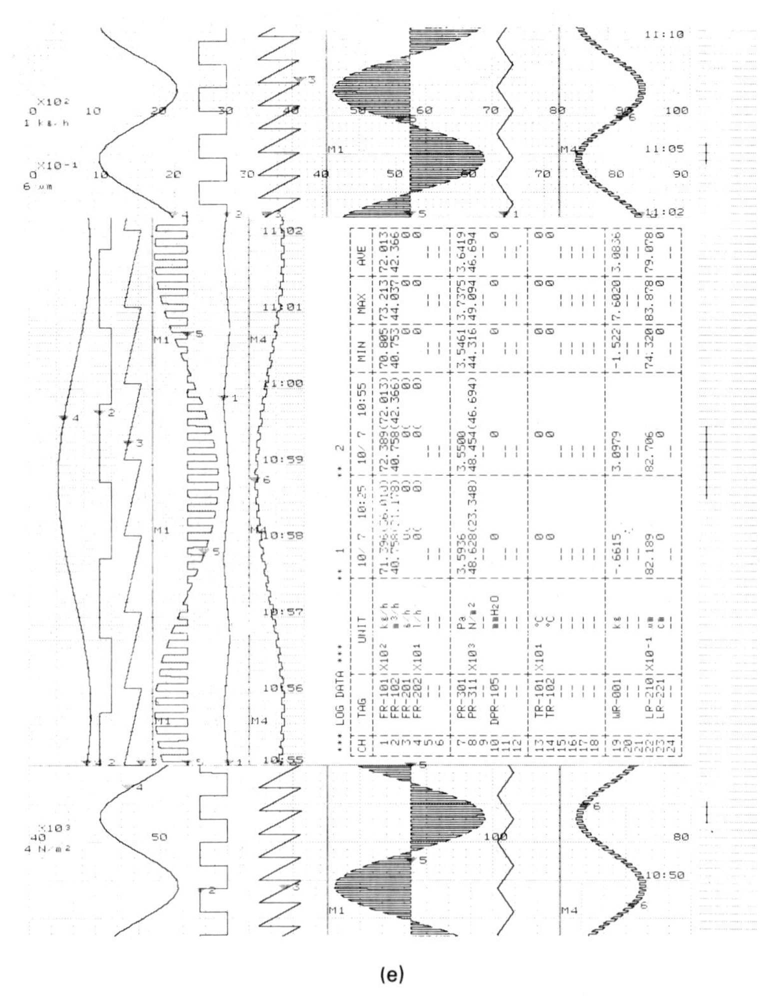

(e)

Fig.7.15 Multivariable chart recorder. (a) Two pen recorder with overlapping pens. The mechanism is simple and cheap, but the two traces are shifted with respect to each other. (b) Two pen recorder with separate pens. Time relationship is maintained at the expense of chart width. (c) Multipen recorder. (d) Typical chart from a dotting recorder. The recorder puts a number alongside each trace, in this example pens 1, 3, 5 are being used. (e) High quality thermal chart with analog trace and report generation facility. (Chart reproduced courtesy of Toshiba and their UK agents Protech Ltd.)

A common approach, useful for slowly changing signals, is the combination of the dotting principle of fig.7.12 with a multiplexor as in fig.7.15c. The pen is replaced by a print head with a different symbol, or ink colour, for each variable. Signals are sequenced in turn to the pen positioning servo, the print head being stepped to the correct symbol or colour each time. A typical scan rate is

5 seconds per point, giving an overall cycle time of 1 minute for twelve channels. Obviously this approach cannot be used for fast changing variables, but it is sufficiently fast for most temperature monitoring applications.

The UV recorder of section 7.3.3 can also handle multichannel signals, but is more of a laboratory or maintenance instrument, and is not really suitable for permanent panel mounted installation.

7.3.6. Event recorders

Full proportional representation is not required where signals such as limit switch or valve operation are to be recorded. Recorders which log on/off signals are known as event recorders, and use far simpler pen mechanisms. Most are simply a solenoid which deflects a pen a small distance (typically 5 to 10 mm) giving a trace similar to fig.7.16. Event recorders can show relationships

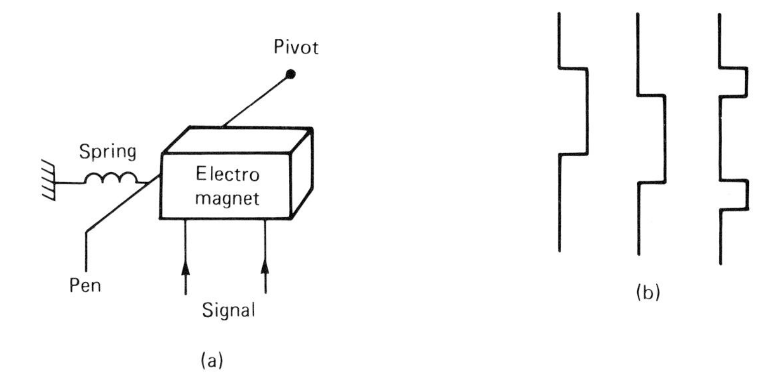

Fig.7.16 The event recorder. (a) Construction. (b) Typical trace.

between signals and be a useful fault finding tool, particularly when intermittent faults are being pursued. Event recorders are available with 24 channels, and a range of input signals from 5 to 240 V AC and DC.

7.3.7. Flat bed (XY) plotter

The flat bed, or XY, plotter has two separate pen position servos, as in fig.7.17a. This allows relationships between two variables to

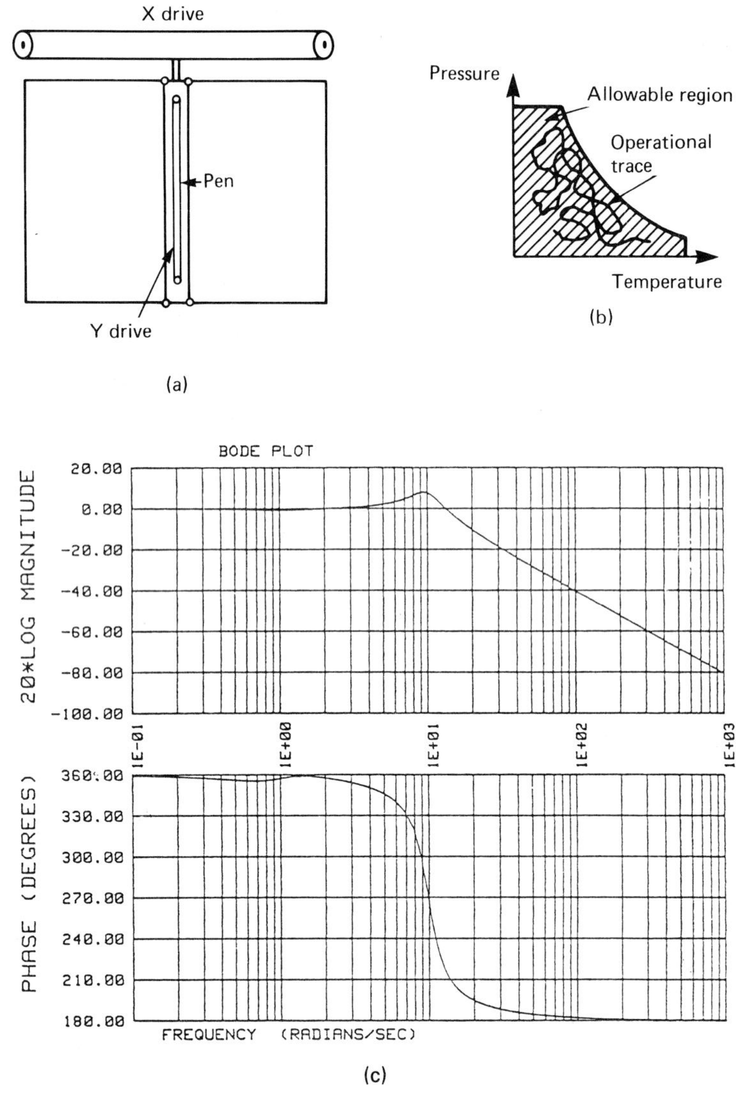

Fig.7.17 XY plotter. (a) Construction. (b) Typical application. (c) Output from a high quality Hewlett Packard XY plotter.

be plotted by connecting one variable to each servo. The plotter is really a laboratory instrument, but can be found in industry where a plant must be operated with certain conditions. Figure 7.17b shows a possible application where a plant must be kept within a certain range of pressure and temperature. An XY plotter with one axis connected to a pressure transducer and the other axis to a temperature transmitter will produce a spider's web trace that can easily be examined for excursions outside the permitted region. Flat bed plotters are also useful output devices for computers, allowing accurate graphs and sketches to be drawn. Computer aided design (CAD) machines use plotters to produce the final drawings.

7.4. Display devices

7.4.1. Introduction

The multidiscipline subject of opto-electronics was discussed in chapter 8 of Volume 1. The present section covers the application of light emitters in various forms from simple indicator lamps to multiline alphanumeric displays for the purpose of displaying plant status to operators. The reader is referred to Volume 1 for a description of the operation of the various devices.

7.4.2. Indicators

A single indicator lamp can convey an on/off indication to an operator, and as such they are used to display the status of plant items such as limit switches, pumps, motors, etc. Indicators come in a variety of shapes, sizes and intensities, some of which are shown in fig.7.18.

LEDs and neons are relatively low intensity devices and consequently best suited to small mimic panels. Most indicators

Fig.7.18 Typical industrial indicators.

are incandescent bulbs which give high intensity and low operating currents when operated on higher voltages (e.g. 110V AC). The use of transformer indicators, taking 110 V AC drives but utilising 6 V bulbs via a transformer integral with the indicator body, gives increased operator safety.

Indicator colours can be a source of contention. The relevant British Standard, BS 4099, defines (somewhat simplified):

Red Warning of potential danger or situation requiring action.
Amber Caution, change, or impending change, of condition.
Green Indication of safe condition. Authority to proceed.
White Any meaning when doubt exists about the application of red, amber or green.
Blue Any meaning not covered by red, amber or green.

but this does not really cope with a simple motor starter using illuminated push buttons, as in fig.7.19a. It is really a matter of site convention as to the colouring of the lenses and the light sequence. At the author's plant, the start button would be green and the stop button red, with the indicator showing which button was pressed last. In many plants both the colours and light sequence would be reversed. The most important factor, however, is consistency.

Plants should not be overindicated; the operator should have adequate status information, but not be swamped by a visual

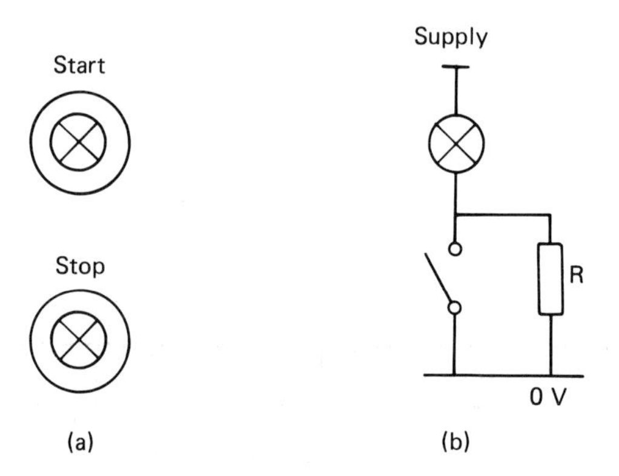

Fig.7.19 Using indicators. (a) The illuminated pushbutton problem. What colour are the buttons? (b) Lamp warming resistor. R should be chosen to give one tenth of normal operating currents.

overkill. Crucial indication should be given in both pass/fail states to cover for bulb failure. The eye is very good at recognising patterns, so a gap in a row of green lights with a red light below is easily spotted. Needless to say, a lamp test push button should be fitted for all but the most non-critical indications.

Bulbs have a limited life, and a large inrush current from cold. Bulb life can be improved and the inrush current reduced by the use of lamp warming resistors, as in fig.7.19b.

7.4.3. Numerical indicators

Numerical indicators, or digital displays, are used as an alternative to the analog instruments of section 7.2. Their primary advantage is improved resolution. A good analog instrument with moving pointer can be read to a resolution of 1% of FSD. A four digit display, showing 0 to 9999, can have a resolution, and accuracy, of 0.01%. Digital indicators are also used to display the state of devices which inherently move in finite steps–tap numbers on multitap transformers, for example.

Digital indicators have some disadvantages. They do not convey the 'feel' of a process that can be obtained with well laid out analog instruments or bar graphs. Compare fig.7.20a, b and c which shows the same temperatures, of cooling water, say,

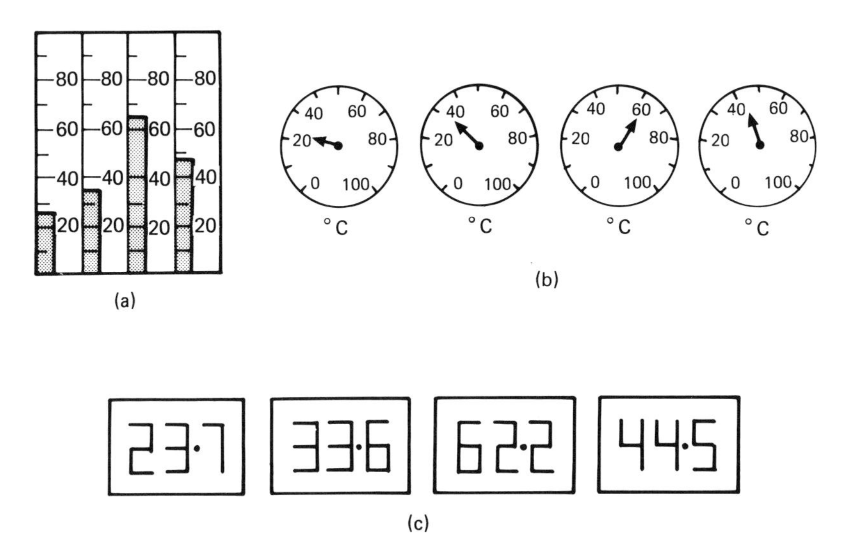

Fig.7.20 Comparison of display methods. (a) Bar graph. (b) Analog dials. (c) Digital displays.

displayed in bar graph, meter and digital form. The two analog displays can be scanned in one glance; the digital display needs careful reading.

Digital displays are also poor when used with process variables which are subject to rapid change. In these circumstances the display becomes a blur of digits from which rate of change, or under extreme conditions even the direction, cannot be inferred. The update rate of digital displays needs to be carefully chosen; too fast and the display flickers in an annoying manner at points such as 999 to 1000, too slow and the display makes large jumps as the process variable changes.

Analog displays are therefore best suited to applications where the process variable can change quickly, is subject to continuous change, needs relatively low accuracy reading or is part of a block of instruments that need to be scanned for an anomalous reading. Digital displays are required where accuracy is important–say, better than 1 or 2%–the variable is changing slowly (e.g. most temperatures) or the variable is inherently digital in nature. The author views with some suspicion the trend towards digital indication in motorcars, particularly for speedometers.

The basic display devices–incandescent, light emitting diodes (LEDs), liquid crystal displays (LCDs), and gas discharge–are described in chapter 8 (Opto-electronics) of Volume 1.

Seven segment displays use seven bars, as in fig.7.21, which can be arranged into the digits 0 to 9 plus some letters as shown. ICs are available to decode a 4 bit BCD number to the seven segments. Obviously several displays can be used to give any required resolution.

LED displays are available with character sizes up to 30 mm. These can be viewed at distances up to several metres in normal

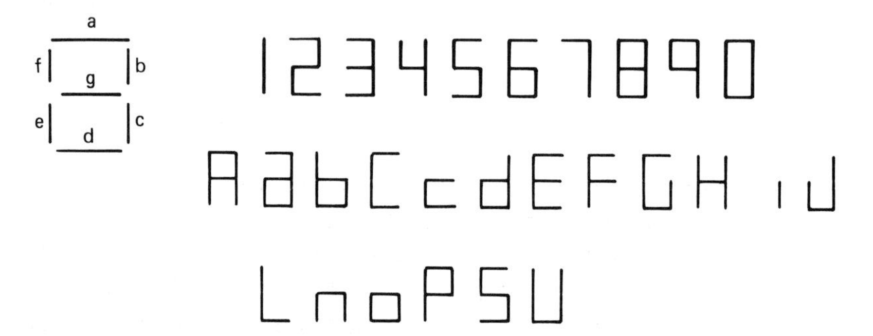

Fig.7.21 The seven segment display and available characters.

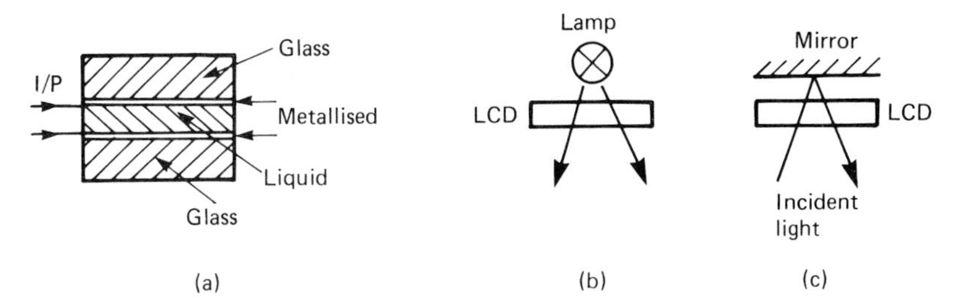

Fig.7.22 Liquid crystal displays. (a) Construction. (b) Transmissive mode. (c) Reflective mode.

room lighting, but are difficult to read in daylight. LCD indicators can operate either in the transmissive mode (with internal lighting) or the reflective modes of fig.7.22. LCDs are not as clear as LEDs in normal room lighting, but the contrast of reflective mode displays improves with increasing light levels. LCD indicators require minimal current, and units operating on 4 to 20 mA signals can derive their power supply from the loop current without degrading the signal. Life expectancy is reduced if LCDs are operated with a DC bias. The circuit of fig.7.23 is often used to give an AC drive.

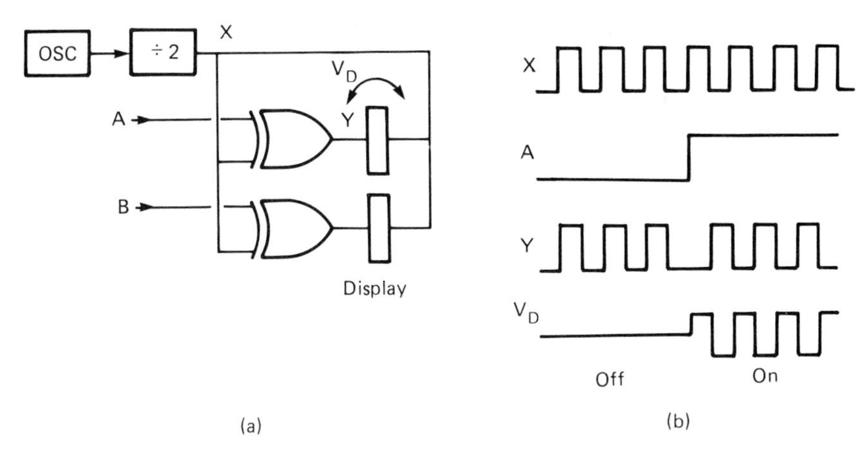

Fig.7.23 Driving LCDs without DC bias. (a) Circuit. (b) Operation.

Gas discharge devices have a bright easy-to-read orange display. In general they are far clearer than LED and LCD indicators and can be read in daylight. A variation is the Nixie tube (a registered trade name of the Burroughs Corporation). This uses shaped cathodes behind a mesh anode, as in fig.7.24, and gives natural shaped digits.

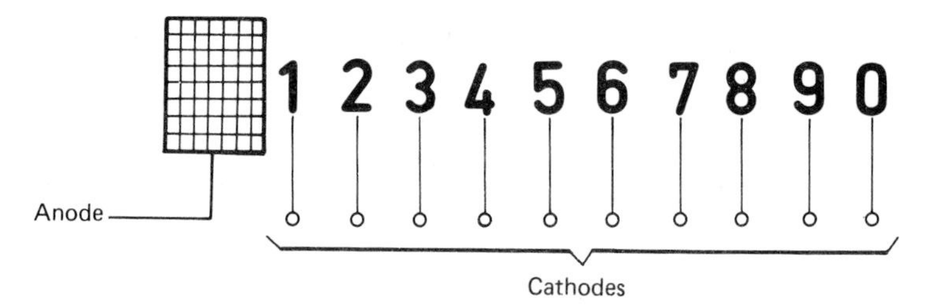

Fig.7.24 The Nixie tube.

Although shaped cathode displays such as the Nixie tube give natural looking digits, they suffer from a varying depth of display, with '1's, say, looking significantly nearer the front of the display than '9's and '0's. The depth also restricts the angle of viewing to around 45° from the normal. The anode voltage in Nixie and other gas discharge displays is around +180V, posing a possible service hazard for technicians used to working on low voltage display devices. The +180V anode supply is normally derived inside the display from a DC to DC inverter operating from the 5V or 12V logic supply. Ready made encapsulated inverters are also readily available. The high voltage anode supply also increases the cost of the units as high voltage driver transistors are required.

Fig.7.25 A large (1500 mm) alphanumeric display based on a 14 segment display. Each display segment is built up from many individual indicators; in this case LEDs. Displays similar to this are used in crane weighing and similar applications where information is to be read at long range. (Photo courtesy of Displait.)

Where large scoreboard displays are needed, incandescent or fluorescent tube based displays are required. Figure 7.25 shows an LED multi segment display in use on a crane weighing application. Mechanical seven segment displays are also available. These rotate coloured bars with solenoids.

Numerical displays can be made easier to read by the use of leading and trailing zero suppression. Without leading zero suppression, the number 27 on a four digit display would appear

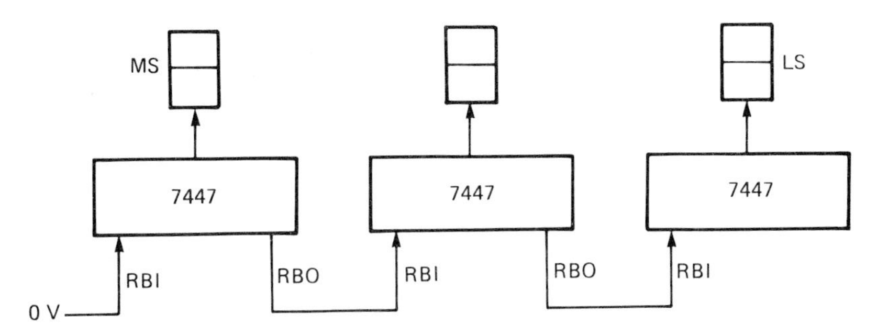

Fig.7.26 Ripple blanking (leading zero suppression).

as 0027. Leading zero suppression blanks the unnecessary zeros. Trailing zero suppression works after the decimal point, so 27.1 on a 99.99 display would appear correctly, and not as 27.10. Zero suppression is provided on most display driver ICs; fig.7.26 shows how the ripple blanking pins can be used to this end on the popular 7447 seven segment driver IC.

7.4.4 Multiplexing

Digital displays require a large number of signal lines: twelve for a three digit BCD display plus two power supply connections. Time division multiplexing can considerably reduce the complexity and the number of lines. Figure 7.27 shows a typical scheme for a four decade display.

The four BCD inputs are applied to four data selectors IC1–4. Note that IC1 takes the A (LSB) inputs, IC2 the B inputs, and so on. An oscillator, running typically at 1 kHz, drives a 2 bit counter which selects the inputs to the data selectors in turn.

The counter is decoded to drive the strobe transistors TR1–TR4 which are connected to the common anodes of the seven segment displays. The seven segment decoder provides the seven segment data for each display in turn, following the strobes from the transistor. Only one digit is illuminated at one time, but the strobing is too fast for the eye to follow and all four digits appear lit.

Multiplexing can reduce cabling significantly. Figure 7.27c shows 4 × 3 digit display driven with multiplexing. This requires twelve data lines and four strobe lines. Driven directly they would need 48 lines. Multiplexing does reduce the intensity of the

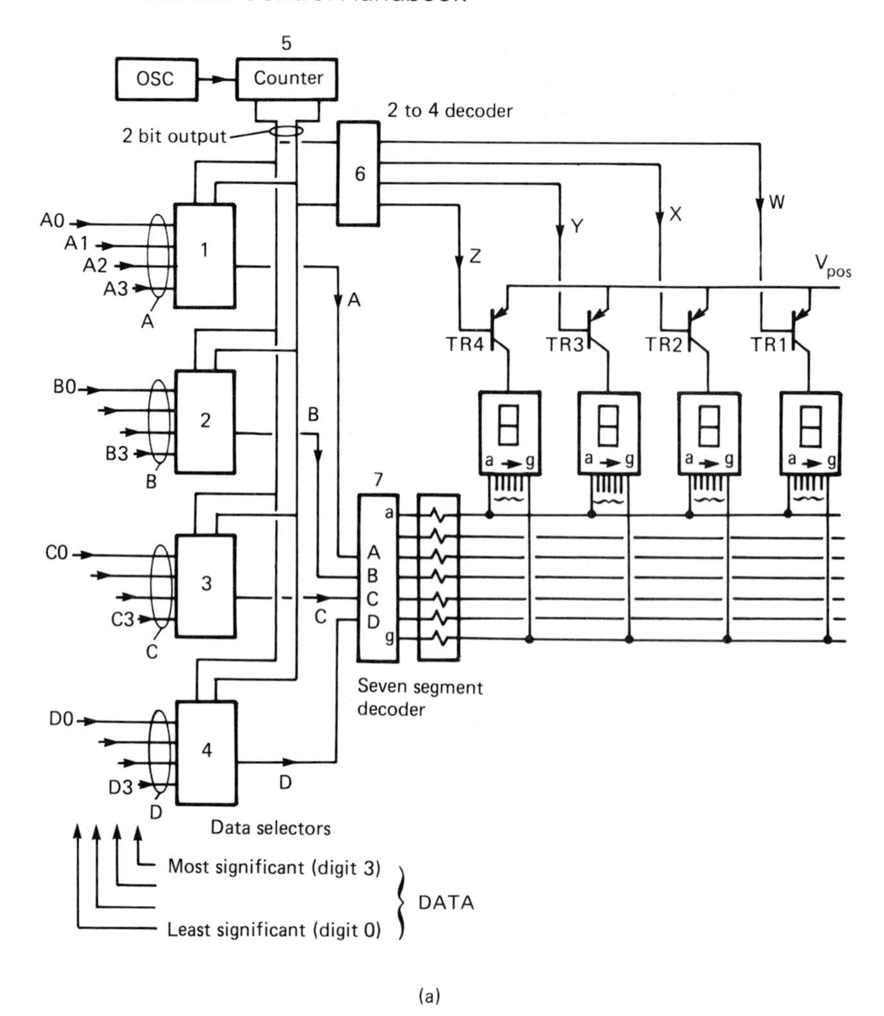
OSC
5
Counter
2 bit output
2 to 4 decoder
6
W
X
Y
Z
V_{pos}
A0
A1
A2
A3
A
1
B0
B3
B
2
C0
C3
C
3
D0
D3
D
4
TR4
TR3
TR2
TR1
a → g
7
a
A
B
C
D
g
Seven segment decoder
Data selectors
Most significant (digit 3)
Least significant (digit 0)
DATA

(a)

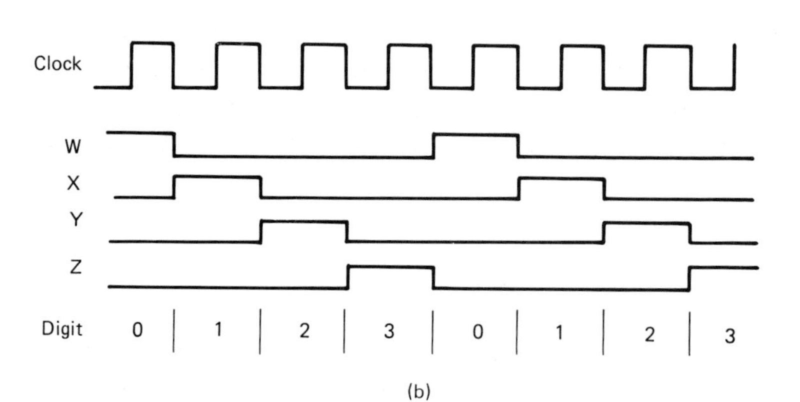
Clock
W
X
Y
Z
Digit
0
1
2
3
0
1
2
3

(b)

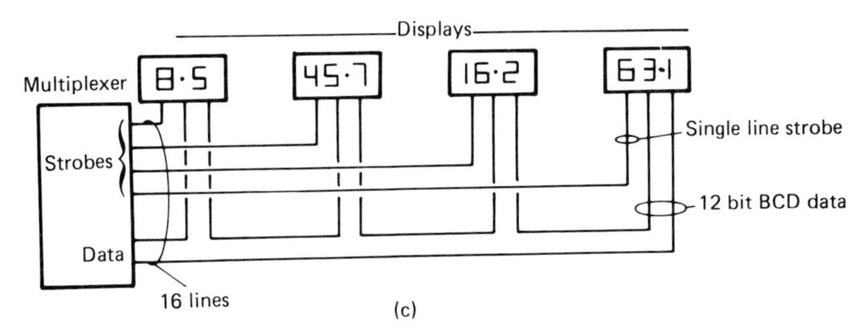

Fig.7.27 Multiplexed outputs to reduce cabling. (a) Multiplexed displays. (b) Waveforms for four digit display. (c) Multiplexed displays on an instrument panel.

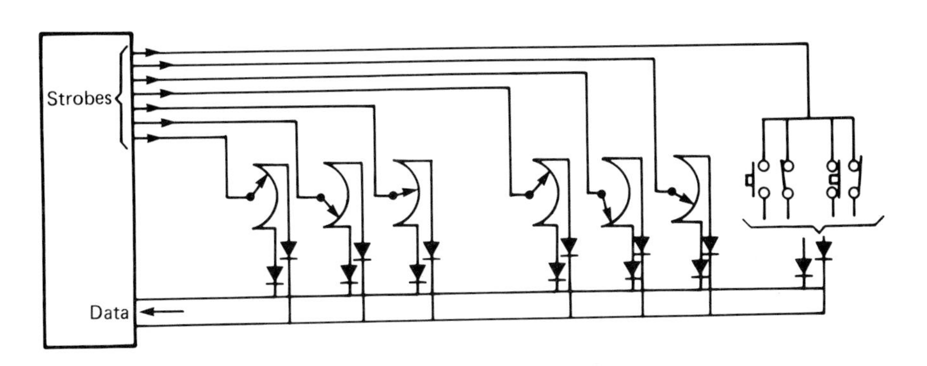

Fig.7.28 Multiplexing of inputs.

displays, and if a large number of devices are being driven, each digit requires a local 4 bit latch to hold the data locally.

The technique of multiplexing can also be used to read push button and decade thumbwheel switches with reduced cabling costs. Figure 7.28 shows a typical scheme. Each decade switch is read sequentially, and the data strobed into storage latches. Note that diodes are necessary to avoid sneak paths through unselected switches.

7.4.5. Alphanumeric displays

A seven segment display can show a limited range of alphabetic characters. A sixteen segment display, shown in fig.7.29, can

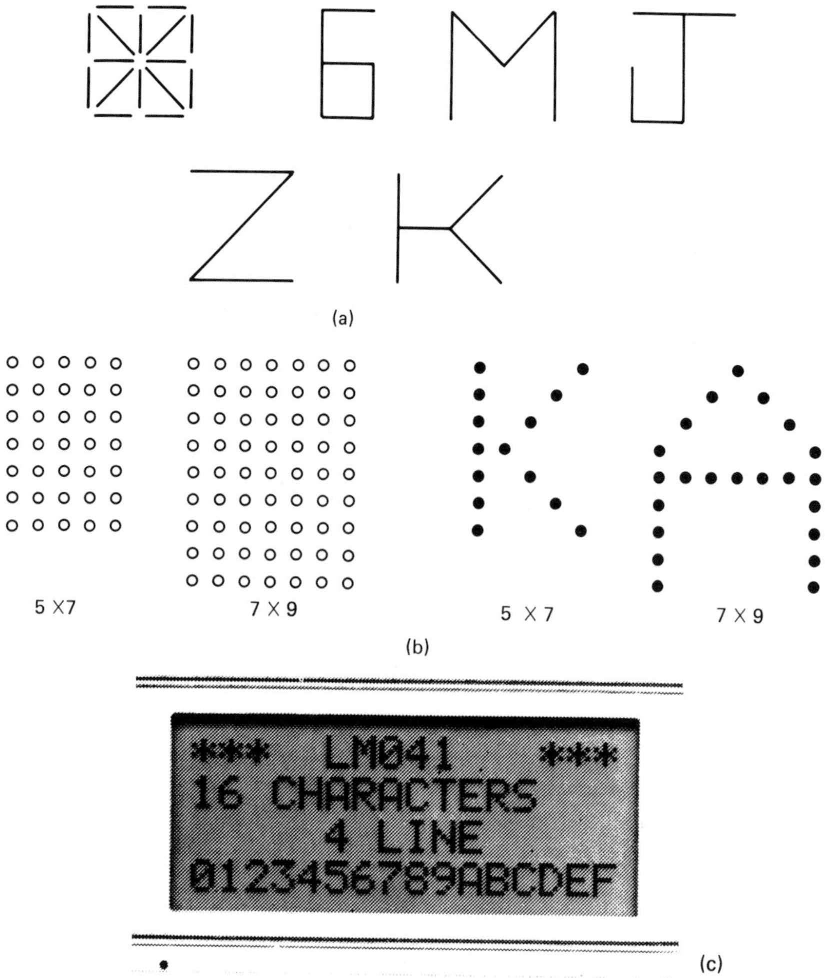

(a)

5 X7 7 X 9 5 X 7 7 X 9

(b)

*** LM041 ***
16 CHARACTERS
4 LINE
0123456789ABCDEF

(c)

Fig.7.29 Alphanumeric displays. (a) 16 segment display. (b) Dot matrix displays. (c) Alphanumeric dot matrix liquid crystal display. The device, with an area of 62 × 25 mm can display 4 lines of 16 characters. (Reproduced from the Hitachi catalogue with permission.)

display all alphanumeric characters. These are usually obtained with internal storage and decoding and accept 7 bit ASCII coded data.

Dot matrix LCD displays similar to fig.7.29 are a convenient way of displaying text in an easily read form. They have, however, a rather narrow viewing angle. Devices displaying four lines of 64 characters, accepting serial data down an RS232 link, provide a method of displaying large amounts of text with minimal installation cost.

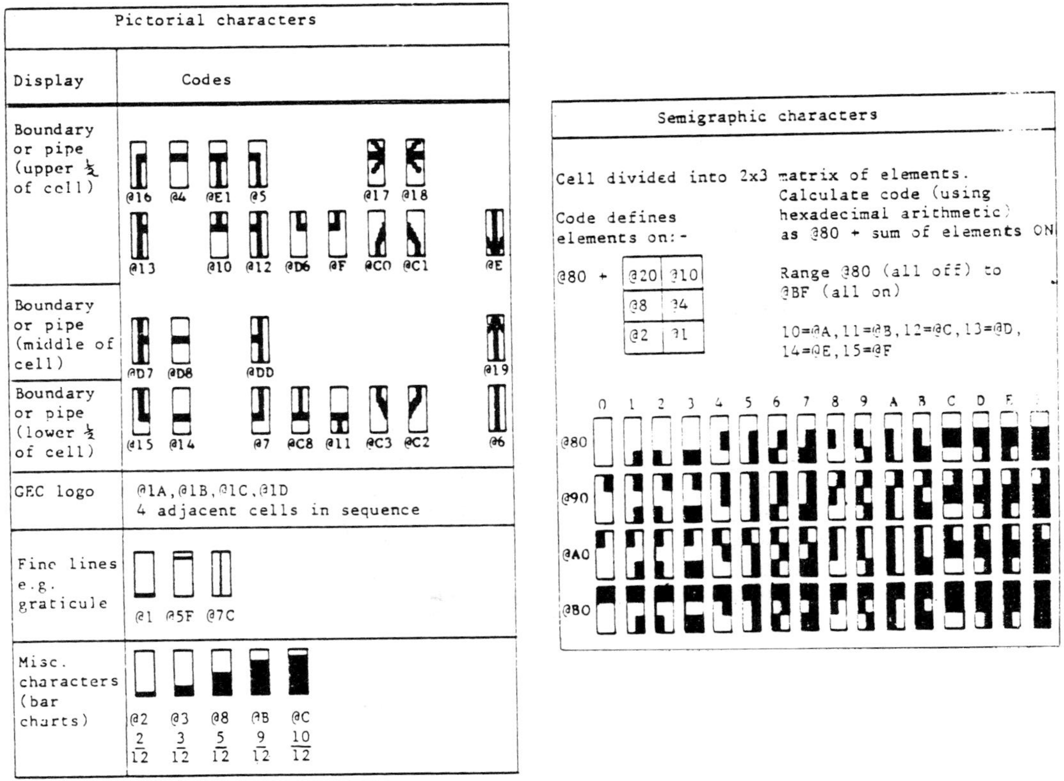

Pictorial characters

Display	Codes
Boundary or pipe (upper ½ of cell)	@16 @4 @E1 @5 @17 @18 @13 @10 @12 @D6 @F @C0 @C1 @E
Boundary or pipe (middle of cell)	@D7 @D8 @DD @19
Boundary or pipe (lower ½ of cell)	@15 @14 @7 @C8 @11 @C3 @C2 @6
GEC logo	@1A, @1B, @1C, @1D 4 adjacent cells in sequence
Fine lines e.g. graticule	@1 @5F @7C
Misc. characters (bar charts)	@2 $\frac{2}{12}$ @3 $\frac{3}{12}$ @8 $\frac{5}{12}$ @B $\frac{9}{12}$ @C $\frac{10}{12}$

Semigraphic characters

Cell divided into 2x3 matrix of elements.

Code defines elements on:-

@80 +

@20	@10
@8	@4
@2	@1

Calculate code (using hexadecimal arithmetic) as @80 + sum of elements ON

Range @80 (all off) to @BF (all on)

10=@A, 11=@B, 12=@C, 13=@D, 14=@E, 15=@F

	0	1	2	3	4	5	6	7	8	9	A	B	C	D	E	F
@80																
@90																
@A0																
@B0																

Fig.7.30 Block graphics set used on the GEC GEM-80 programmable controller. The GEM-80 family also includes a pixel addressed graphics system called IMAGEM. (Illustration courtesy of GEC Electrical Projects.)

7.5. Visual display units

Section 4.8 described the operating principles of computer driven visual display units, or VDUs, as they are more commonly called. These can be an effective way of displaying plant data to the operators. There are effectively two types of VDU display. A block graphics display uses a fixed grid (typically 80 cells wide by 40 cells high) on which one text character or one graphics character can be displayed. Figure 7.30 shows a typical set for the block graphics display for a GEC GEM-80. Quite complex displays can be built up, as shown in fig.7.31.

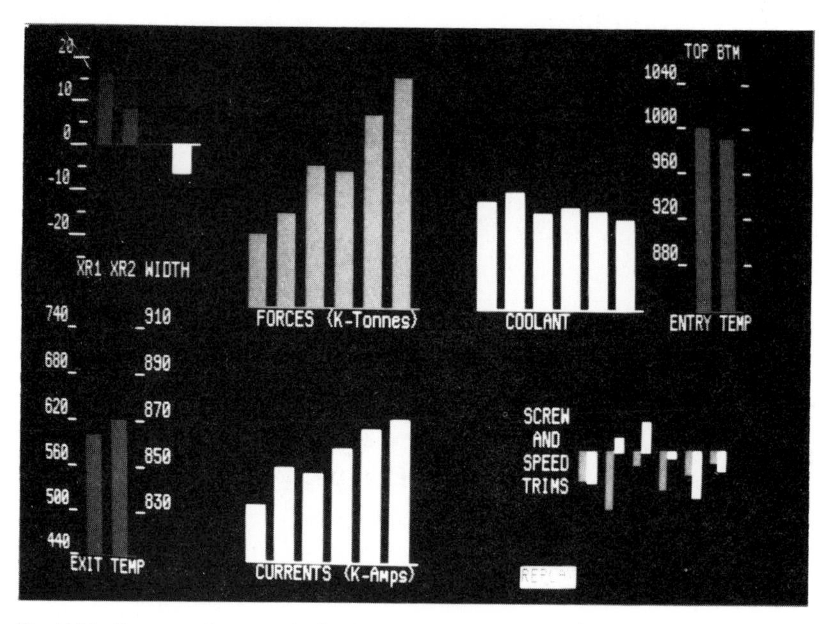

Fig.7.31 Operator bar graph display constructed from the elements of fig.7.30. Bar graph heights change dynamically with plant data. (Photo courtesy of GEC Electrical Projects, Rugby.)

Block graphic displays cannot display graphs and trend displays (although bar graphs can be constructed), and cannot cope with drawing circles other than those prefixed in the graphics set. Pixel addressed graphics allow the user to control each and every point, called pixels, on the screen (typically a resolution of 1000 by 600 points). This gives impressive results, as fig.7.32 demonstrates. The penalty is a much larger screen storage memory. A block graphics screen requires a few kilobytes locations–just 1K for a 40 × 25 standard Teletext display. A 1000 by 600 pixel display will

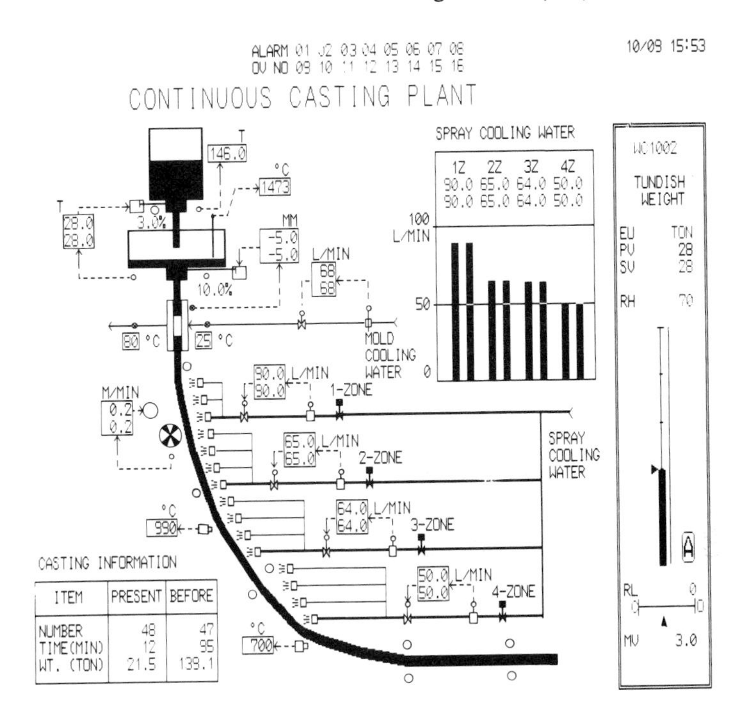

Fig.7.32 High quality pixel addressed VDU screen. (Photo courtesy of Toshiba and their UK agents Protech Ltd.)

require about 128K of memory. Pixel graphics also require more detailed programming, probably via a graphics related language with words such as MOVE, FILL, DRAW, CIRCLE, etc., and need relatively complex drive electronics.

VDUs can display plant mimics. The technology of video games has not, as yet, really reached industrial VDUs, and the plant status is usually shown by colour and intensity/flashing changes (called attribute changes in the jargon). A motor symbol, for example, can be shown green for running, red for stopped and flashed for tripped. Attribute changes can be achieved quickly without the time consuming redrawing of the whole screen. Dynamic events, such as the removal and insertion of the lance in fig.7.33a, can be achieved by drawing two lances, one in foreground colour and one invisible in background colour. Similarly the level in the tank of fig.7.33b can be shown by 'filling' the symbol with foreground colour up to the surface of the liquid, and background from the surface to the top.

VDUs can only display a limited amount of detail before the screen appears cluttered. Displays are therefore usually arranged

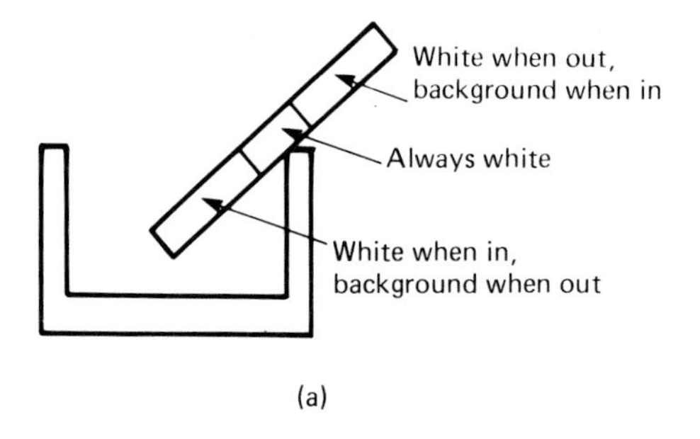

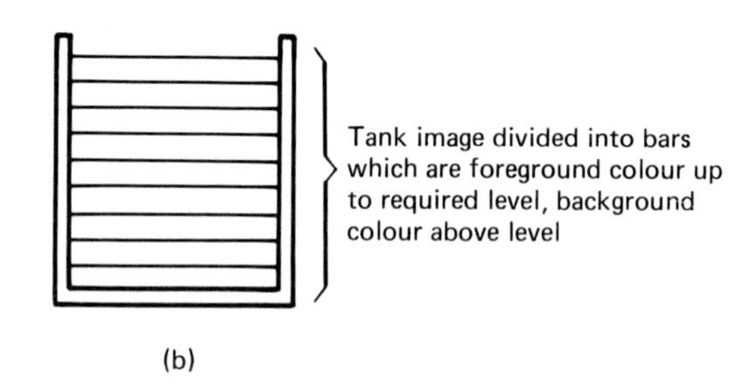

Fig.7.33 Dynamic mimics with block graphics. (a) Position of an object. (b) Level indication in a tank.

as layered pages, with a top level overview page leading down to pages which show more detail on smaller areas.

Normally a VDU connects to a computer via a character generator unit and a high speed serial link, as in fig.7.34. Usually an operator's keyboard will be associated with the VDU. The use of serial links brings both advantage and disadvantage.

On the credit side there is simplicity of installation, commissioning and maintenance. A very complex control desk with

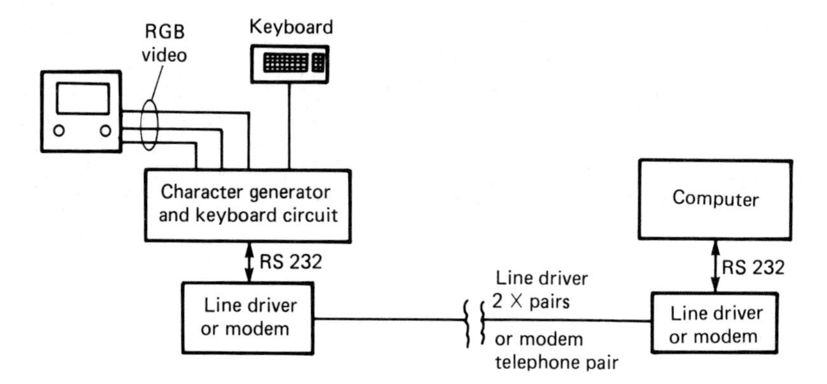

Fig.7.34 Cabling for a colour VDU.

instruments, buttons, indicator lights, etc., can be replaced by a table on which rests a VDU and keyboard with just two serial link cables and a power supply lead to connect.

The major problem with VDUs is speed of response. The maximum practical data transmission speed in most industrial applications is 19.2 kbaud (19,200 bits per second), although fibre optic links show promise of significantly higher speed. At this speed, it takes about 0.5 seconds to transmit one kilobyte of data. A total screen re-draw can take a few seconds.

The normal way to achieve a reasonable speed of response is to write dynamic data on to a fixed background skeleton picture which only changes when the user selects a new VDU 'page'. Some character generator units allow pre-definitions and storage of skeleton pages at the character generator itself. Page changes can then be achieved at high speed via page number codes transmitted from the computer.

The weak links of VDU based systems are usually the colour monitors which, because they employ relatively delicate tubes and EHT power supplies, have much lower mean time between

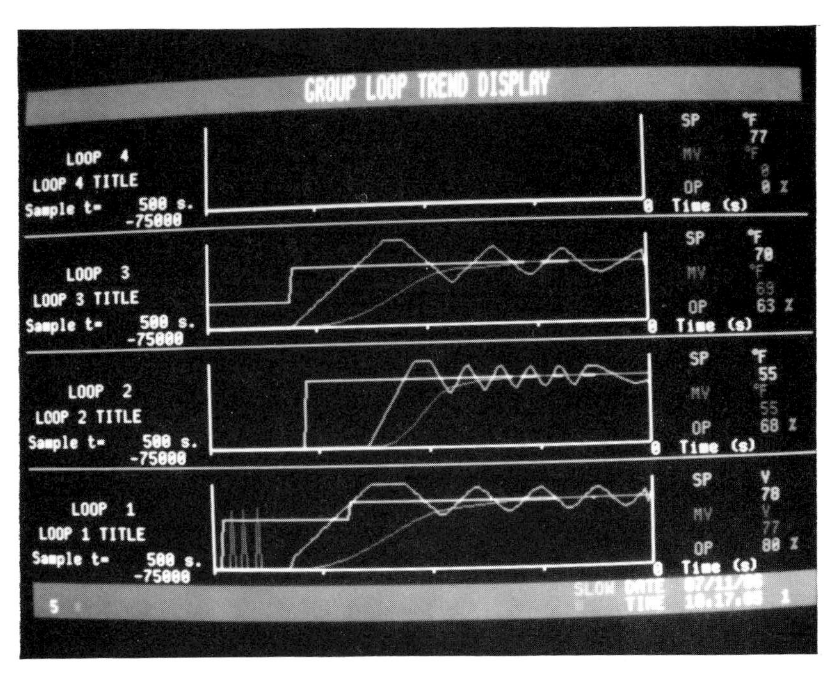

Fig.7.35 IMAGEM trend display generated by a GEC GEM-80. (Photo courtesy of GEC Industrial Controls.)

failure (MTBF) than the rest of the computer system. It follows that it is advisable to design a system based on VDUs with redundancy of displays, any of which can display any page, unless the plant can be run 'blind' for the 15 to 30 minutes it might take to identify a fault, obtain a spare from store and effect the change.

Chart recorders were described in section 7.3. VDUs can give clear and graphic trend displays, as in fig.7.35. These scroll automatically and require little, if any, maintenance. Data can be archived to computer disk and the video 'paper' wound back, or archived to floppy disks for long term storage and analysis off plant. Small VDUs purely dedicated to trend recording are

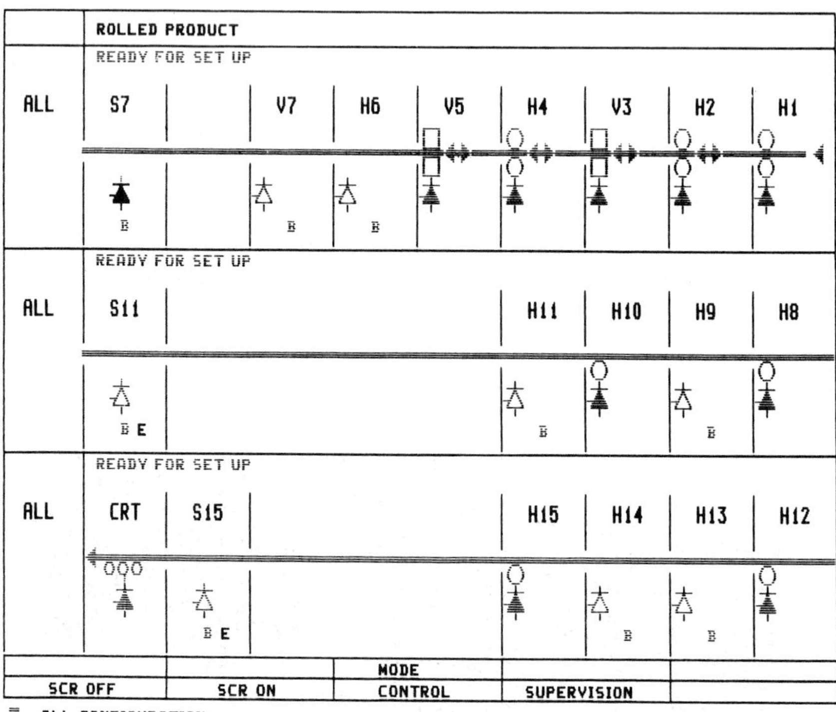

Fig.7.36 Soft keys. The 10 rectangles at the bottom of the screen correspond to 10 pushbuttons on the operator's keypad. The meaning of these buttons can thus change according to the VDU page displayed, allowing simple keyboards to be used. Soft keys in conjunction with touch screen VDUs are particularly versatile. (Illustration courtesy of ASEA, Vasteras, Sweden.)

available as direct chart recorder replacements, interfacing direct to plant transducers.

Figure 7.36 illustrates a useful technique called soft keys. The VDU is associated with a keyboard on which are situated, say, ten unlabelled keys. These are mimiced, and identified, at the bottom of each VDU page. The meaning of the keys therefore changes according to the selected screen page. This might seem rather confusing to the user, but in practice the very simple keyboard layout and the direct presentation of the options available at any stage make soft keys easy to use.

It is possible to dispense completely with the keyboard by means of a touch screen VDU. This takes soft keys a stage further by having the operator point to an area of the screen to select the required action. A pump can be started, for example, merely by pointing to it on the screen, or a new page selected by pointing to a soft key. Touch sensitive screens convert the position of the operator's finger to X and Y coordinates. Common methods are a matrix of fine wire mesh or small optical transmitters and photocells across the screen, or detecting the change in electron beam current as it scans past the operator's finger.

7.6. Alarm annunciators

Faults occur in all process control systems. Fault conditions need to be brought to the operator's attention clearly and concisely so

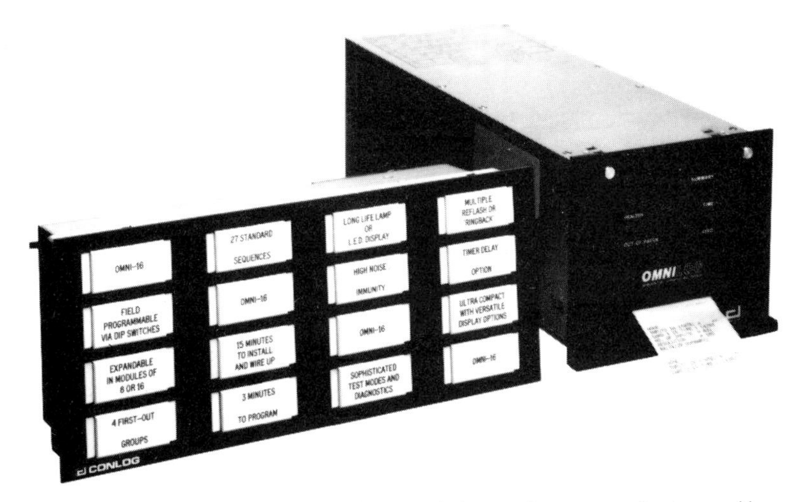

Fig.7.37 Conlog alarm annunciator and alarm printer manufacturered by Bowthorpe Controls. (Photo courtesy of Bowthorpe Controls & Wm McGeoch & Co.)

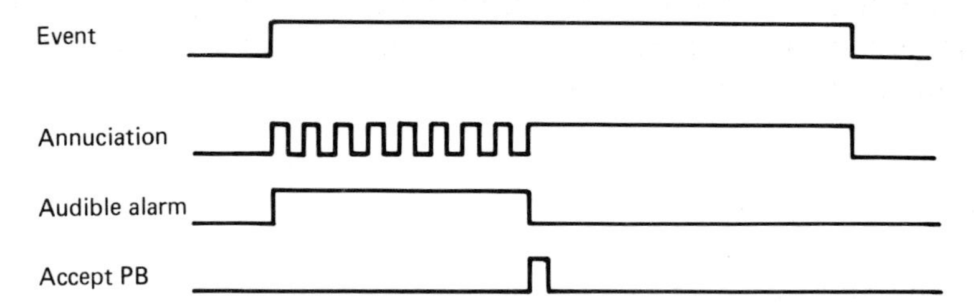

Fig.7.38 Typical annunciator action.

that rectifying action can be taken. This is usually achieved with alarm annunciator panels, such as in fig.7.37. Each indicator identifies a fault condition which requires attention.

Figure 7.38 shows a typical response for one indicator. The light flashes on the occurrence of a fault, and an audible alarm sounds. When the alarm is acknowledged, the light goes steady if the alarm condition is still present and the alarm ceases. The steady alarm clears when the fault condition is removed.

Annunciator panels can display a myriad of fault indications which arise from a single cause. An overkill of fault annunciation was apparently a contributory factor of the Three Mile Island nuclear incident. Alarm indication can be simplified by the use of first up groups.

Consider a simple hydraulic system. This could annunciate on pump stopped or low hydraulic pressure or low tank level. These will all interact. A pump stopped alarm will also cause a low hydraulic pressure alarm; a low pressure alarm will necessitate stopping the pump to prevent damage or an oil spill. Similarly low tank level may cause low pressure, and should stop the pump to prevent air being drawn in. If these alarms are linked into a first up group, the initial alarm condition will be displayed and the consequential alarms ignored or queued. The operator can thus determine the source of the problem quickly and without ambiguity.

VDUs, described in the previous section, usually incorporate one or more alarm pages. Alarm status can be shown by changing the colour attributes of alarm text: flashing red for unaccepted alarms, say, steady yellow for accepted, but still present, alarms and blanking the text (attribute the same as the background) for a healthy state.

Alarms can be generated from contact closure or opening, or direct from analog values which go above, or below, a preset value. In general, normally closed contacts which open on a fault are preferred as these will give a fault indication under cable fault, supply failure or transducer failure conditions.

7.7. Data logging and recording

7.7.1. Data loggers

A working plant produces a vast amount of data which can be useful for subsequent analysis of performance, efficiency, running costs, etc. Section 4.3.2 described logging as one of the roles of an industrial computer, where plant data is archived for later examination.

If a computer is not thought necessary, or desirable, for control, similar archiving can be achieved by means of a commercial data logger. These aim to provide a record of plant operation by:

(1) Recording plant analog values at regular timed intervals.
(2) Performing continuous checking for alarm conditions, and recording plant state and time when an alarm occurs.
(3) Recording digital events, with time, as they occur.

The resulting data is usually printed out as it occurs to give a continuous event log, and archived to disk or magnetic tape for later analysis.

Figure 7.39 shows a block diagram of a typical data logger, and it can be seen that this is effectively an industrial computer without an output unit.

Analog values are scanned by a multiplexer, and frozen by a sample-and-hold unit prior to digitisation by an ADC. Signals are normally presented to the data logger in a standard form (e.g. 4 to 20 mA) and converted to engineering units (degrees C, psi, etc.) in the control unit program. This is commonly achieved by specifying high and low range engineering values and any linearisation routines (e.g. square root) for each channel as part of the set-up procedures.

Input channels need not be scanned sequentially; flows, for example, may need to be scanned more frequently than temperatures. The set-up procedure will usually allow a logging scan rate and an alarm scan rate (with alarm limits) to be specified for each channel.

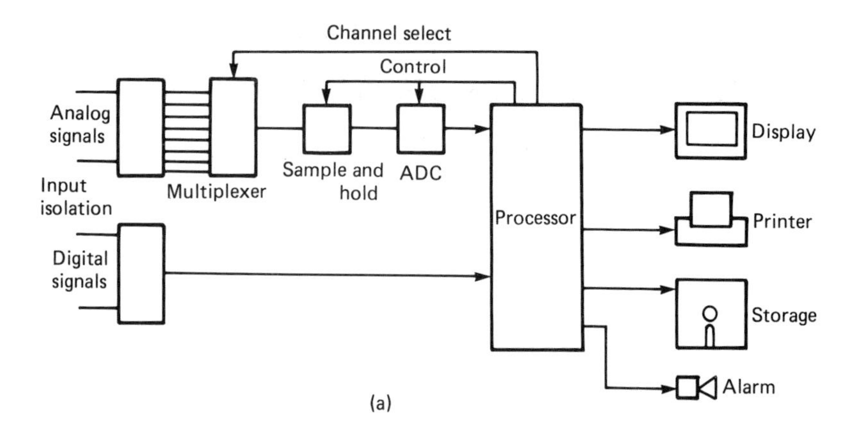

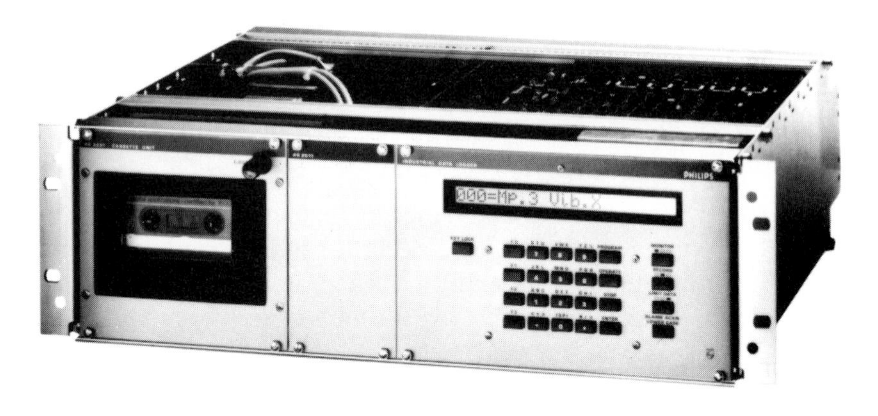

(b)

Fig.7.39 Data logger. (a) Block diagram. (b) The Philips PR2011 data logger. This unit can handle up to 256 channels for display, recording and alarm annunciation. (Photo courtesy of Philips Scientific & Industrial Equipment.)

Digital inputs will have an event text associated with them. In simple units this can be as terse as 'input 27 on'. More sophisticated units allow the user to define that input 7 corresponds, say, to 'reactor B auto mode selected', and choose if the digital event is to be recorded for contact closure, contact opening or both.

7.7.2. Instrumentation tape recorders

The magnetic tape recorder is an alternative way of archiving plant data for later analysis. The falling costs of computer systems

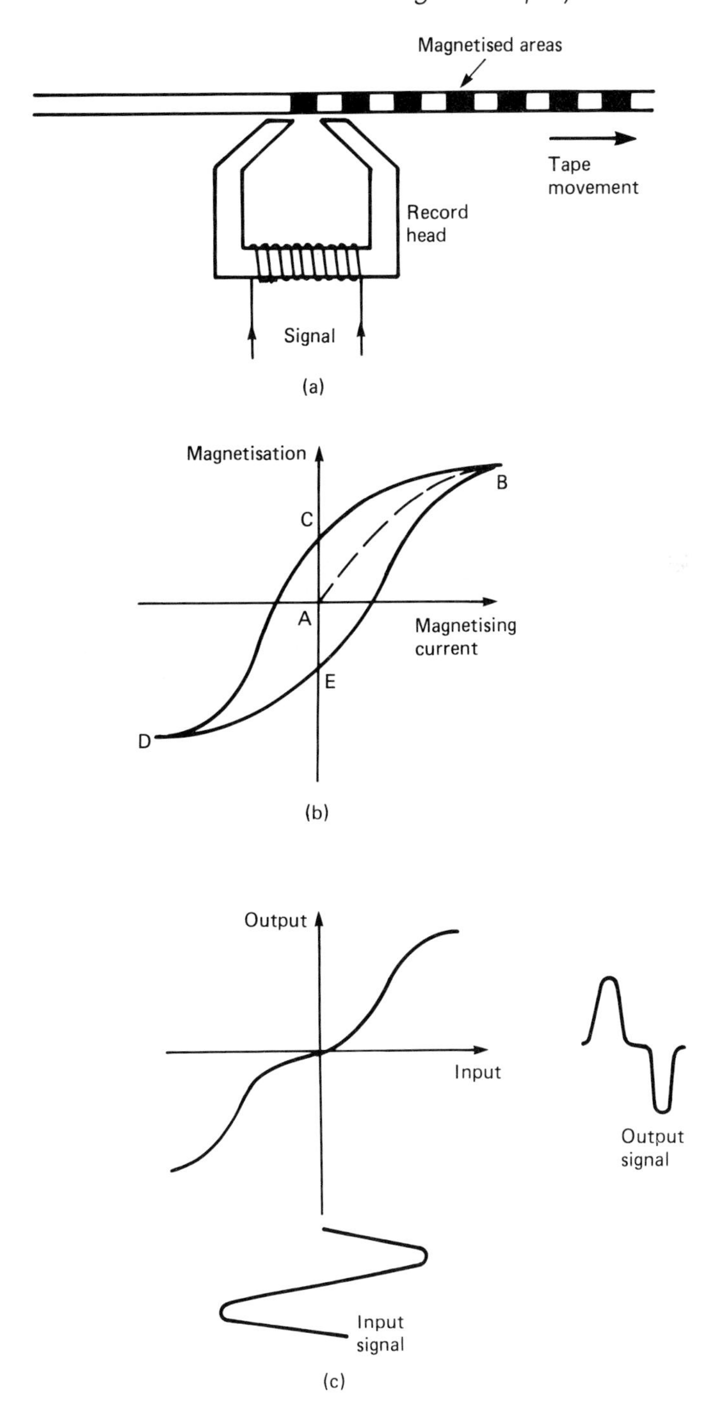

Fig.7.40 Magnetic recording. (a) Principle of tape recording. (b) Hysteresis curve. (c) Non-linear response.

and disk storage (particularly Winchester disks), however, make instrumentation recorders less attractive for industrial users, and tape recorders are more likely to be found in laboratory or experimental environments.

An instrumentation recorder works on the same principle as a domestic tape recorder, although many simultaneous tracks (typically sixteen) are recorded. Plastic tape, coated with ferromagnetic material, is passed over an electromagnetic record head, as shown in fig.7.40a. The current passed through the record head coil induces a high magnetic field across the record head gap, and magnetises the ferromagnetic material on the tape. To retrieve the information, the tape is passed over a playback head, similar in construction to the record head, and the variations in magnetisation of the tape induce a voltage in the playback head coil.

Although simple in theory, the process is fraught with many difficulties. The first of these is hysteresis, illustrated in fig.7.40b. If a magnetic material is magnetised by an external field, it will follow a curve similar to AB. If the field strength is reduced to

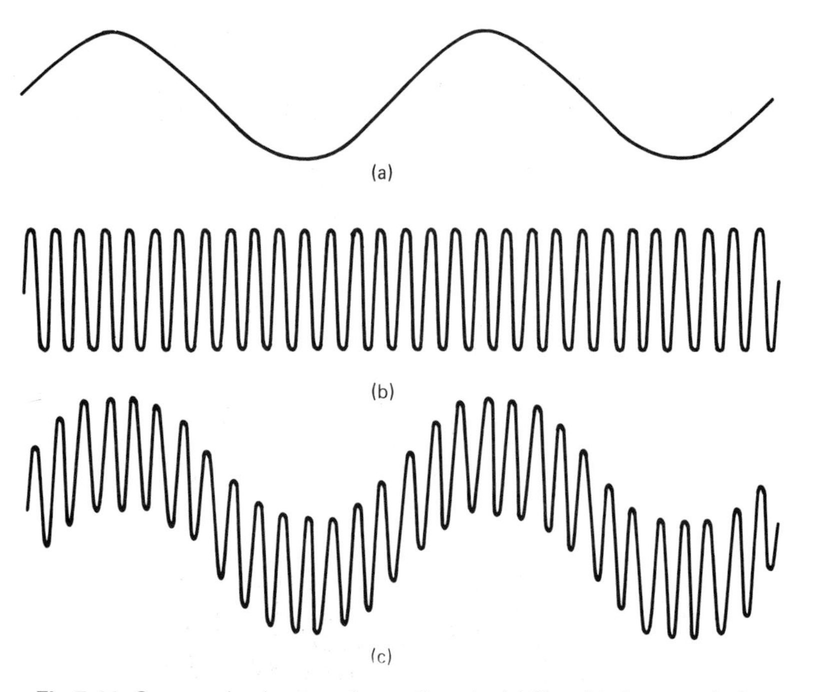

Fig.7.41 Overcoming hysteresis non-linearly. (a) Signal to be recorded. (b) Modulating waveform. (c) Resultant.

zero, the magnetisation of the material does not fall to zero but remains magnetised at point C, called remnant magnetisation. Field reversals cause the material to traverse the curve CDEB.

Hysteresis produces a very non-linear response between applied field and resulting magnetisation, as shown in fig.7.40c. This can be overcome by the addition of an AC bias signal, as in fig.7.41. The bias signal is typically 30 to 50 kHz (dependent on tape speed) and is around 3 to 5 times the amplitude of the maximum input signal amplitude. It should be noted that the bias signal is *added* to the input signal, i.e. it is not a form of modulation. The bias signal causes the magnetic material on the tape to traverse both sides of the hysteresis curve, and the resulting non-linearities cancel. The bias signal is removed at playback by a simple low pass filter.

The response of the tape is also highly frequency dependent, as both the tape magnetisation and the play back signal are proportional to the rate of change of the magnetic flux across the head gap. Almost all process variables are essentially low frequency; very few can change faster than 50 Hz. It follows that the process signals must be encoded in some way before they can be recorded.

Frequency modulation (FM, described further in Volume 3), modulates the frequency of a carrier wave according to the level of input signal, as illustrated in fig.7.42. Because the information

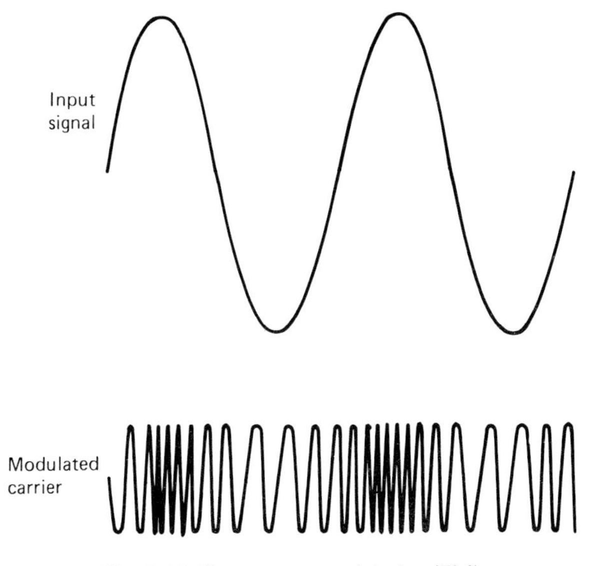

Fig.7.42 Frequency modulation (FM).

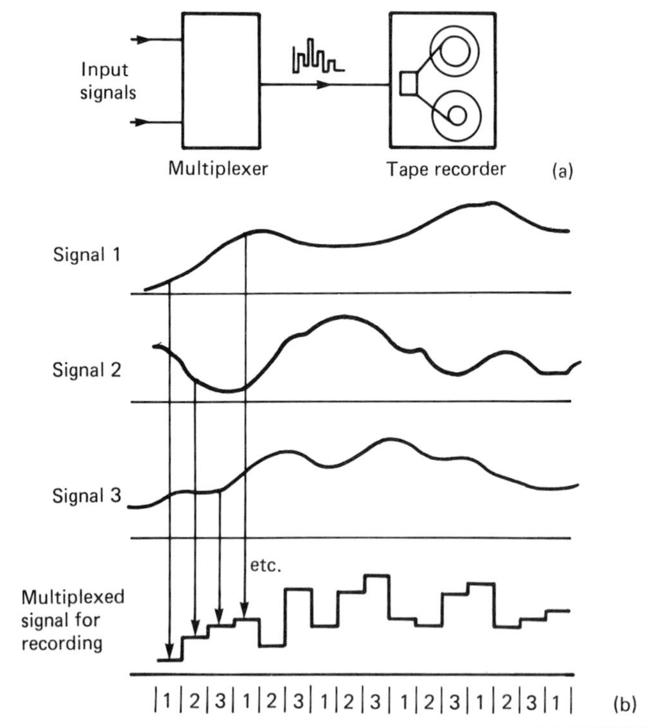

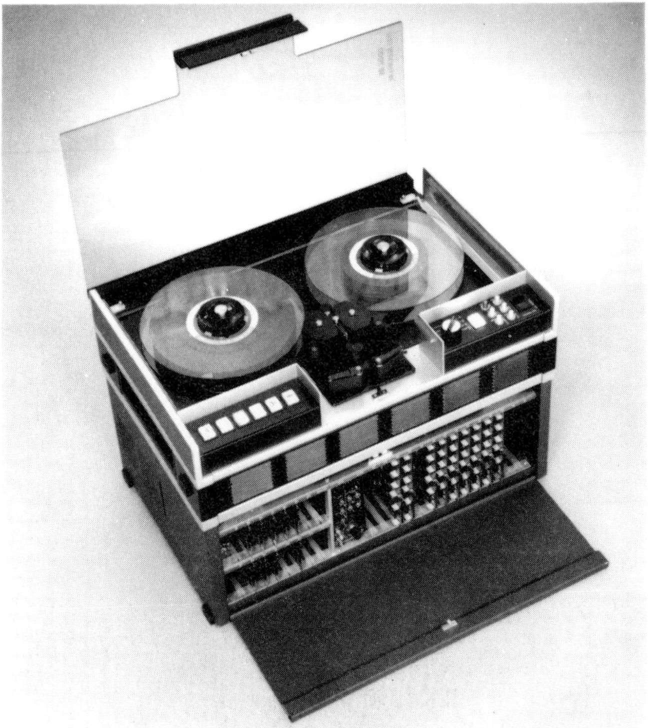

Fig.7.43 Pulse height recording. (a) Multiplexed recording system. (b) Typical signals. (c) Instrumentation tape recorder, showing input scaling cards. (Photo courtesy of Thorn EMI Datatech.)

about the signal is conveyed in the carrier frequency and not its amplitude, errors from hysteresis and the tape/head frequency response are overcome.

Pulse height recording, as in fig.7.43, is a way of encoding data and multiplexing signals on to one recorder track. Each signal is sampled in turn and converted to a pulse whose height is proportional to the input signal level. A related technique uses pulses whose height is fixed, but whose width is proportional to signal level.

Increasingly, variables are digitised by a multiplexer and ADC similar to the input section of fig.7.39 to give a digital representation before recording. Typically 12 bit representation will give 0.01% resolution. Sample rates must be chosen according to Shannon's sampling theorem (see section 4.8) to give a faithful representation of the measured variable. Digital representation gives almost total immunity from distortion at the cost of greater tape usage. Typically a recording density of 63 bits per millimetre can be achieved with nine tracks across the 12.7 mm tape.

Digital data can be recorded serially along one track per signal, or in parallel across, typically, nine tracks. With parallel

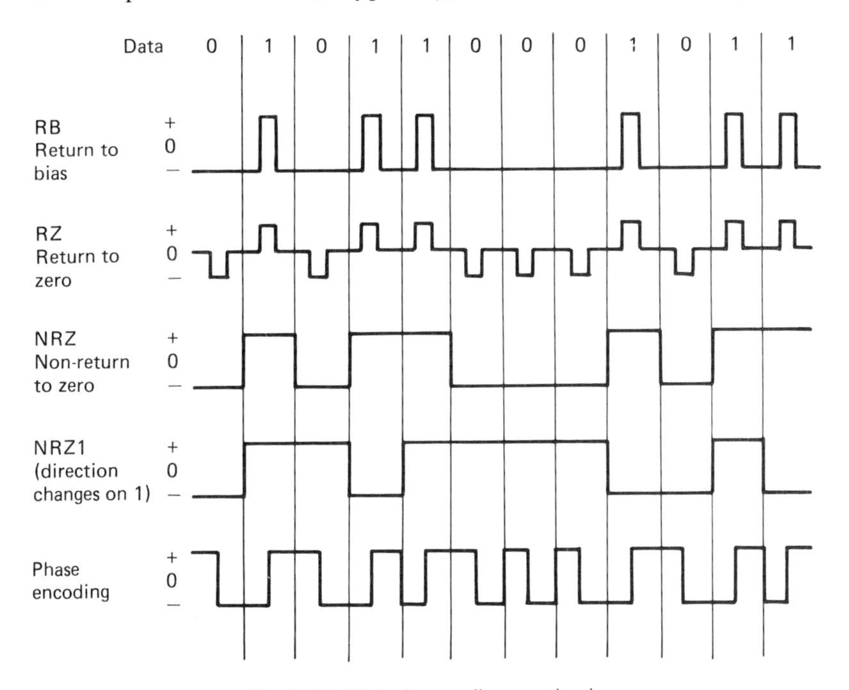

Fig.7.44 Digital recording methods.

recording, one track is used as a parity check to give error indication.

Various digital encoding methods are used; fig.7.44 shows the more common. Digital signals can be FM modulated, or used directly to drive the tape material into saturation–point B and D on fig.7.40b.

Tape speed accuracy and consistency are important particularly for FM recording where speed variations (called wow for low frequency variations and flutter for high frequency variations) will appear directly as signal changes. A reference signal track is often used to control tape speed. This is a very stable high frequency signal recorded on the tape which is compared with a crystal oscillator on playback to correct tape speed. On digital systems a clock track is sometimes used to synchronise the tracks and control tape speed.

7.8. Ergonomics

Ultimately, every plant can be represented by fig.7.45 which shows what is often called, rather grandly, the MMI, or man machine interface. Ergonomics is the study and design of this interface such that the operator can perform his duties efficiently, in comfort and with minimum error.

Display of plant data has been the topic of most of this chapter and is crucial to the operator's effectiveness. An overkill of data which has to be painstakingly scanned for crucial information can

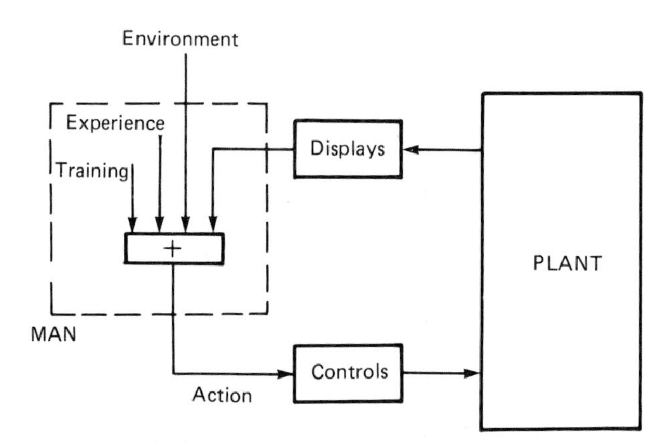

Fig.7.45 Man as part of the process.

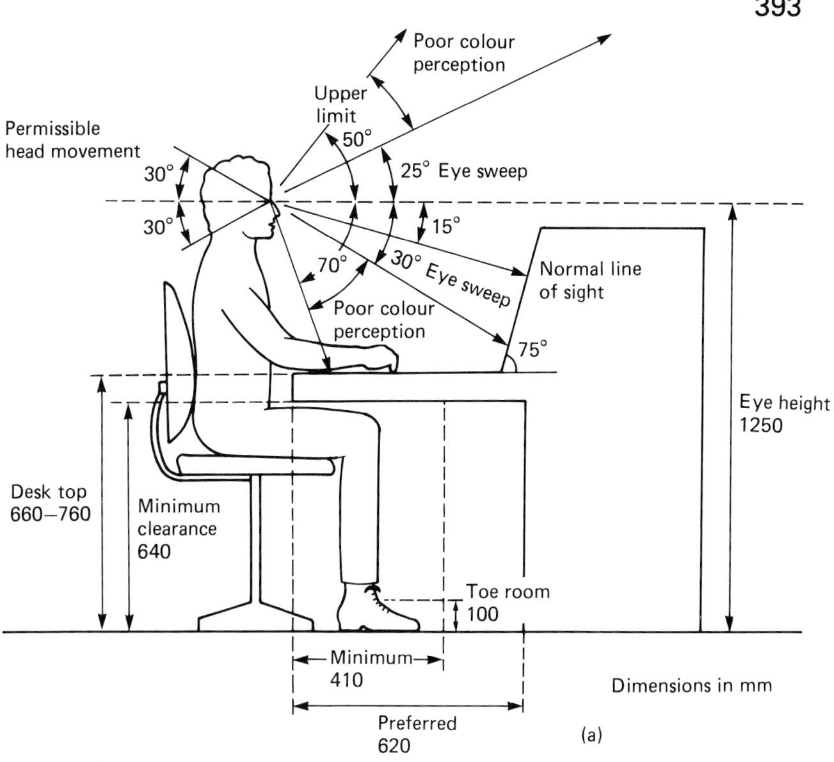
Poor colour perception
Upper limit
50°
Permissible head movement
30°
30°
25° Eye sweep
15°
70°
30° Eye sweep
Normal line of sight
Poor colour perception
75°
Eye height 1250
Desk top 660–760
Minimum clearance 640
Toe room 100
Minimum 410
Preferred 620
Dimensions in mm

(a)

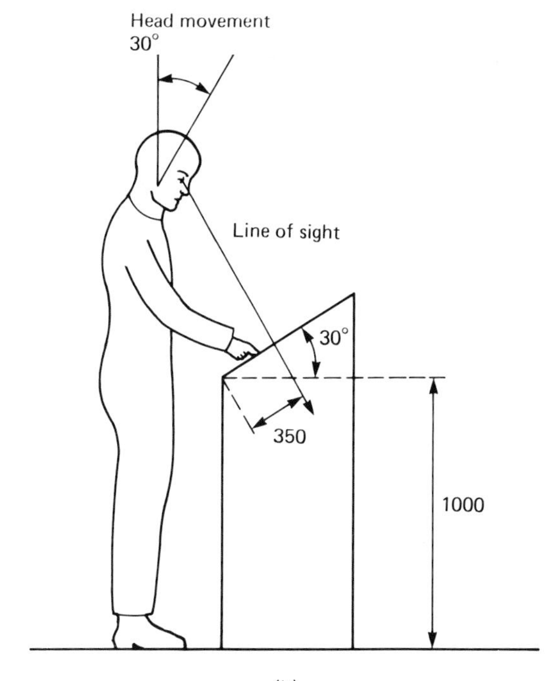
Head movement 30°
Line of sight
30°
350
1000

(b)

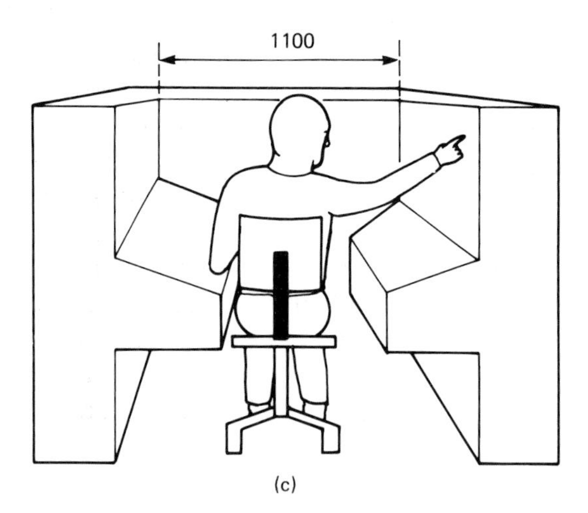

Fig.7.46 Control panel design. (a) Desk design. (b) Standing panel. (c) Multisection panel.

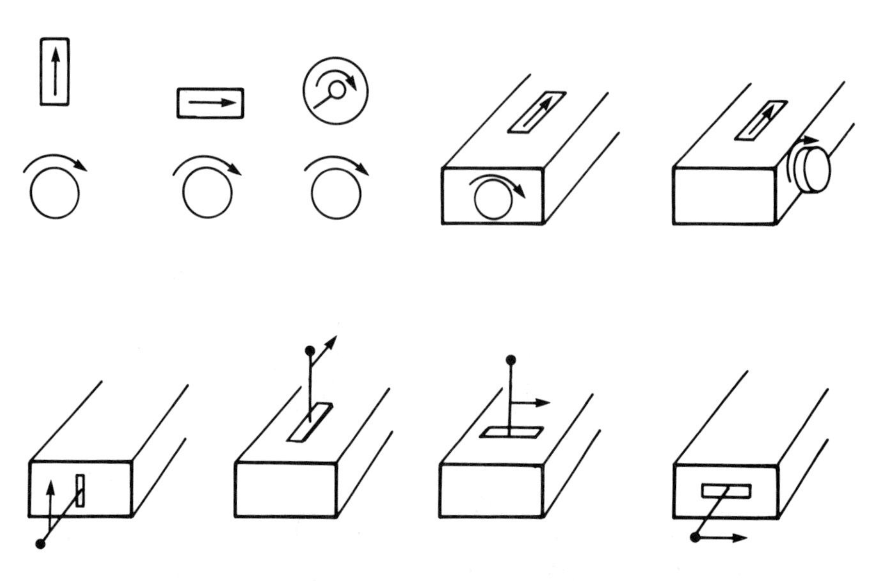

Fig.7.47 Human expectations of controls and indications. Arrows show direction for increase (of flow, power, speed, temperature, etc.). It is odd to note that UK light and power outlet switches are reversed from normal expectations. Controls relating to motions (e.g. crane controls) should follow plant movement.

be as detrimental as too little. Grouping of displays and consistency in the displays are important in allowing the operator to form a mental picture of what is actually happening in the plant under his control. The displays should be arranged so they can be scanned with minimum effort. Figure 7.46 shows boundaries of human perception, and preferred angles for desk tops.

Consistency is important in the action of control devices; we all expect to turn something on, to increase some variable, by turning a knob clockwise. Similar expectations are shown in fig.7.47. Relative positioning of control devices and displays is important; people expect to see a reaction to an action, even if it is only a light that says the action is being acted upon (the lift call-button effect).

Possibly the most important aspect is the worker's immediate work space and environment. Reliable work cannot reasonably be expected from an operator who has a headache or a sore back within an hour of starting work. Factors such as noise, dust, smell, vibration, temperature (and temperature changes), lighting levels (and glare) and humidity all contribute to a worker's ability to concentrate. Psychological factors such as the degree of concentration necessary, and the ability to mentally rest and 'coast' for a short period are also important.

The layout of controls, displays and seating for convenience of operation are often overlooked. It is unfortunately true, for

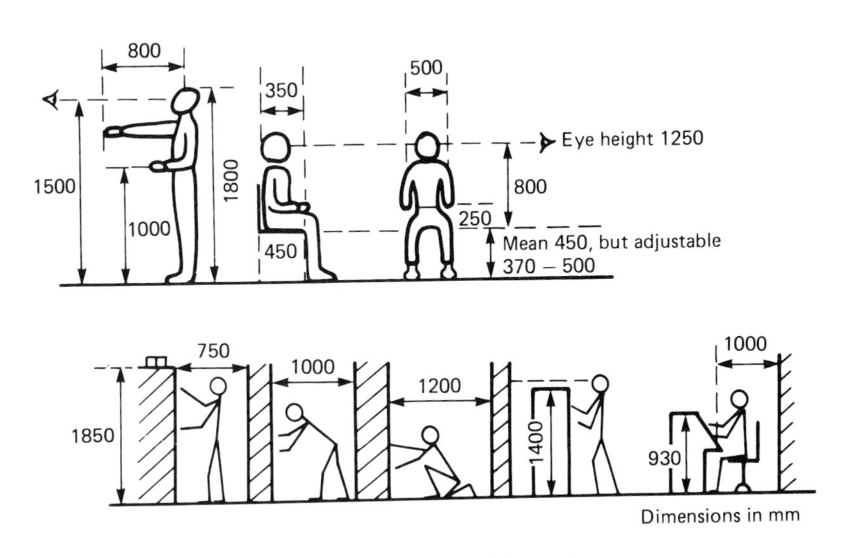

Fig.7.48 The human body and the environment.

example, that most machine tool lathes and drilling machines are designed for operation by workers 4 feet tall with 8-foot-long arms, as can be seen by the postures that have to be adopted in use. Figure 7.48 shows comfortable working spaces and reaches which suit the majority of the population.

Chapter 8
Maintenance, fault finding and safety

8.1. Introduction

It is the duty of the design engineer to ensure that in any new system:

(a) At least one item is obsolete.
(b) At least one item is on twelve months delivery.
(c) At least one item is experimental.
(d) The drawings arrive six months after the equipment and do not include commissioning modifications.
(e) The instruction manuals are written in a confusing mixture of the banal and the impossibly complex which assumes that the user is a moron with an engineering doctorate.

8.2. System failures

8.2.1. Introduction

Equipment inevitably fails, and there is no such thing as a totally reliable system. It is not possible to predict when an item will fail; it is not even possible to say, with certainty, than an item will not fail in the next 30 seconds. Discussions of reliability are consequently based on statistical analysis, rather than predictions for one specific piece of equipment. In a similar way, one can talk with great accuracy of the life expectancy of a human being, but this cannot be used to predict the lifespan of one person. It is similarly possible to say, in general terms, what are the effects of smoking, alcohol, lack of exercise, overeating, etc., but everyone knows someone who indulged in all the vices and lived to a ripe old age. In industrial terms, it is not feasible to say that this system, operated under these conditions, will run for 1000 hours, plus or minus 5 hours, without failure. It is, however, realistic to say that if a very large number of systems are studied, the average lifetime will be 1000 hours.

Production management often expresses the wish for fault free systems. This is unachievable, but by using techniques such as redundancy (described later) any desired level of reliability can be achieved. High reliability may not, surprisingly, be what is really required.

Low reliability is achieved at low cost, but brings additional costs in terms of repair effort and lost production. As reliability increases, production losses decrease, and maintenance costs decrease to a point where reliability improvements require more staff to 'stand guard' and continuously check the plant. High reliability techniques such as redundancy are also expensive to install. The overall reliability cost curve thus has a shape similar to fig.8.1, with the optimum operating point being the minimum overall cost. Identifying this point is the art of plant maintenance.

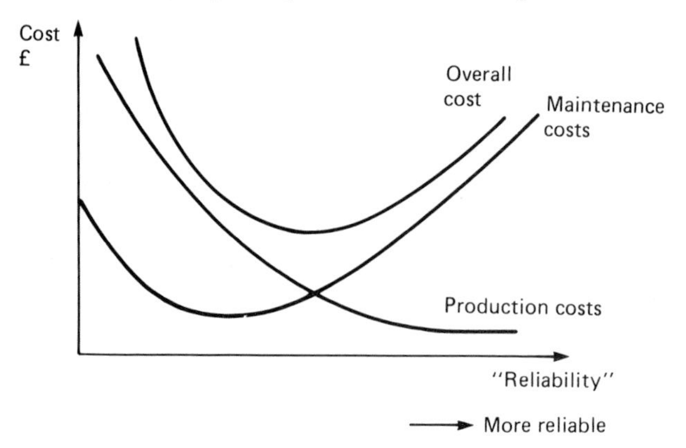

Fig.8.1 The financial implications of reliability.

Most plants are, to some degree, failure tolerant and tend to operate in some form of failure mode for most of the time. Good plant design considers the effect of failures and the ability of the plant to continue operating safely and reasonably economically for a period of time whilst a fault is identified and rectified.

8.2.2. Reliability

Because it is not realistic to predict when one item will fail, statistical methods are used to discuss reliability. The reliability of an item or system is the probability that it will perform correctly under the defined operating conditions for a specified time period. For example, a transducer may have a 95% probability of operating without failure for two years when used in accordance with the manufacturer's data sheets.

Reliability measurements are based on tests done on a large number of items. If N items are run for a time t, and N_s are still working at the end of the test and N_f have failed, the reliability for time t is:

$$R_t = \frac{N_s}{N} = \frac{N - N_f}{N} \tag{8.1}$$

The unreliability, Q_t, is defined as:

$$Q_t = \frac{N_f}{N} = \frac{N - N_s}{N} \tag{8.2}$$

Note that $0<R_t<1$, $0<Q_t<1$ and $R_t+Q_t = 1$.

8.2.3. MTTF and MTBF

Reliability is related to a specified time period (e.g. 1000 hours, one year), but an estimate of life expectancy is more useful in most circumstances. This is given by mean time to failure (MTTF) for non-repairable items (e.g. lamp bulbs and disposable equipment) and mean time between failure (MTBF) for repairable items. Both MTTF and MTBF are obtained from tests based on a large number of items.

Table 8.1 shows results of a typical test to determine MTTF for, say, light bulbs. A 1000 bulbs are lit and observed daily; 25 fail during the first day, 17 during the second, and so on. The mean number of bulbs alight each day multiplied by 24 gives the total operational time, in hours, each day. At the end of 10 days, 145 bulbs have failed, and 220596 operational hours have accumulated. The MTTF is therefore 220596/145 or 1521 hours.

MTBF is a measure of the time between failures for repairable items. A MTBF test for a lighting system would set up 1000 bulbs as before, but bulbs would be replaced as they failed and a record kept of the number of failures over a given time. Strictly speaking, the MTBF is the operational time/number of failures, with the operational time being the length of the test multiplied

Table 8.1 *MTTF calculation (outer letters/numerals are for spreadsheet, section 4.9 and fig. 4.42).*

Column	*A*	*B*	*C*	*D*	*E*	*F*	*G*
Row	*Day*	*Hours*	*Failures*	*Cumulative failures*	*Survivors*	*Mean survivors*	*Total time operational*
1	0	0			1000		
2			25			987.5	23700
3	1	24		25	975		
4			17			966.5	23196
5	2	48		42	958		
6			15			950.5	22812
7	3	72		57	943		
8			11			937.5	22500
9	4	96		68	932		
10			13			925.5	22212
11	5	120		81	919		
12			16			911.0	21864
13	6	144		97	903		
14			14			896.0	21504
15	7	168		111	889		
16			12			883.0	21192
17	8	192		123	877		
18			9			872.5	20940
19	9	216		132	868		
20			13			861.5	20676
21	10	240		145	855		
22							———
23							
24						Total time	220596

$$\text{MTTF} = \frac{\text{total time}}{\text{cum. failures}} = \frac{220596}{145} = 1521 \text{ hours}$$

by the number of items under test less the repair time; but as all realistic systems will have MTBFs far larger than the repair time, it is usual to define MTBF as the length of the test multiplied by the number of items divided by the number of failures.

Suppose 1000 bulbs are kept alight for 10 days, with bulbs being replaced as they fail. At the end of 10 days 136 bulbs have failed, so the MTBF is 240×1000/136 or 1765 hours. In general, for most items MTTF and MTBF will be similar and the two terms are (incorrectly) used indiscriminately.

Note that both MTTF and MTBF are average values, not a prediction for one bulb. Table 8.1 shows that some bulbs had a life of less than 24 hours, and some would, of course, have a life far greater than the MTTF or MTBF.

8.2.4. Maintainability

When equipment fails, and all equipment can, it is important that it is returned to an operational state as soon as possible. The term maintainability describes the ease with which an item can be repaired, and is defined as the probability that a piece of faulty equipment can be returned to an operational state within a specified time. A DC drive with a thyristor bridge fault, for example, could have a 0.85 probability of being operational within 30 minutes of the fault occurring.

Mean time to repair (MTTR) is another measure of maintainability, and is defined as the mean time taken to return a failed piece of equipment to an operational state. Like MTTF and MTBF, it is a statistical figure derived from a large number of observations. The DC drive mentioned above, for example, might have a MTTR of 20 minutes.

Maintainability is determined partly by the designer and partly by the user. Important factors are:

(1) The design and use of the equipment should be such that the failure is immediately apparent and can be quickly localised to a changeable item. This requires good documentation, readily identifiable test points and indication, and modular construction. Simple and cheap techniques, such as running the neutral out to the limit switch junction box in fig.8.2 to allow input signals to be checked, can make significant impact on plant maintainability.

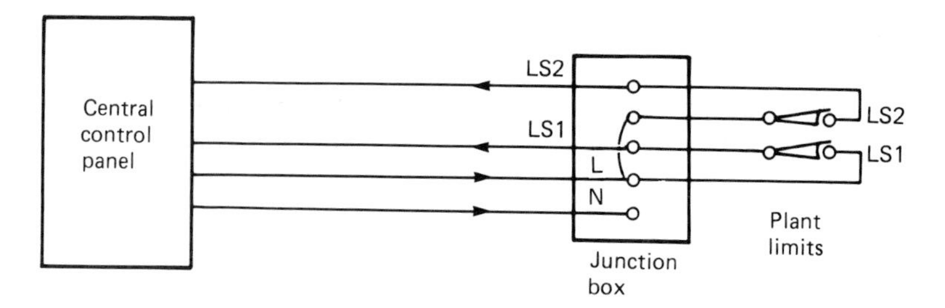

Fig.8.2 Designed in maintainability. The neutral in the junction box serves no functional purpose, but allows plant limit states to be checked with a multimeter at the box.

(2) Vulnerable components should be easily accessible. On a DC drive, for example, a high proportion of faults may lead to the rupturing of the HRC protection fuses, so these should be provided with indicator flags and the mounting designed so they can be replaced quickly.

(3) The maintenance personnel should be competent, well trained and have suitable tools and test equipment. MTTR will obviously depend on how long it takes for maintenance personnel to respond to a fault call.

(4) Adequate spares should be carried and be accessible. MTTR will usually be reduced if a policy of unit replacement rather than unit repair on site is adopted. In the case of the DC thyristor drive referred to earlier, it may be advantageous to treat the whole unit as a replaceable item to be repaired later in the workshop.

Items (1) and (2) are the responsibility of the designer, items (3) and (4) the responsibility of the user.

Plant availability is the percentage of the time that equipment is functional, i.e.

$$\text{availability} = \frac{\text{uptime}}{\text{uptime} + \text{downtime}} \tag{8.3}$$

$$= \frac{\text{MTBF}}{\text{MTBF} + \text{MTTR} + \text{PMT}} \tag{8.4}$$

where PMT is the planned maintenance time, during which the plant is off line for essential scheduled servicing. Normal availability of a plant will be well over 95%.

8.2.5. Failure rate

If N_s components are in operation, and ΔN_f components fail over time Δt, the failure rate $\lambda(t)$ (also called the hazard rate) is defined as:

$$\lambda(t) = \frac{1}{N_s} \frac{\Delta N_f}{\Delta t} \qquad (8.5)$$

As Δt tends towards zero we can write:

$$\lambda(t) = \frac{1}{N_s} \frac{dN_f}{dt} \qquad (8.6)$$

This is, of course, a theoretical model, as in reality we are dealing with a finite number of components and finite time intervals.

The failure rate for most systems follows the bathtub curve of fig.8.3. This falls into three distinct regions. The first, called 'burn in' or 'infant mortality', lasts at most a few weeks and exhibits a high failure rate as faulty components, bad soldering, etc., become apparent. Manufacturers generally employ heat cycling tests to provoke infant mortality before equipment is shipped to the user. In complete instrumentation and process control systems an initial 'problem period' is not uncommon as the designer's mistakes and computer software bugs become apparent.

The centre portion, called 'maturity', exhibits a low constant failure rate. During this period failures are random. The final period, often called 'senility', is characterised by a rising failure rate. Generally this increased unreliability is caused by structural

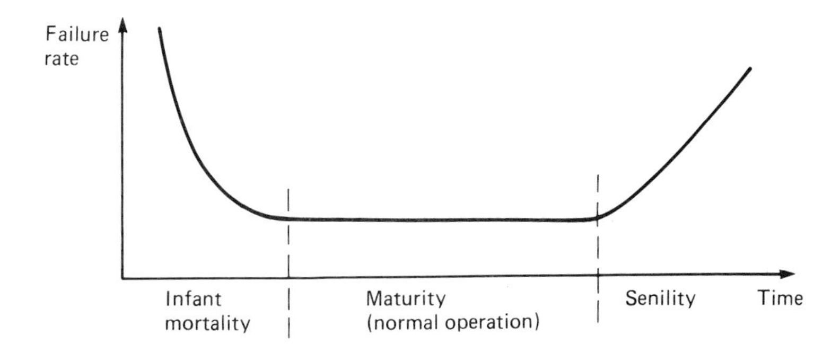

Fig.8.3 The bathtub curve.

old age: connectors oxidising and losing their spring, electrolytic capacitors drying out, open circuits resulting from temperature cycling induced strain, and so on.

In the centre portion, which is where plant operates continuously, it is possible to predict the probability that a piece of equipment will operate without failure for a given time.

If the failure rate is constant, we can write:

$$\lambda = \frac{1}{N_s}\frac{dN_f}{dt} \tag{8.7}$$

or

$$\frac{dN_f}{dt} = \lambda N_s \tag{8.8}$$

From equation 8.1, the reliability is

$$R = \frac{N_s}{N} = \frac{N - N_f}{N} = 1 - \frac{N_f}{N} \tag{8.9}$$

Differentiating gives:

$$\frac{dR}{dt} = -\frac{1}{N}\frac{dN_f}{dt} \tag{8.10}$$

Substituting from equation 8.8:

$$\frac{dR}{dt} = -\lambda\frac{N_s}{N} \tag{8.11}$$

or from equation 8.9:

$$\frac{dR}{dt} = -\lambda R \tag{8.12}$$

Integrating from t=0 to t:

$$R = Ae^{-\lambda t} \tag{8.13}$$

where A is an integration constant. However, at t=0, R must be 1 because the equipment must work for zero time, hence:

$$R = e^{-\lambda t} \tag{8.14}$$

For the centre ‘maturity’ period, the failure rate is simply:

$$\lambda = \frac{1}{\text{MTBF}} \text{ or } \lambda = \frac{1}{\text{MTTF}} \tag{8.15}$$

dependent on whether the item under consideration is repairable or not.

For example, a transducer has an MTBF of 17,500 hours (approximately two years). The probability that it will run at least 8750 hours (approximately one year) without failure (and hence its reliability for a six month period) is:

$$\begin{aligned} R = e^{-\lambda t} &= \exp(-t/\text{MTBF}) \\ &= \exp(-8750/17{,}500) \\ &\simeq 0.6 \end{aligned}$$

The probability that an item will run for at least its MTTF or MTBF is simply (but rather surprisingly):

$$\begin{aligned} R &= e^{-1} \\ &= 0.37 \end{aligned}$$

The 0.5 reliability time (i.e. the time for which the probability of no failure is 0.5) is approximately 0.7 MTBF (or 0.7 MTTF).

8.2.6. Series and parallel reliability models

Figure 8.4 shows a typical instrumentation system, where a power supply provides 24 V DC for a transducer whose output is shown

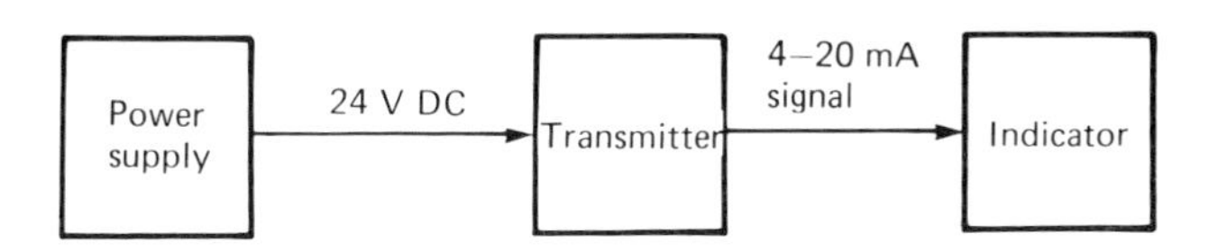

Fig.8.4 Series reliability model.

on a bar graph display. The failure of any component will result in the loss of the display and a 'system failure'. Given the reliability, or MTBF, of each unit, it is possible to predict the reliability, and MTBF, of the whole system.

Suppose the MTBFs, and the corresponding λs, are:

	MTBF	λ
PSU	10000	0.0001
Transmitter	15000	0.000067
Display	20000	0.00005

The reliability of a component for time T is the probability that it will run for time T without failure. The probability of three components running for time T is the product of their three reliabilities (assuming there is no interaction and no common influence such as external damage):

$$R = R_{1T}.R_{2T}.R_{3T} \tag{8.16}$$
$$= e^{-\lambda_1 T}.e^{-\lambda_2 T}.e^{-\lambda_3 T} \tag{8.17}$$
$$= e^{-(\lambda_1 + \lambda_2 + \lambda_3)T} \tag{8.18}$$

i.e. the resulting λ is the sum of the individual component λs. For the example above:

$$\lambda = 0.0001 + 0.000067 + 0.00005$$
$$= 0.000217$$

so the system MTBF:

$$= \frac{1}{0.000217}$$
$$= 4600 \text{ hours}$$

i.e. much lower than the MTBF of any individual unit.

The arrangement of fig.8.4 is called a series reliability model, because the failure of any unit results in the failure of the whole system.

In fig.8.5, the temperature of a vat is measured by two independent temperature measuring systems. Such an arrangement is called a parallel reliability model, because it requires a

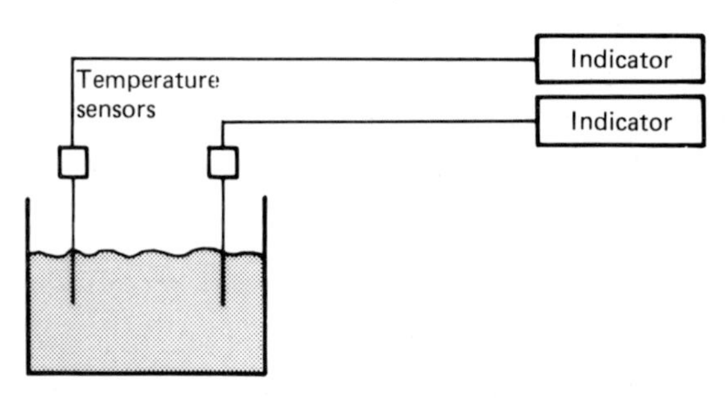

Fig.8.5 Parallel reliability model.

failure of both sensors to lead to a loss of temperature display. It is assumed that there is no interaction, and there is no common failure route. (The loss of mains supply is an obvious potential common failure; the analysis below assumes a totally reliable supply.)

For a failure to occur within a specified time, both systems must fail. The probability of this is the product of the unreliability of each unit, i.e. the system unreliability is given by:

$$Q = Q_{1T}.Q_{2T} \tag{8.19}$$

$$(1 - R) = (1 - R_{1T}).(1 - R_{2T}) \tag{8.20}$$

$$R = R_{1T} + R_{2T} - R_{1T}.R_{2T} \tag{8.21}$$

Suppose that the two systems of fig.8.5 have an MTBF of 15,000 hours. The reliability for 8000 hours for one sensor is:

$$\begin{aligned} R_{1T} &= e^{-\lambda T} \\ &= \exp(-8000/15{,}000) \\ &= 0.59 \end{aligned}$$

The reliability for the two in parallel from equation 8.21 is:

$$\begin{aligned} R &= 0.59 + 0.59 - 0.59.0.59 \\ &= 0.83 \end{aligned}$$

i.e. the parallel arrangement exhibits a higher reliability than the individual units. The above reliability corresponds to an MTBF of over 40,000 hours.

The parallel arrangement is often called 'redundancy' and is used in critical applications (e.g. nuclear reactors, petrochemical plants) where very high reliability is required. If three parallel paths are used, with each path having a reliability of 0.85, for example, a system reliability of 0.997 is achieved.

It is important for systems employing redundancy that protection against common failure routes (called common mode failures) is included. Separate and independent power supplies, for example, are essential. Less obvious precautions are different sensor positions and cable routes to avoid common mechanical damage, different sensor types (e.g. thermocouples and RTDs) to avoid problems affecting one type of sensor, and precautions against outside events affecting all channels simultaneously. In fig.8.6, for example, duplicate temperature sensors are used to indicate the exit water temperature from a cooling jacket. If the

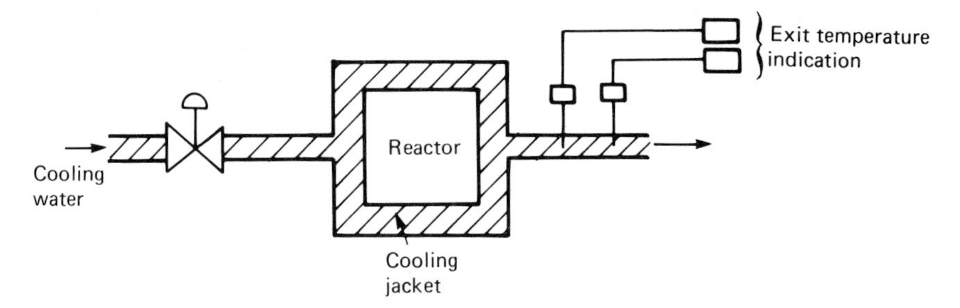

Fig.8.6 Potential common mode failure caused by incorrect sensor positioning.

water flow fails–caused by a pipe blockage, say–both indicators will show the same, probably incorrect, temperature. A common mode failure similar to fig.8.6 was a contributing factor to the operators' confusion during the Three Mile Island nuclear incident.

Variations on redundancy are the majority vote and highest (or lowest) voting systems. With two parallel systems, it can be problematical establishing which unit is working, and which has failed, in a fault condition. Majority voting uses 3, 5 or 7 systems, and takes the majority signal (2 of 3, 3 of 5, etc.) as being correct. Highest (or lowest) voting is normally employed where safety is the prime consideration (e.g. shutdown systems). The most unsafe signal (e.g. the highest temperature reading) is assumed to be correct from a control viewpoint. Majority voting gives the high reliability of redundancy; high voting gives very high safety, but the operational reliability is worse than a single system as any failure can lead to a potential shutdown.

Other approaches employing redundancy are arithmetic averaging of several signals, or discounting readings from the highest and lowest of three sensors (called median redundancy). All redundancy techniques are expensive in initial installation costs.

8.3. Maintenance philosophies

Even with the best planned maintenance and preventative maintenance procedures, faults will occur in all equipment at some time. Maintenance departments, responsible for the servicing, repair and operational improvement of plant, are often seen as a necessary evil, but are in reality a crucial part of a production team. Plant downtime is invariably expensive, and a maintenance department should not just 'fix it when it breaks' but consider the whole economic cost of maintaining the equipment in their care. Admittedly such a broad view is difficult to take at 3 am with a shift manager asking the three inevitable questions 'What's wrong?', 'Do you think you can fix it?' and 'How long will it take?'

There is a fundamental difference between most process control/electronics problems and, say, mechanical faults. In the latter case the fault is usually obvious, often to non-technical persons, but a repair is lengthy. Typical examples are seized bearings, broken couplings and so on. Process control problems tend to be more subtle. Symptoms are noticed by production staff, and a logical fault finding procedure is needed to locate the fault. Once diagnosed correctly, the repair is usually straightforward and quick.

The reliability of modern equipment can create problems for the maintenance staff. With MTBFs measured in years, it is possible that the maintenance technician really sees a piece of equipment for the first time when it fails (and the instruction manuals are collecting dust in a cupboard). More reliable equipment also means that a maintenance technician can cover, and hence needs to know about, much more plant. It is therefore essential that maintenance staff are involved in the installation and commissioning of all new plant, as this is the best time to observe test procedures and learn how it works. Refresher training at regular intervals is also helpful. Design staff can assist by standardising on types of equipment in all areas, so the maintenance staff only have to learn the details of, say, one type of programmable controller. Small price differences between competitive equipment can be totally wiped out by production losses at the first fault.

A fault manifests itself as a symptom; a gas fired burner is stuck on low fire, say. Investigation shows that the air valve is almost closed and the gas valve, correctly following the air flow, is also on a low setting. The non-operable air valve is also a symptom. Further investigation reveals the diaphragm in the pneumatic actuator has ruptured; this is the fault which is causing the symptoms. A repair is effected by replacing the diaphragm or, more probably, by changing the actuator. The fault finding procedure splits naturally into realisation that there is a fault, location of the fault and rectification of the fault. Symptoms and faults should not be confused; a blown fuse, for example, is a symptom, not a fault.

There is, however, an often overlooked fourth stage which is the analysis of the fault to see if it was a random occurrence or was due to some underlying cause. A continuing failure of the diaphragm in the example above would suggest a possible environmental or application problem. Faults are repaired usually by shift staff who inevitably only see one quarter of the faults and are naturally mainly concerned with restoring production. A managerial fault recording and analysis system will allow a broader view of problems and possible solutions.

Apart from revealing reasons for higher than expected fault rates, such studies may suggest a need for better diagnostic aids and instruments or lack of knowledge at the first line maintenance level, or may show that the level of repair being carried out on site is too deep (and it would be more cost effective to use replaceable units with the repairs being effected off line in the workshop). The fault finding procedure does not stop when the plant goes back into production.

Fault finding can be considered as first line maintenance (repairs carried out on plant) and second line maintenance (repairs carried out in a workshop). In both cases it is a logical procedure which homes in on the fault along the lines of fig.8.7. Symptoms are studied, and from the available information possible causes are postulated. Test are conducted to confirm, or refute, these possibilities. Further information gained from these tests allows the probable causes to be narrowed down and further tests made. The procedure is repeated until the fault is found.

One of the arts of fault finding is the balancing of the probabilities of the various possible causes of a fault against time, effort and equipment necessary to perform the tests required to confirm or refute them. (This manifests itself in the sensible old rule ‘check the power supply first’. The probability of a supply problem may be less than 10%, but it usually is a possible cause that can be checked out in a few seconds.) Good equipment design should provide diagnostic aids such that tests for the most probable faults can be performed quickly without specialised test instruments. It does not help the MTTR if the most probable faults can only be located with a 50 kg test box which has to be signed out of the stores and carried 20 metres up a vertical ladder. Such problems should be revealed by the managerial fault analysis system.

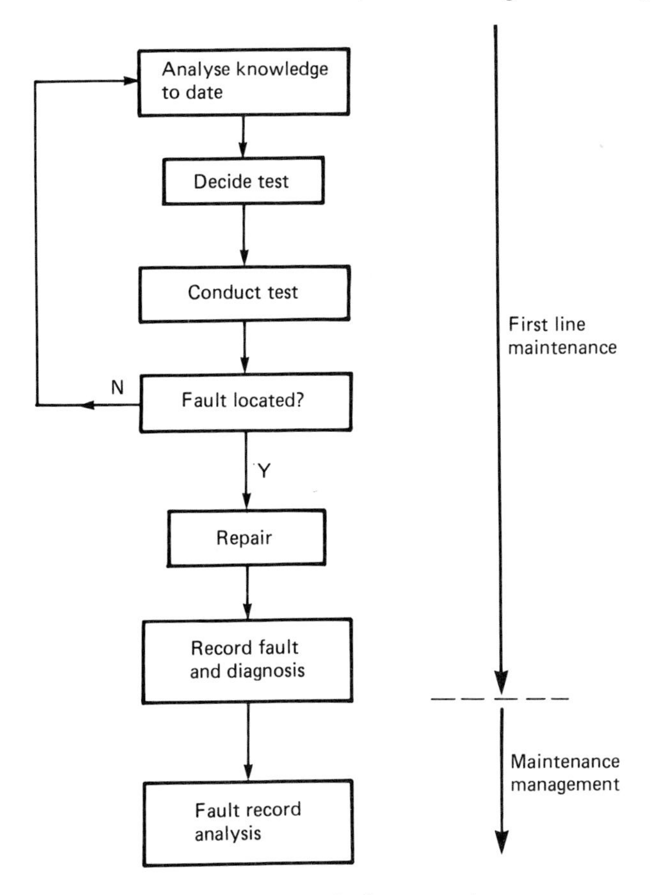

Fig.8.7 The fault finding procedure.

Fault finding is simplified if equipment is conceptually modularised, a topic discussed further in section 8.6. Figure 8.8 shows a flow control loop which can be broken down to six modules, each of which can be tested independently. With care, most systems can be considered as a series chain of modules, as in fig.8.8b. The loop of fig.8.8a could be broken, for example, by putting the controller in manual and considering the controller output as the input to the chain and the feedback signal as the output.

Fault finding is based on signal injection and signal tracing. Signal injection, shown in fig.8.9a, introduces test signals whose effects are observed. Signal tracing, shown in fig.8.9b, follows normal signals through the chain. In both cases, the so-called half split method of fig.8.9c, where the possible fault area is reduced by half with each test, is the best approach.

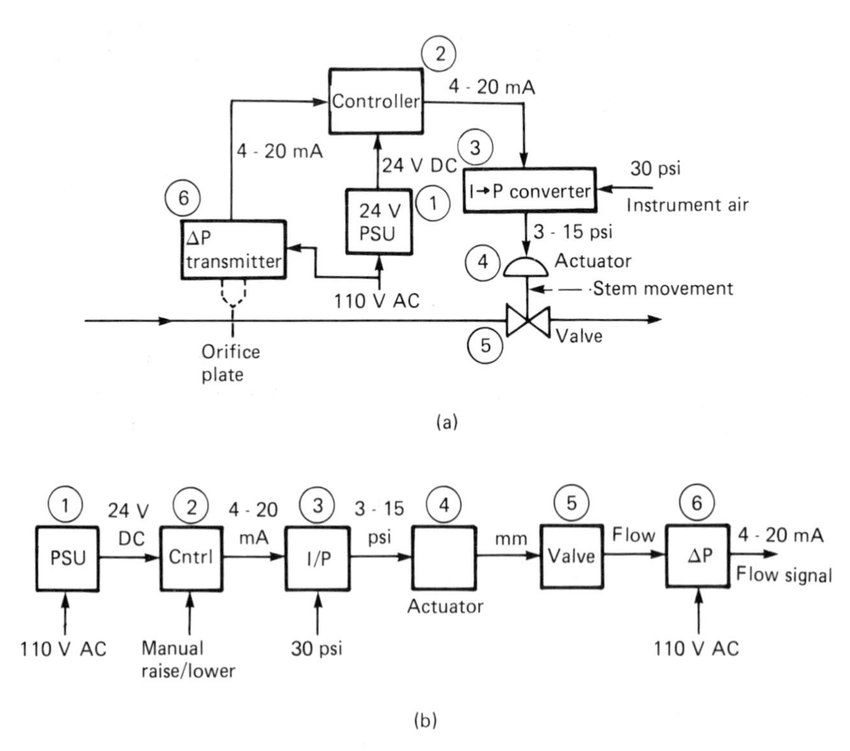

Fig.8.8 Fault finding methodology. (a) Flow control system. (b) Flow control system modularised for fault finding.

Plant reliability can be improved by a sensible planned maintenance programme. It is possible to identify the life of essentially mechanical plant items (e.g. filters, oils, seals, etc.) with a fair degree of accuracy from manufacturers' data sheets or from plant records accumulated from the fault recording process described earlier. For such items a routine schedule of replacement can easily be devised.

Most process control equipment, however, does not 'wear out' but exhibits the bathtub reliability curve of fig.8.3. Once the maturity period is reached, routine replacement will not improve reliability and may even take the equipment back into the infant mortality region. Planned maintenance therefore usually takes the form of regular calibration checks (such as the setting of span and zero on transducers).

The most important factor in determining reliability is probably the competency and experience of the maintenance personnel. Training is often overlooked, but a well trained staff is the best

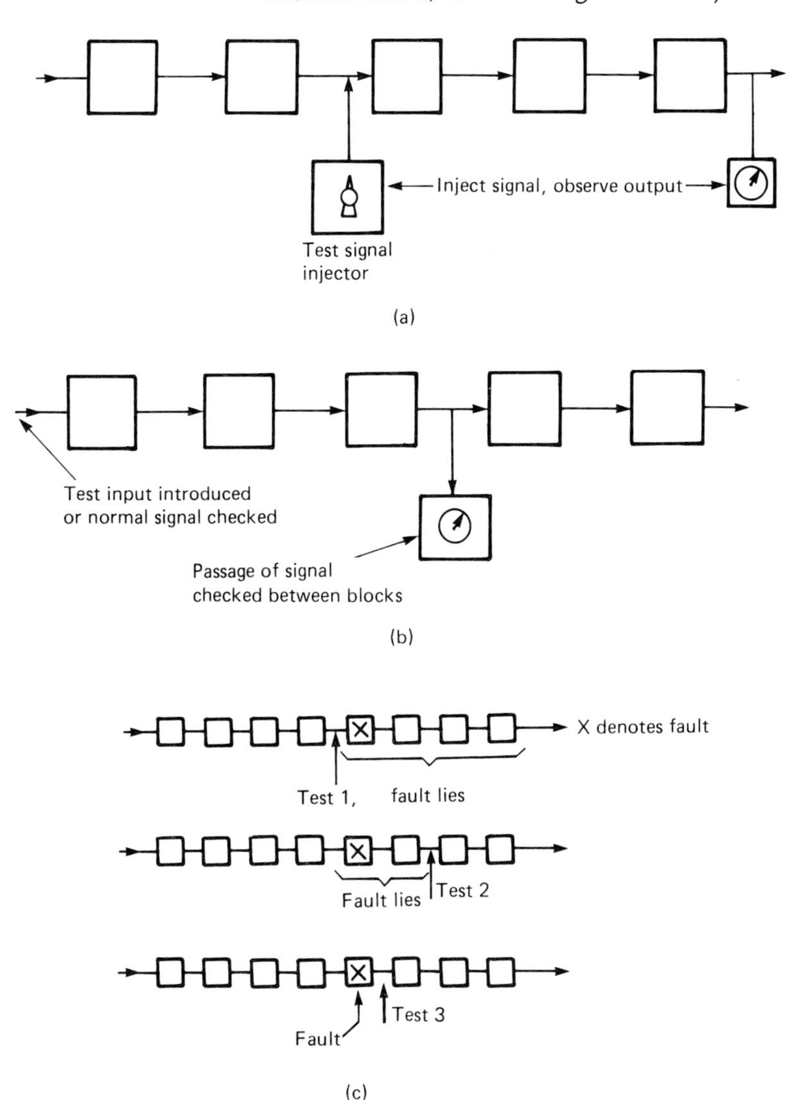

Fig.8.9 Fault finding techniques. (a) Signal injection. (b) Signal monitoring. (c) The half split method.

investment a company can make. Often, however, the training is of the wrong type. If a technician, say, goes on a manufacturer's course on a specific PLC there is a natural tendency to assume thereafter that plant faults can only exist inside the PLC. There is also a defensive tendency for first line personnel, when in doubt, to do what they can do well. When the ideas run out, a technician freshly trained in stripping down and servicing the instrument air compressor will invariably strip down and service the instrument air compressor. Training should therefore deal with a broad view of a plant; what the various items do and how they interact. On large plants it is useful to have a course on the physical locations of plant equipment. Many wrongly identified plant mounted limit switches and transducers have been changed in error!

8.4. Fault finding instruments

The previous section showed that testing is essentially of a signal injection or signal tracing nature. Instruments for assisting in the tracing of faults therefore tend to be devices for injecting test signals or displaying the value of signals at strategic points.

The classical (and in many organisations the only) fault finding aid is the multimeter, used to measure voltages and currents. Digital multimeters are useful where accurate readings are needed but can be confusing where values change quickly, for reasons outlined in section 7.4.3. Auto-ranging meters can be disturbing if the signal goes across several range changes. Ideally auto-ranging meters should have optional manual range selection or a hold high range option.

In the author's experience, the high impedance of digital meters (normally a desirable characteristic) can cause misleading indications in some quite common circumstances. A typical problem is shown in fig.8.10 where a triac output from a programmable controller is used to operate a pneumatic

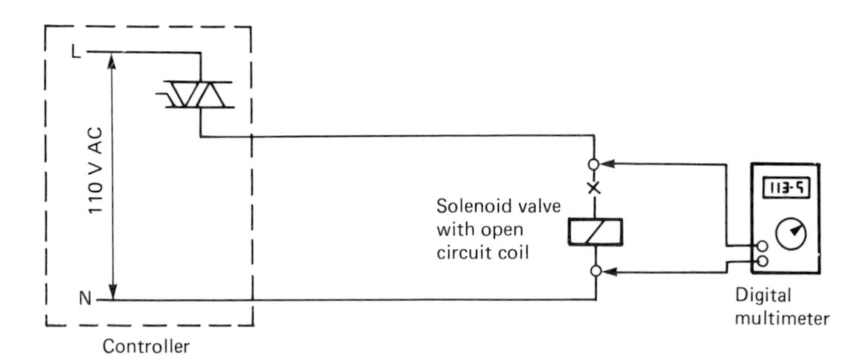

Fig.8.10 The need for care with digital multimeters.

solenoid. The coil of the solenoid has gone open circuit, and a digital meter has been connected across the coil. A triac has a small leakage current in the off state; this is insufficient to pick up the coil or displace a moving coil meter, but *will* be enough for the digital multimeter which will register full line volts regardless of whether the PLC output is on or off. The technicians could thus be misled into suspecting a PLC or output card fault.

The other classical fault finding device is the oscilloscope, but its bulk and relative delicacy tends to limit its use to control rooms and other centralised points. Storage scopes are useful for capturing transients, a role also filled by the UV recorder described in section 7.3.3.

A large proportion of instrumentation signals use 4 to 20 mA and similar signals, so a test device for the injection and monitoring of current signals is essential for the maintenance and calibration of process control equipment. A typical device is shown in fig.8.11.

Fig.8.11 An instrumentation injection/monitor/calibration unit; the SUPERCAL manufactured by Rochester Instruments. This unit provides and monitors mA, mV, V for thermocouple and transducer testing, and can simulate and monitor resistance temperature transducers. Additional facilities include a calculator function, frequency monitoring and injection plus displays in engineering units. (Photo courtesy of Rochester Instruments.)

Monitoring and injection of millivolt signals is required where thermocouples are used. Devices such as fig.8.9 usually allow the monitoring and injection of both voltage and current signals. The use of millivolt sources/display with thermocouples requires thermocouple tables and a knowledge of the cold junction temperature. Thermocouple principles are described in section 2.5 of Volume 1, but the procedures are:

Injection

(1) Measure the cold junction temperature (usually ambient) and note the voltage V_a corresponding to this from the tables.
(2) Note the voltage V_t from the tables corresponding to the required temperature.
(3) Set the injection voltage to $V_t - V_a$. The system under test should indicate the required temperature.

Monitoring:

(1) Find the ambient temperature voltage V_a from the tables, as above.
(2) Measure the thermocouple voltage V_t at the place where the ambient temperature measurement was made.
(3) Find the temperature corresponding to $(V_a + V_t)$ from the tables; this is the thermocouple temperature.

Direct reading temperature indicators are, of course, readily available but these are not as versatile for test purposes as a millivolt source/monitor.

Pneumatic test devices are also available for the injection and monitoring of pneumatic signals. A typical device is shown in fig.8.12. Similar devices are available for hydraulic systems, but hydraulic monitoring (i.e. pressure gauges) should be built into the system at the design stage as most hydraulic test sets are very bulky and difficult to transport.

Faults on sequencing systems can be difficult to trace particularly if (as is usual) many events happen in a short period of time. So-called 'signature analysers' are a useful, if somewhat expensive, approach to this problem. Essentially signals from the

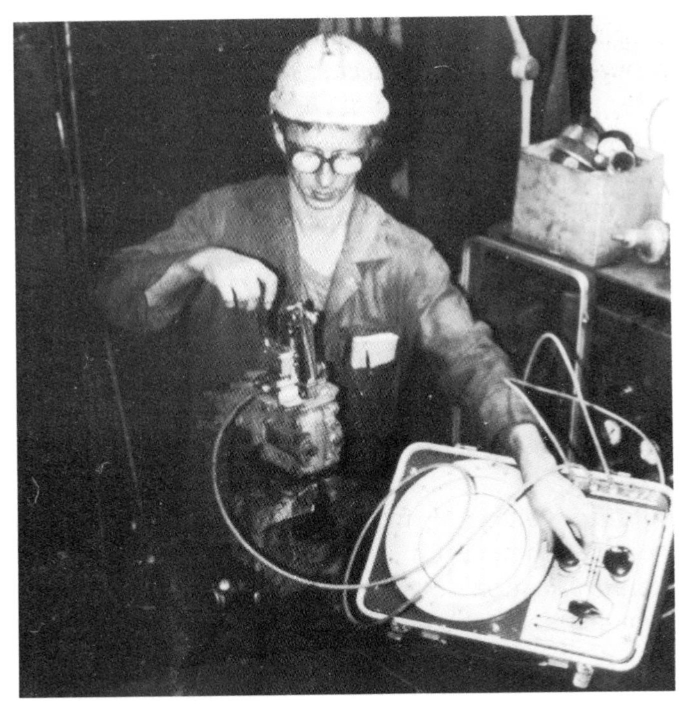

Fig.8.12 A pneumatic injection/monitoring unit being used to calibrate a differential pressure transmitter.

plant are brought back to the analyser which has been pre-programmed with acceptable patterns of plant behaviour. The analyser compares the plant operation with its internal model and flags any deviation (e.g. input 27 late coming on). Like expert systems, the use of signature analysers is best 'built into' a plant at the design stage rather than grafted on to an existing plant.

8.5. Noise problems

8.5.1. Introduction

Electrical interference, usually called 'noise', is often the cause of poor or erratic performance of a process control system. Many so-called intermittent faults are found to originate with noise problems, particularly in digital sequencing systems. Chapter 7 in Volume 3 covers the theoretical aspects of data transmission through a noisy environment; the present section is primarily concerned with the practical aspects of noise elimination.

The tracing and removal of noise problems can be more of an art than a science, and requires an almost detective-like instinct. Is the noise always present, or occurring randomly? Is the noise related to events occurring on a plant (e.g. a compressor starting)? Has any new plant been added and tied into the grounding system? Does the noise have a recognisable waveform? These and similar questions should be asked when noise problems occur. Good design with screening, isolation and sensible grounding should, however, prevent most noise problems.

8.5.2. Types of noise

It is useful to categorise types of noise. There are, in general, three types of noise that can be encountered, as summarised by fig.8.13. Random, or white, noise is a wideband random fluctuation in a signal. In audio circuits this appears as a 'hiss'. All electrical components generate white noise with resistors, thermionic valves and zener diodes being particularly effective noise generators. Internally generated white noise should not be

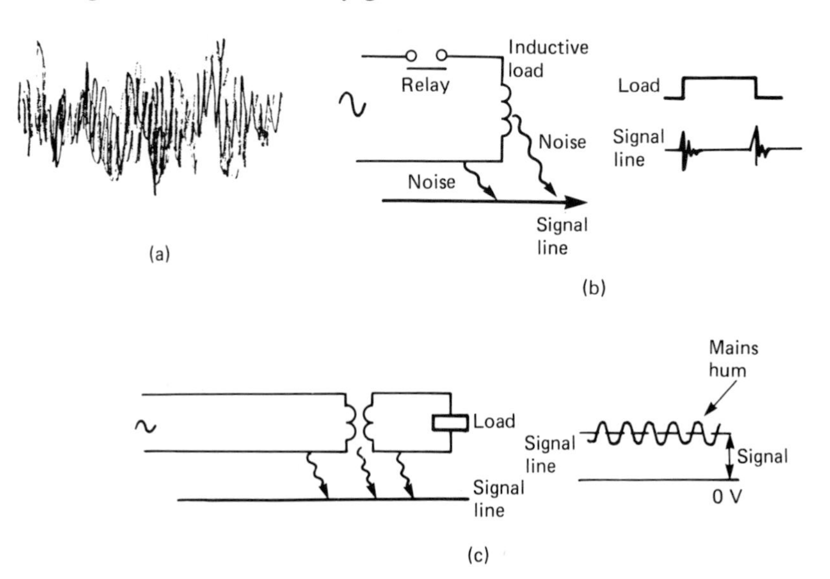

Fig.8.13 Types of noise. (a) Random noise. (b) Impulsive noise. (c) Cross coupled noise.

a problem in process control except with high gain instrumentation amplifiers. Poor connections and badly soldered joints can, however, produce random noise.

Impulsive noise appears as random spikes on an otherwise steady signal. In process control these spikes are invariably associated with the switching of adjacent heavy or inductive loads. Poor or dirty connections can again generate impulsive noise.

Cross coupled noise appears from capacitive or inductive common impedance Z3 causes interaction between the two items. (UK)/60 Hz (USA) interference from power cables to adjacent signal cables. Noise spikes can occur from cables switching power to inductive loads.

8.5.3. Noise sources

Figure 8.14 shows the common routes by which noise can be introduced into a system. For simplicity no noise sources are

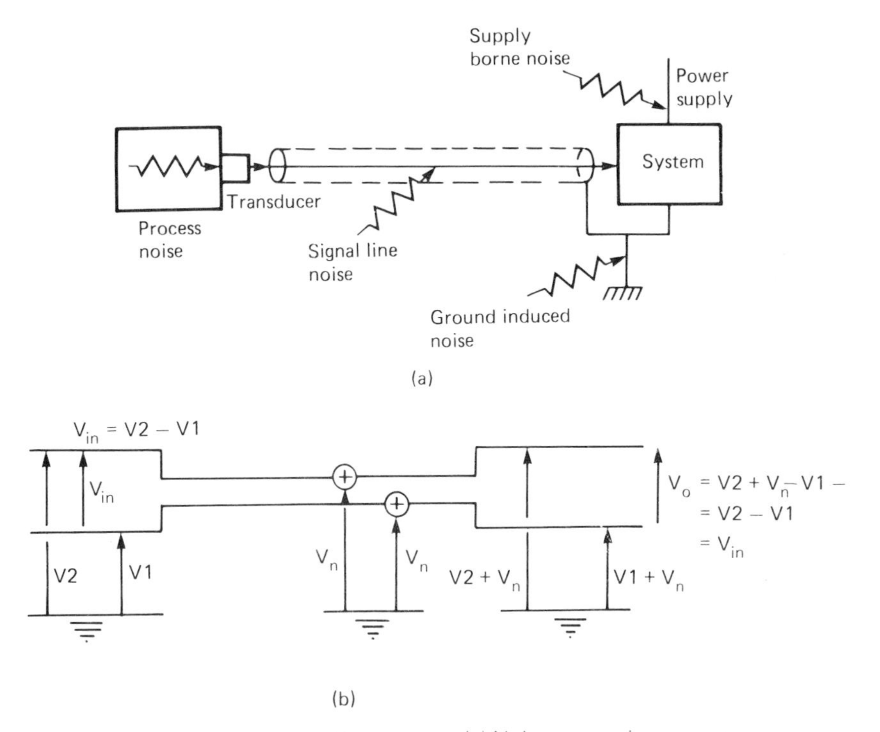

Fig.8.14 Noise and real life systems. (a) Noise routes into a system. (b) Common mode noise.

shown affecting the primary sensor, but these are, of course, subject to the same noise sources as the rest of the system. A noise problem will usually arise as a combination of the effects of fig.8.14.

Many process variables are inherently noisy; level measurement is particularly difficult, for example, as the surface is affected by ripple and turbulence. Other problems arise from pressure spikes in compressor systems, flame flicker in temperature measurement and suspended solids or entrapped gases in flow measurement. All these, and similar effects, contribute to a noisy primary signal.

Noise coupled on to signal lines is a common problem. This noise usually affects all signal lines in a particular cable equally, and is called common mode noise. The effect is shown in fig.8.14b. Series mode noise occurs when circulating currents are induced around signal pairs, and is far less common (but more difficult to deal with). Signal line induced noise, often called pickup noise, can be reduced by utilising sensible spacing between signal and power cables (typically 1 to 2 metres minimum), the avoidance of long parallel runs, and the use of screening on signal cables or running signal cables in conduit. Electrical currents are far less affected by common mode noise than are voltages and this is one reason why standards such as the 4 to 20 mA current loop are widely used in instrumentation. The best solution, though, is to keep a reasonable separation between power and signal cables.

Ground induced noise is often overlooked. The earthing arrangements in instrumentation systems require special care, and the topic is treated further in section 8.5.4.

Some systems generate internal noise, TTL based logic systems being particularly vulnerable. Internal noise prevention is largely a matter of sensible PCB layout and power supply rail decoupling. This subject is treated further in section 3.10.

Noise can also be introduced via the supply lines. The switching of large loads, the use of thyristor drives and random lighting strikes on the supply authority's overhead power lines all cause erratic system behaviour. The use of line LC filters or constant voltage transformers (CVTs) is highly recommended for line sensitive equipment.

Noise problems can be overcome by removing the cause(s) of the noise at source, or by making the system less sensitive to the noise. The former method should be used wherever possible. The

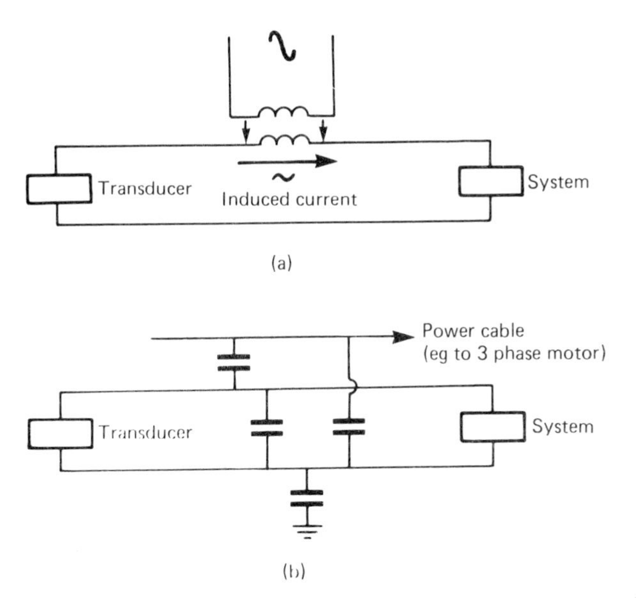

Fig.8.15 Methods by which noise enters system. (a) Electromagnetic coupling. (b) Capacitive coupling.

suppression of an inductive load or the construction of a stillwell around a level transducer are examples of noise removal. The use of filters is an example of making a system noise tolerant.

8.5.4. Grounding

Ill conceived grounding arrangements can be the cause of particularly elusive noise problems and intermittent faults. Problems arise when different equipment or parts of the same equipment, share a common route to ground, as in fig.8.16a. The common impedance Z3 causes interaction between the two items. Generally the resistance of the connections is small, and it is the inductive impedance that causes the problem on fast changing currents.

One common example of poor grounding called an earth loop occurs when screened cable is used to connect equipment on different earthing circuits, as in fig.8.16b. There will inevitably be a significant AC potential (as high as a few volts in extreme cases) between the two earths, causing a large current to flow down the screen and any other 0 V link. Screened cable must have the screen connected to ground at one place only. Particular care

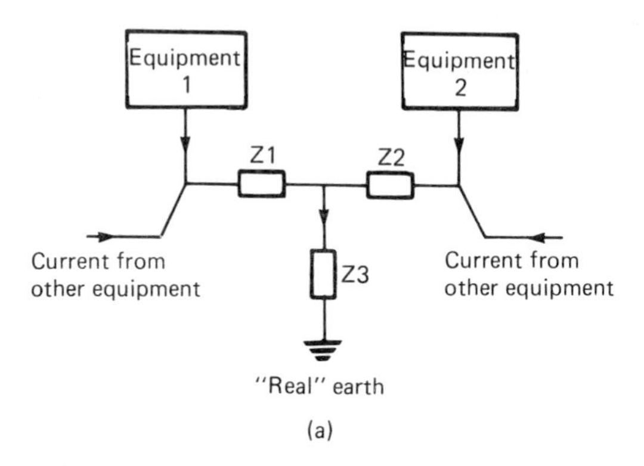

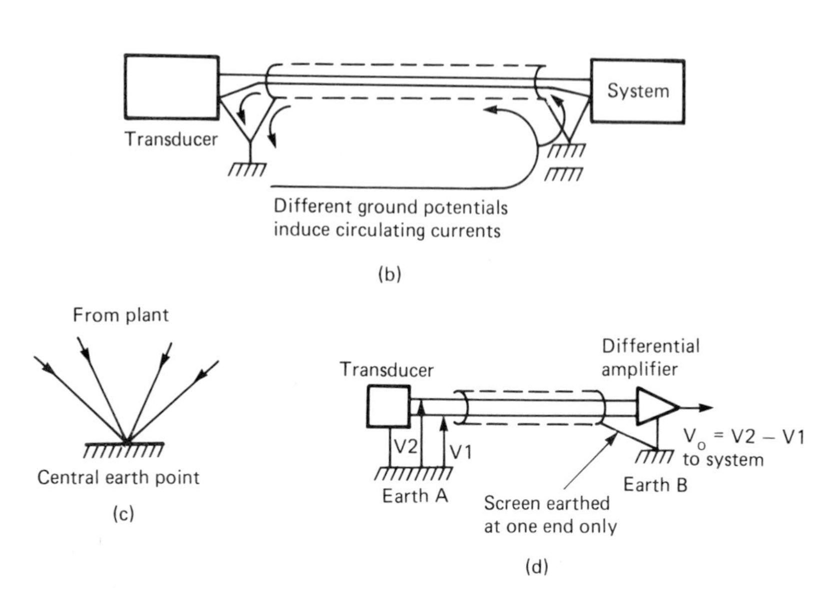

Fig.8.16 Noise induced via ground lines. (a) Noise from shared ground routes. (b) Problems with multiple earths. (c) The CEP. (d) Earthing problems minimised by differential amplifier.

should be taken to ensure screen continuity through junction boxes whilst avoiding multiple earths.

Ideally, it is possible to identify at least three different 'grounds' in a system, namely a safety ground for earthing cases, cubicles, instruments, etc., for the prevention of electrical shock, a power ground for current returns from power supplies and high current loads such as relays and solenoids, and signal grounds from low current instrumentation. These should all meet at one, and only one, central earthing bus which has a direct low impedance route to ground, as in fig.8.16c.

This ideal state of affairs cannot exist on a widely separated plant, and in these circumstances differential or isolation amplifiers can be used to remove the common mode noise arising from different earth potentials, as in fig.8.16d. Particular care should again be taken to avoid the creation of earth loops.

8.5.5. Suppression of inductive loads

In fig.8.17a a current is flowing through an inductor L. The voltage across the indicator is given by:

$$V = L\frac{dI}{dt} \tag{8.22}$$

This predicts that the voltage induced is proportional to the rate of change of current. The switching off of inductive loads results in large transient voltages, arcing across contacts and a very real source of interference, as shown in fig.8.17b.

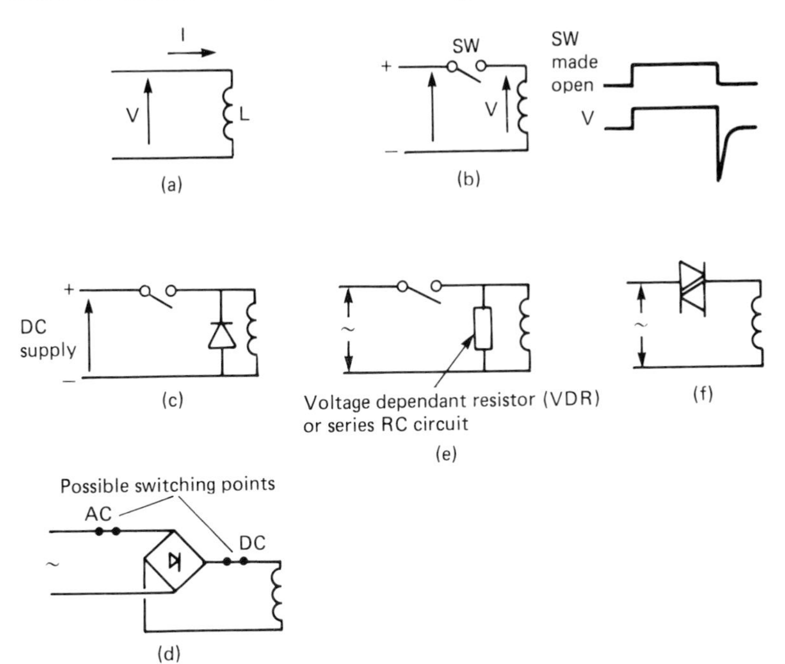

Fig.8.17 Noise from inductive loads. (a) Inductive circuit. (b) DC circuit. (c) Spike suppression diode. (d) AC/DC circuit. (e) AC circuit. (f) Triac circuit.

DC switching can be made noise free by the use of a spike suppression diode, as in fig.8.17c. This provides a route for the current in the inductor to decay gracefully and without an excessive transient. If a DC load is being fed from a rectified AC supply, as in fig.8.17d, AC side switching is preferred as the bridge rectifier provides a route for circulating currents and acts as a suppression diode.

The switching of AC loads presents more problems as diodes cannot be used. For small loads, series RC snubbers can be connected across the load or switching contact. For larger loads, voltage dependent resistors (VDRs, also known as metrosils) can be used. The ideal solution is to use triacs (see section 2.5.3), with zero voltage crossing control circuits. These will turn on at the zero voltage point of the AC supply and inherently turn off when the AC load current is zero. The induced voltage, and hence the interference with other equipment, is minimal.

8.5.6. Differential amplifiers

The differential amplifier of fig.8.18a (analysed previously in section 1.3.3) amplifies the difference between its two input voltages, i.e.:

$$V_o = V1 - V2 \qquad (8.23)$$

assuming all the resistors have equal values. This feature makes it particularly useful for separating out a signal from large amounts of common mode noise, as in fig.8.18b.

Differential amplifiers are commonly employed where signals are to be passed between equipment with different earth potential. Figure 8.18c shows a particularly useful technique where the signal is converted to a current for transmission (as explained earlier, current signals are less affected by noise). The current signal is converted back to a voltage at the receiver by a burden resistor and a differential amplifier. The voltage/current circuit is described in section 1.6.1.

The performance of a differential amplifier is defined by the common mode rejection ratio (CMRR). This is tested as shown in fig.8.18d. The inputs are linked and driven from an oscillator

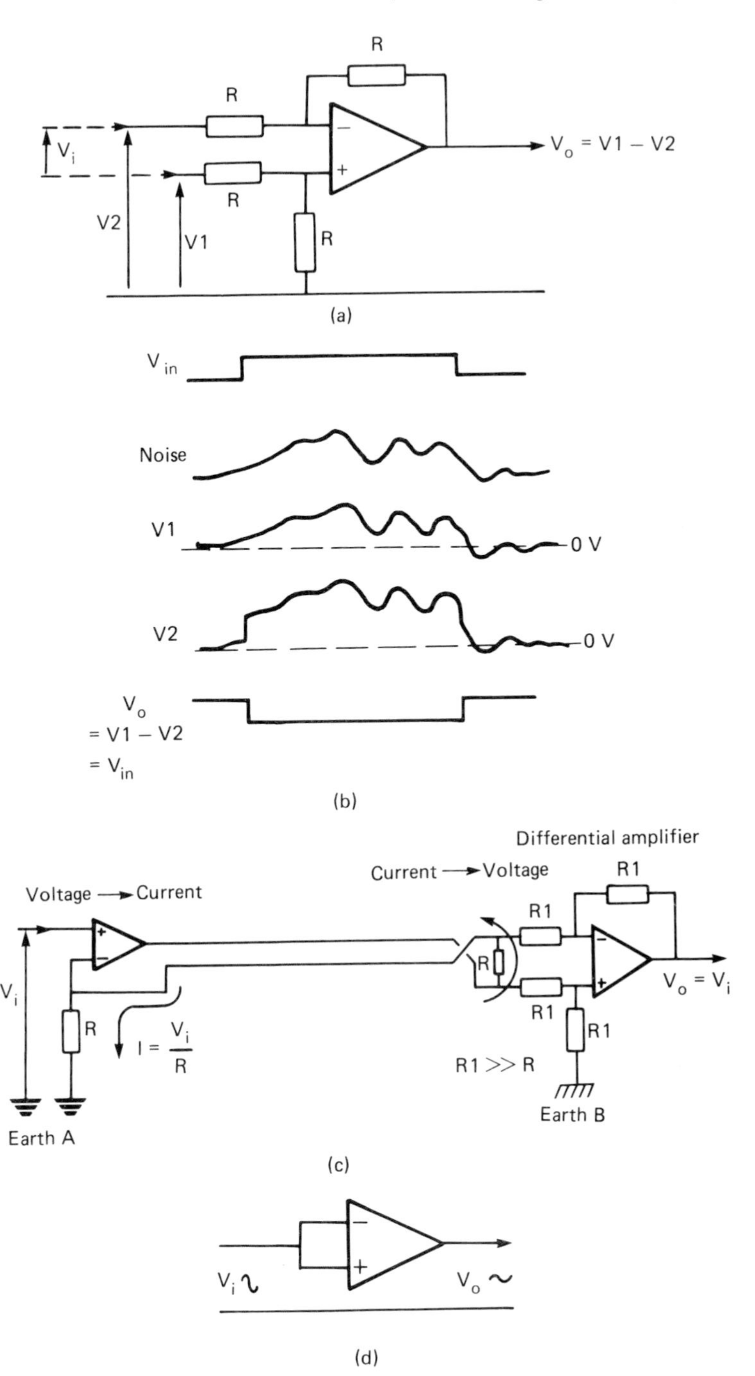

Fig.8.18 Removal of common mode noise. (a) The differential amplifier. (b) Typical signals. (c) Current transmission of signal. (d) Definition of common mode rejection ratio.

simulating common mode noise. The output should, theoretically, be zero, but in practice will follow V_{in} to some extent. If the amplifier gain is A, the CMRR is defined as:

$$CMRR = A\frac{V_i}{V_o} \tag{8.24}$$

The CMRR is very large, and is usually expressed in decibels (dB). A typical value for an instrumentation amplifier is around 140 dB.

8.5.7. Isolation techniques

Differential amplifiers can handle common mode voltages of a few tens of volts. Beyond this, or where faults can induce large

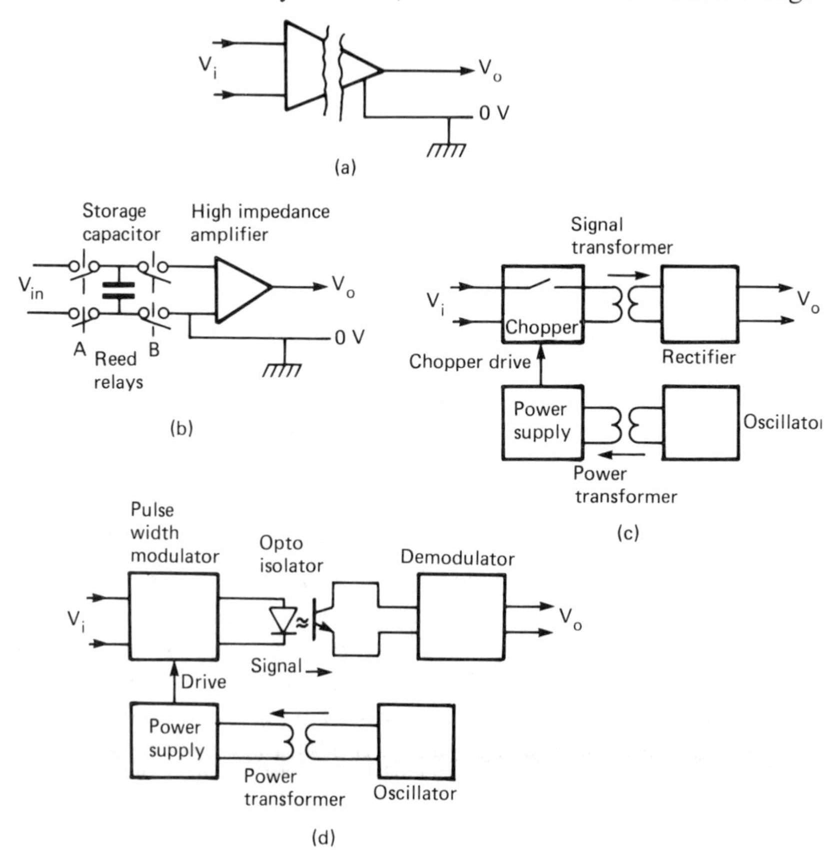

Fig.8.19 Isolation amplifiers. (a) Symbol for isolation amplifier. (b) Flying capacitor. (c) Transformer coupled isolation. (d) Opto isolation.

voltages on to inputs, or equipment is to be used in a particularly electrically noisy environment, devices called isolation amplifiers should be used. These have no direct electrical connections between the input and output terminals (called four port isolation). A typical device can withstand 1 kV between input and output and reject common mode noise of a similar value.

An early technique, but one still used because of its simplicity, is the flying capacitor of fig.8.19b. Reed relay pairs A and B close alternately; with A closed, the capacitor charges to V_{in}. With B closed, the capacitor voltage is presented to the differential amplifier input.

A more modern technique, shown in fig.8.19c, uses modulation of an AC waveform with transformer isolation. A similar technique uses pulse width modulation of a square wave and opto isolators, as in fig.8.19d.

Figure 8.19 shows isolation techniques for analog signals. Isolation of digital signals is, if anything, more important. Usually this is provided by opto isolation, as described in section 3.10.

8.5.8. Filtering

If noise cannot be removed at source, usually the best solution is to apply filtering to the signals before it is used by the control system. Before suitable filtering can be chosen, it is first necessary to determine the bandwidth of the measured signal so that the filtering will remove the noise but not degrade the signal information content. Loop stability can be affected by filtering; an often overlooked fact is the phase shift introduced by the simple low pass filter at frequencies well below the corner frequency, as in fig.8.20.

Filtering normally takes the form of low pass filters (to remove random high frequency noise and impulsive noise) or notch filters to remove noise of a specific frequency (e.g. mains induced hum or level resonance–see section 7.1 in Volume 1). Low pass filters are described in section 1.5.1 and notch filters in section 1.5.3.

Digital filtering is essentially an averaging procedure, and consists of taking a rolling average of the last 'N' samples. The effective time constant is determined by the number of samples and the sample time:

$$V_n = V_{n-1} + \frac{\Delta t}{T}(V - V_{n-1}) \qquad (8.25)$$

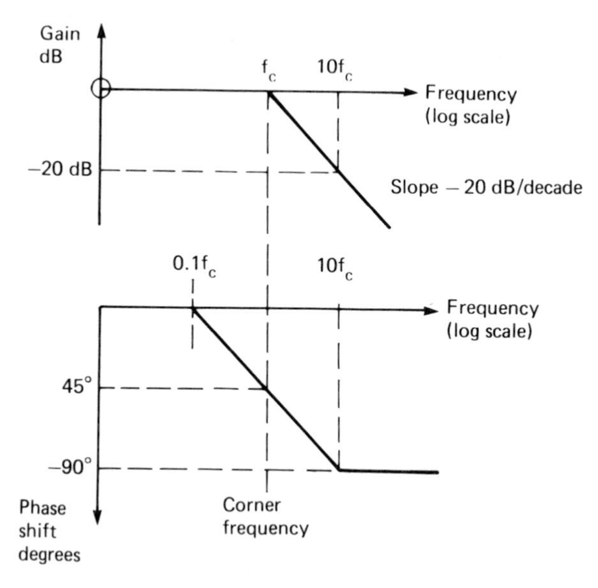

Fig.8.20 Straight line approximation Bode diagram of first order filter.

Where V_n is the current filtered value, V_{n-1} the last filtered value, t the sample time, T the time constant and V the raw unfiltered sample.

8.6. Documentation

Modern plants are both complex and reliable. Together, these characteristics create problems for maintenance staff, as it is not usually possible to build up experience of faults on specific items. (If it is, the fault recording and analysis procedures should have flagged a possible design, application or environmental problem!) The plant documentation is therefore crucial in reducing repair time. It also follows that the documentation should be available, and not sitting in the bookcase in the chief engineer's office!

Figure 8.21 is a common example, familiar to most people, of a car wiring diagram. This, like a politician's answer, is completely factual and truthful but of little use. The drawing has been produced for constructional purposes and not for fault finding. The circuit has been redrawn in a style suitable for fault finding in fig.8.22, with a logical flow of signals from left to right.

Figures 8.21 and 8.22 illustrate a common failing of documentation. There are two types of plant drawing. The first is

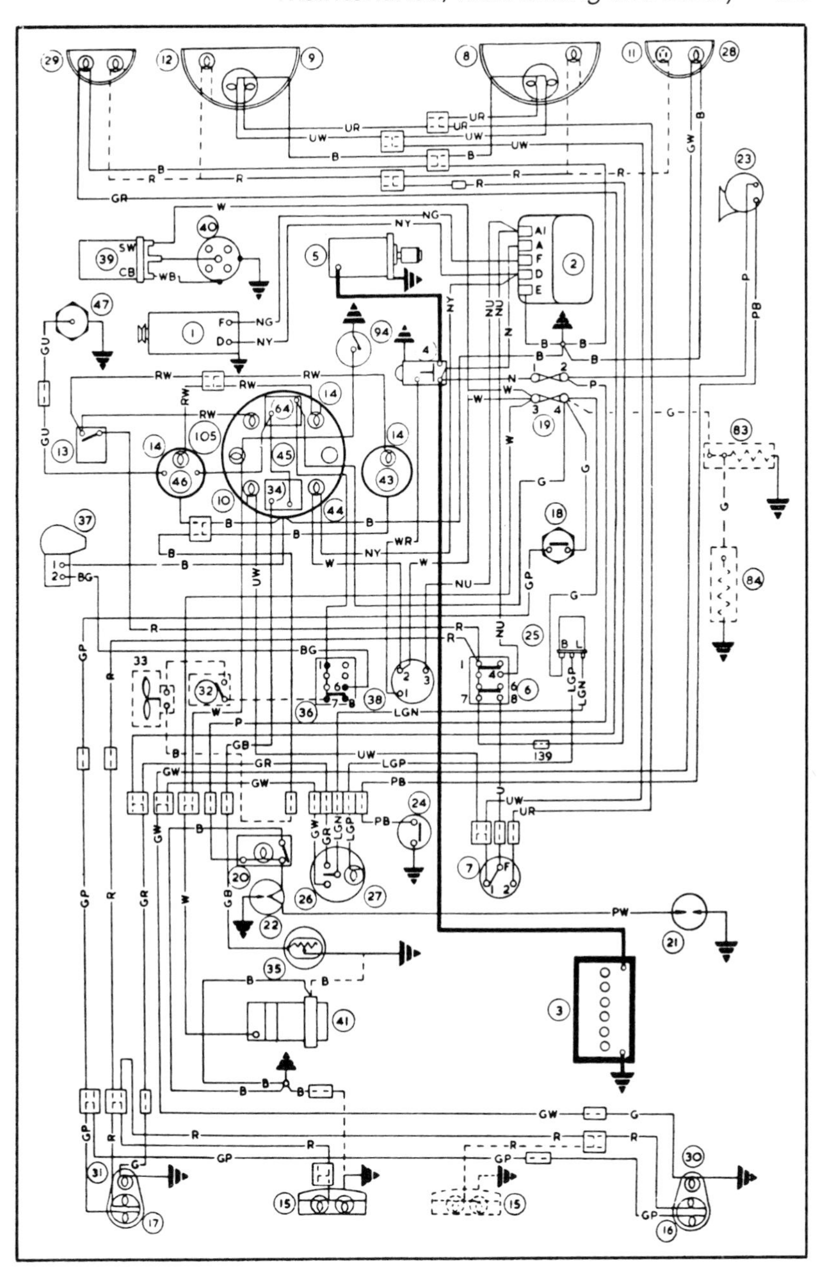

Fig.8.21 A typical car wiring diagram with which most people are familiar. It emphasises spatial relationships in that the layout of components follows, to some extent, the physical arrangement in the car. This results in the diagram having a large number of wiring crossovers and parallel runs, and a lack of any 'direction' or functional flow. In consequence it appears cluttered and is difficult to follow. (Reproduced from IBA Technical Review with permission. The concept of fig.8.21 and fig.8.22 is based on material published by the IBA.)

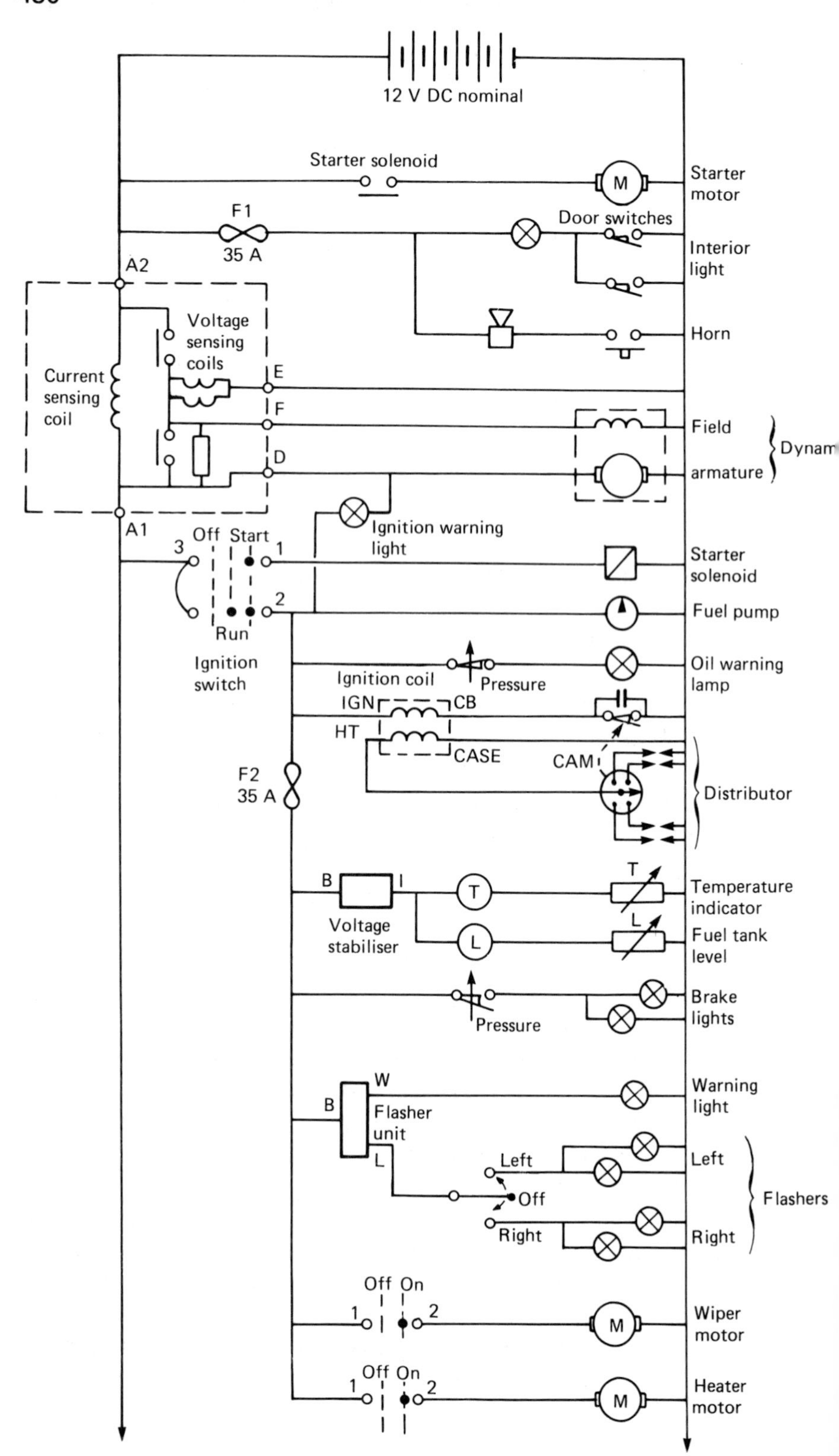

12 V DC nominal
Starter solenoid
Starter motor
M
F1
35 A
Door switches
Interior light
A2
Voltage sensing coils
Horn
Current sensing coil
E
F
Field
D
armature
Dynam
A1
Ignition warning light
Off
Start
3
1
Starter solenoid
2
Fuel pump
Run
Ignition switch
Oil warning lamp
Ignition coil
Pressure
IGN
CB
HT
CASE
CAM
F2
35 A
Distributor
B
I
T
Temperature indicator
Voltage stabiliser
L
Fuel tank level
Brake lights
Pressure
W
Warning light
B
Flasher unit
L
Left
Left
Off
Flashers
Right
Right
Off
On
1
2
Wiper motor
Off
On
1
2
Heater motor

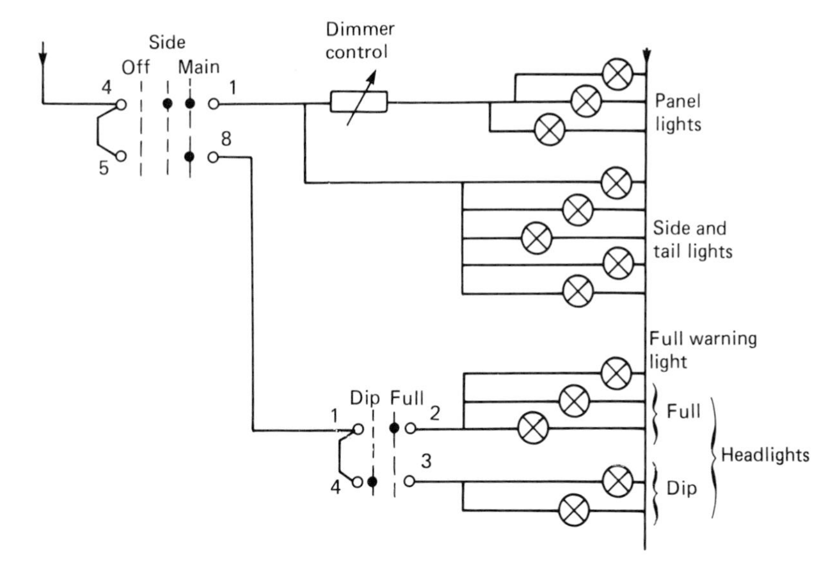

Fig.8.22 Car wiring diagram redrawn for ease of understanding.

produced by the manufacturer to construct and interconnect the plant. Such drawings are essential, but of little use for subsequent fault finding unless there is a major fire or similar disaster. These construction diagrams tend to be of a locational nature (such as fig.8.21) or panel orientated (such as fig.8.23, which is an extract from the drawings for a PLC system). The latter drawing would not be much use in finding out why a particular motor will not start.

Day to day maintenance requires drawings which are functionally orientated whilst retaining locational information. Figure 8.24 shows part of a functional drawing of the same PLC system as fig.8.23, showing the plant signals and where they can be found and traced for fault finding purposes.

Unfortunately, many manufacturers and designers only provide constructional and locational orientated drawings, making the task of maintenance personnel more difficult than it need be.

In the 1960s, the Royal Navy were concerned about the increasing complexity of ship borne equipment and the problems of maintenance. A team at HMS *Collingwood* devised an approach called FIMs, for functionally identified maintenance system. FIMs is diagnostic documentation which is supplementary to the main constructional or functional drawings. It is based on functional blocks whose inputs and outputs can be identified and tested. These blocks are arranged in a hierarchy, as

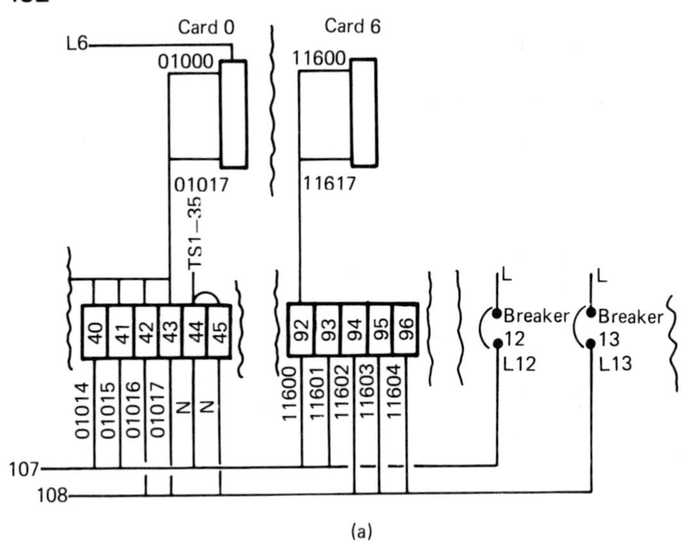

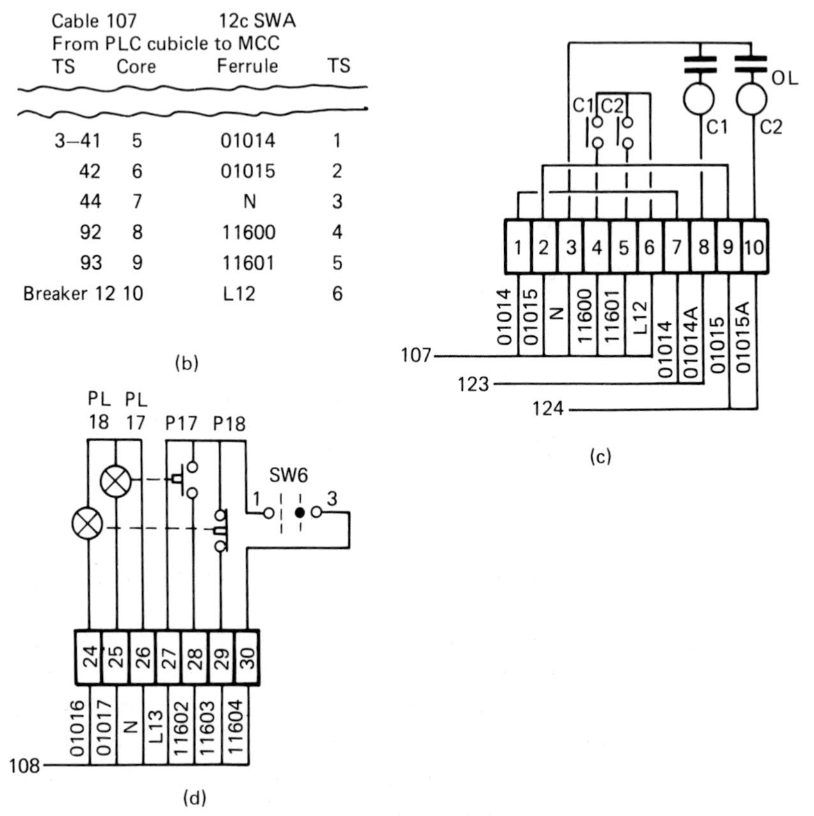

Cable 107 12c SWA
From PLC cubicle to MCC

TS	Core	Ferrule	TS
3–41	5	01014	1
42	6	01015	2
44	7	N	3
92	8	11600	4
93	9	11601	5
Breaker 12	10	L12	6

(b)

Fig.8.23 Plant documentation as usually supplied for first line maintenance. (a) Part of PLC cubicle constructional drawing. (b) Part of cable schedule. (c) Part of MCC constructional drawing. (d) Part of panel constructional drawing.

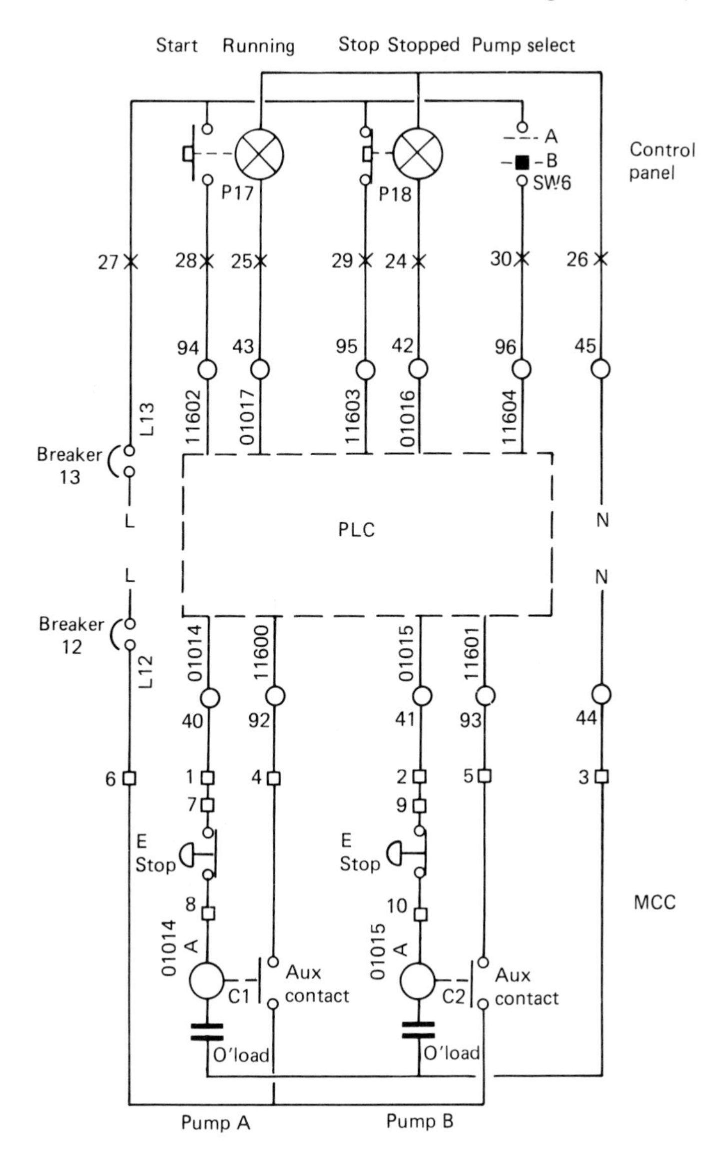

Fig.8.24 The information in fig.8.23 redrawn to show functional relationships.

in fig.8.25. By identifying blocks in which a fault lies, a technician will naturally follow the half split method of section 8.3. FIMs works well with equipment that is modular in nature and has been designed with FIMs in mind, but can be applied to any plant.

Figure 8.26 shows a complete FIMs documentation for a simple modular power supply, and fig.8.27 part of the documentation for a thyristor drive. In each case the FIMs charts will lead the

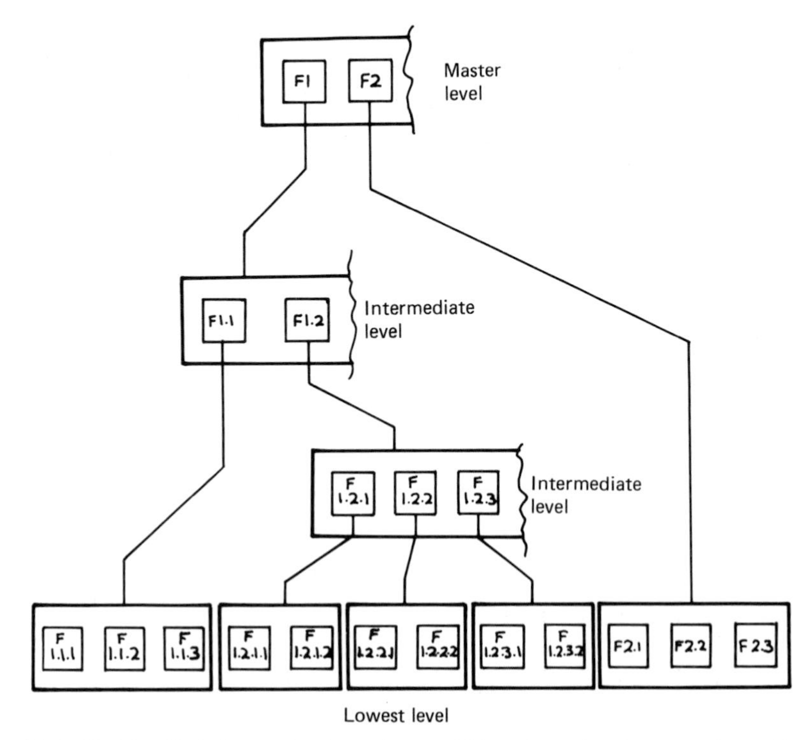

Fig.8.25 The FIMs hierarchy (based on an IBA technical review drawing with permission).

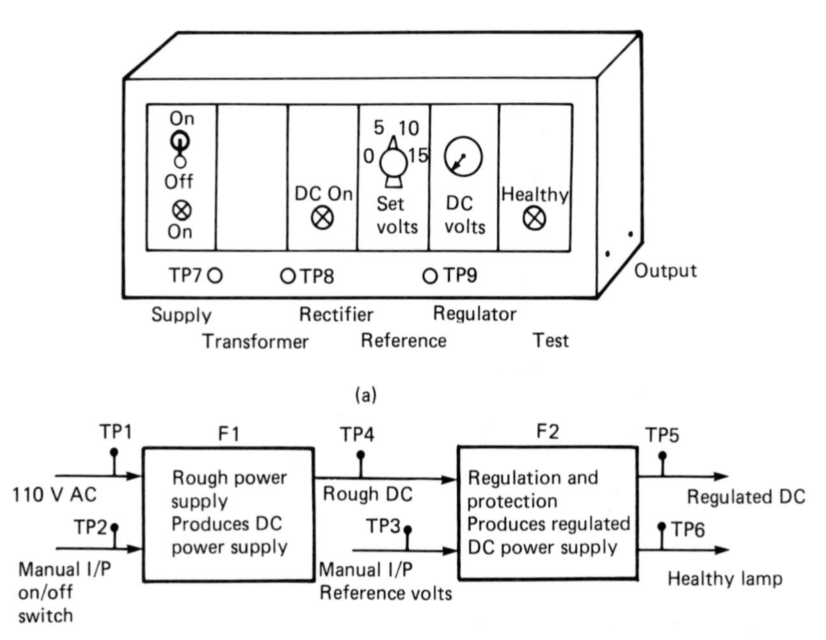

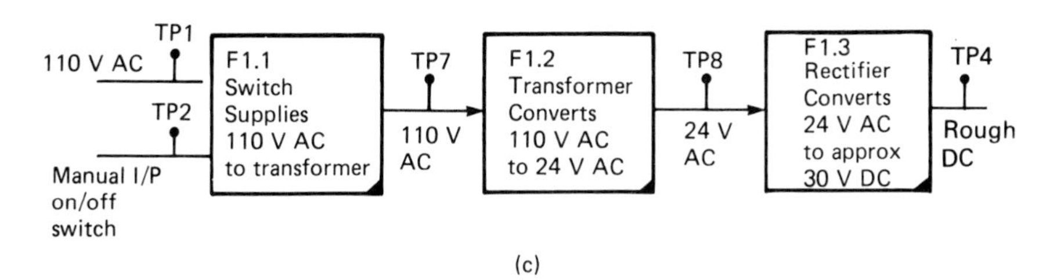

(c)

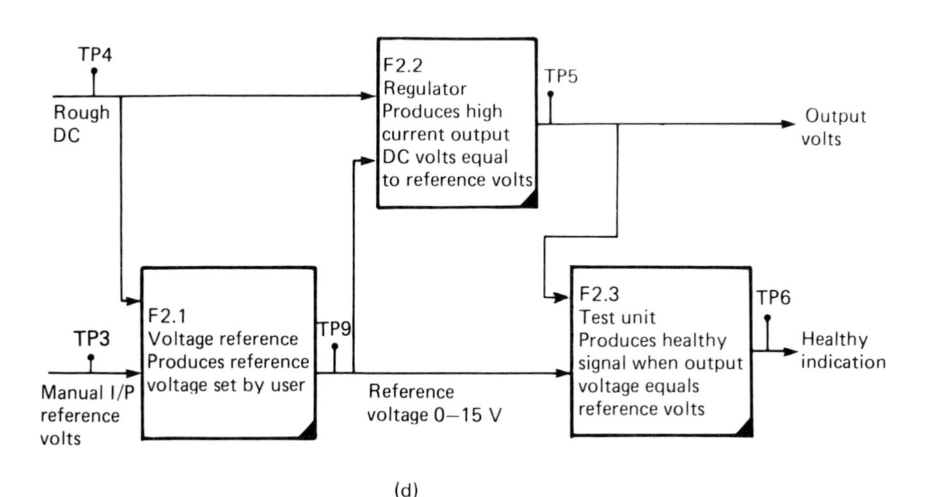

(d)

Test point	Description	Normal indication
1	Neon lamp LP1	ON
2	Supply switch	ON position
3	Reference voltage knob	Pointer shows required voltage
4	Rough DC lamp LP2	ON
5	Output volts	0 – 15 V according to TP3
6	Healthy lamp LP3	ON
7	TP7 multimeter on 250 V AC range	100 – 130 V AC
8	TP8 multimeter on 50 V AC range	27 – 36 V AC
9	TP9 multimeter on 30 V DC range	0 – 15 V according to TP3

(e)

Fig.8.26 Complete FIMs chart for a power supply. (a) Physical construction. (b) Master level. (c) F1 lowest level. (d) F2 lowest level. (e) Test point table.

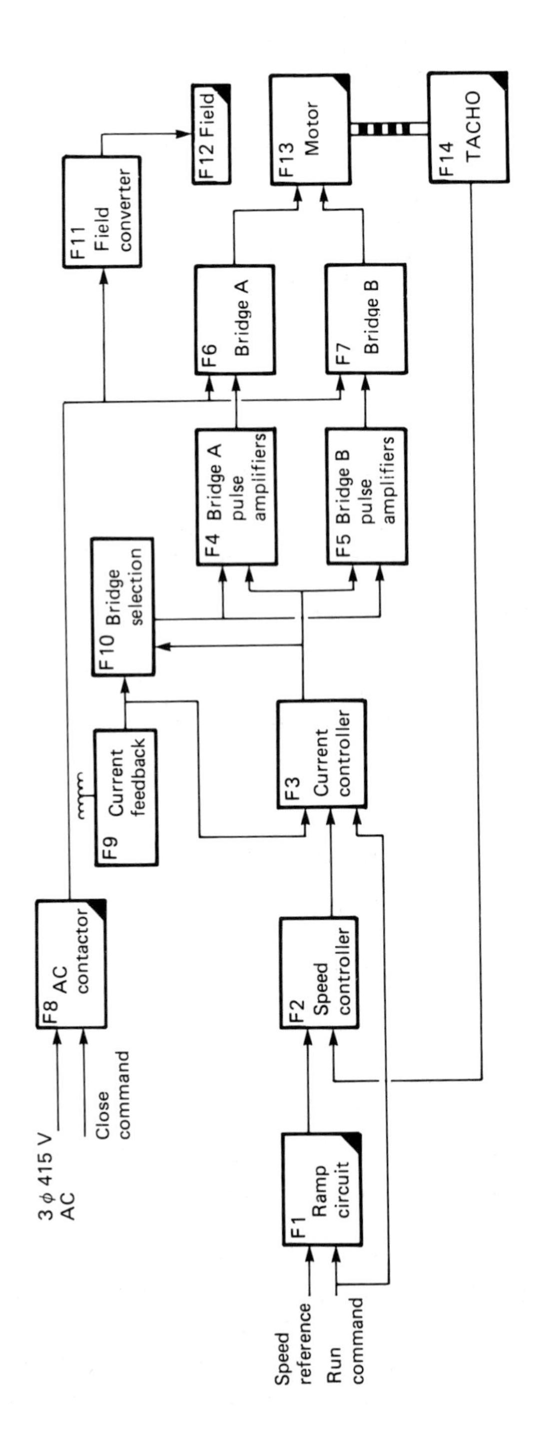

Fig.8.27 Top level FIMs chart for a thyristor drive.

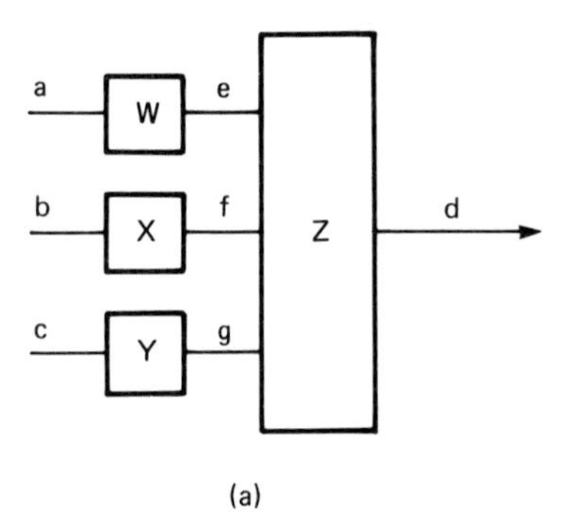

(a)

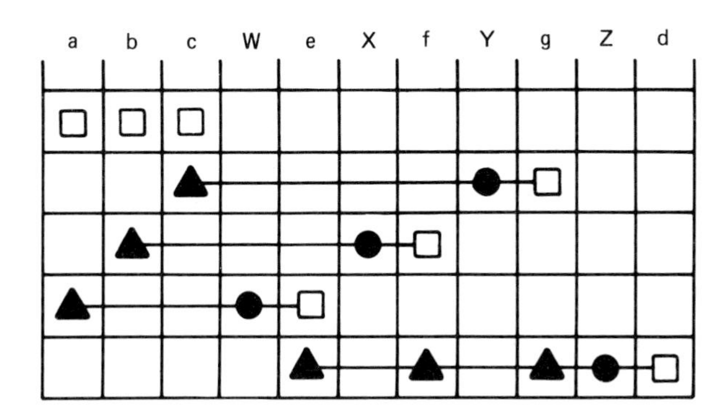

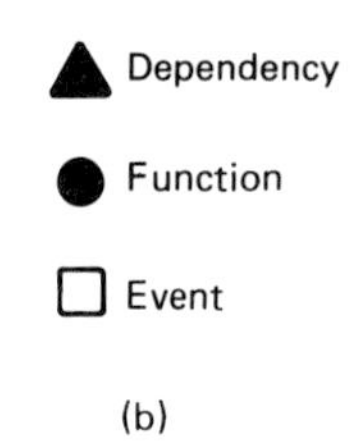

(b)

Fig.8.28 The dependency chart. (a) A simple system. (b) Chart for simple system.

technician to a replaceable unit (denoted by a triangle in the bottom right-hand corner of the block) or a simple circuit diagram.

The FIMs team also devised the concept of a maintenance dependency chart. This shows relationships between items, using the symbols of fig.8.28. The charts are used for fault finding by following the signals back to their origin. On fig.8.28, for example, output d requires unit Z, which in turn requires signals e, f, g, etc.

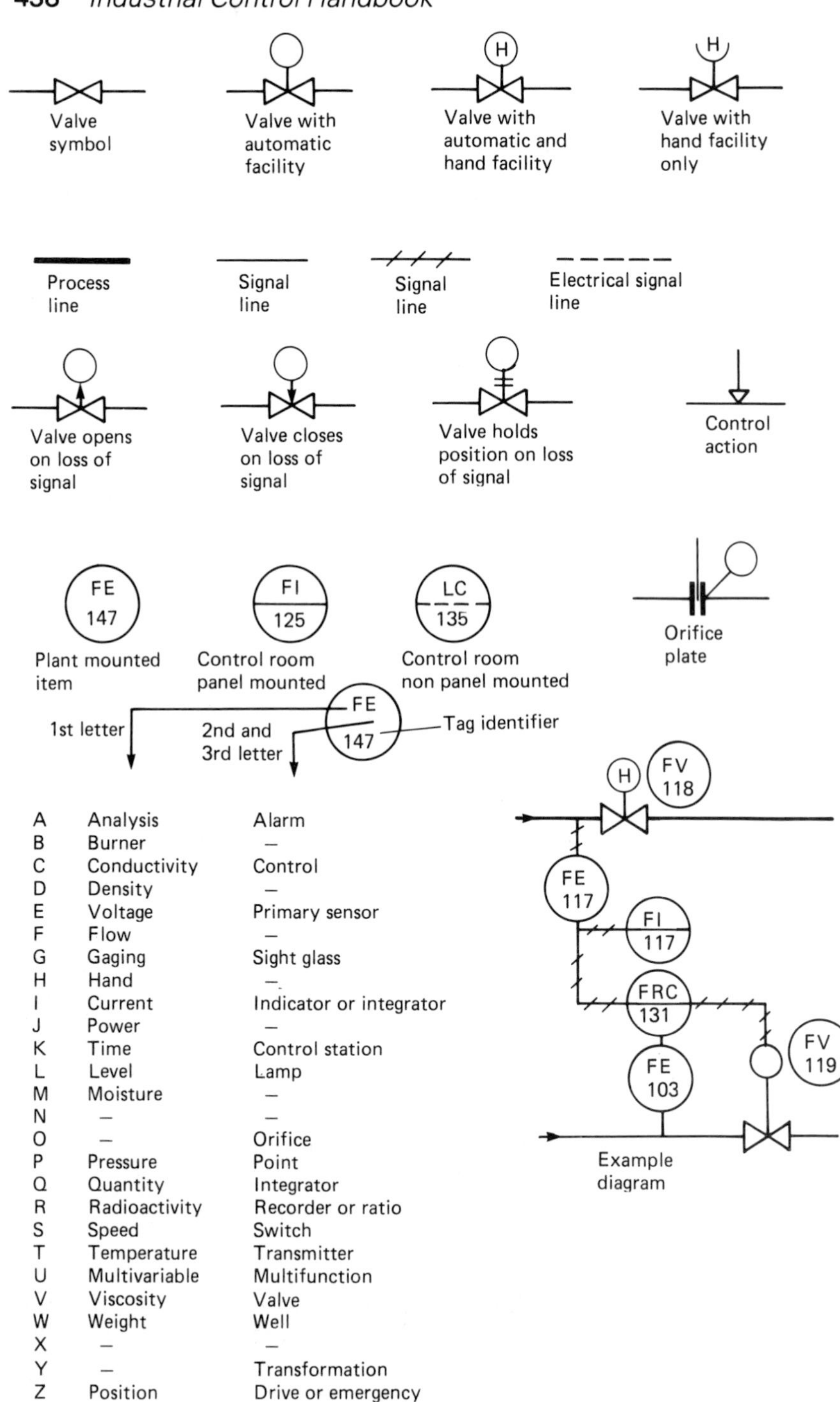

Fig.8.29 P and ID symbols.

A common form of fault finding documentation is the flow chart (also known variously as a symptom analysis chart or, rather grandly, an algorithmically based diagnostic chart). Flow charts aim to lead the technician to the fault via a series of predetermined tests. This approach is well suited to a computer based maintenance system. It is important that all faults are covered (even the subtle ones). Far too often flow charts deal only with the simple, obvious faults such as 'check the fuses' and leave the technician just when he needs most help. Flow charts linked with FIMs can be a powerful tool.

Documentation should include plant and item descriptions and operating parameters at plant commissioning. (Knowing that a feed line is operating at 6 bar is no use if the normal operating pressure is not known.) These should be given by the manufacturers, but it is good practice for maintenance staff to record normal operating conditions before the first fault occurs. The documentation should also include a full spares equipment list with manufacturers' names, addresses and manufacturers' part numbers.

Process control systems can be very complex, and are best described by process drawings. The symbols of fig.8.29 have evolved (Instruments Society of America (ISA), BS 1646 and IS 03511) to show the operation of complex systems. These are often called P and IDs, for piping and instrumentation drawings (not to be confused with PID controllers). Plant mounted equipment should be tagged with the drawing reference identifications (e.g. FE107).

Standard graphical symbols are covered by the following international standards: DIN 40700-40717 (West Germany), BS 3939, ANSI 32.2 (USA), NEMA ICS (USA), CEMA ICS (Canada) and International Electrotechnical Commission (IEC) publication 117. Unfortunately most manufacturers seem to derive their own. Large organisations can probably impose standards on suppliers, but the small purchaser usually does not have sufficient influence.

8.7. Environmental effects

The environment in which equipment operates has a large impact on its reliability. Pressure, humidity, temperature (particularly temperature cycling), corrosive atmospheres and vibration all have an adverse effect on equipment.

Equipment protection is defined in BS 5420 (IEC 144) by the letters IP (for ingress protection) followed by two digits (e.g. IP54). The first digit refers to the degree of protection against solid objects, the second to the protection against liquids, as in Table 8.2. Some IP numbers have commonly used names, but these have no official standing:

Drip proof	IP22
Rain proof	IP23
Splash proof	IP34
Dustproof	IP54
Weatherproof	IP55
Watertight	IP57
Dust tight	IP65

Note the subtle difference between 'shall have no harmful effect' and 'shall not enter' between digits 5 and 6 in Table 8.2. Contrary to some beliefs, water and dust can enter IP55 enclosures in small amounts. It should also be noted that the IP rating is only achieved with the seals in good condition and the door closed.

The temperature range over which equipment will operate needs to be investigated. Most equipment will have ranges specified for storage, operation (with reduced accuracy) outside of which damage will occur. Equipment in the open air in the UK can expect to experience a range of −10°C to +40°C, but this can obviously vary considerably according to local conditions and the proximity of heat sources. Operating conditions for process control equipment is defined in BS 5967, 1980. This defines, for example, temperature and humidity ranges for control rooms (18°C to 27°C) and sheltered outside locations (−25°C to 55°C or −40°C to +70°C), for example, along with humidity and rate of change of temperature limits.

IEC publication 364 gives a more extensive classification. This utilises a three character code (e.g. AD6) to define external influences. The first character defines:

A environment
B utilisation
C construction of buildings

The second character (A...Z) defines the nature of the influence. For environmental effects, A is temperature, B is

Table 8.2

Solid Bodies			Liquids		
First number			Second number		
0	/	No protection	0	/	No protection
1	50 mm	Protection against large solid bodies Hand cannot come into contact with live parts	1		Drops of condensed water falling on enclosure shall cause no harm
2	12 mm	Protection against medium solid bodies Fingers cannot come into contact with live parts	2	leak/roof	Falling liquid shall have no harmful effect up to 15° from vertical
3	2.5 mm	Protection against objects > 2.5 mm diam. Tools (e.g. screwdrivers) cannot contact live parts	3		Falling liquid shall have no harmful effect up to 60° from vertical
4	BICC WIRE 1 mm	Protection against objects > 1 mm	4		Protection against splashing from any direction
5	Talc	Totally enclosed Dust may enter but not in harmful quantities	5		Protection against hose pipe water from any direction. Water may not enter in harmful quantities
6		Dust may not enter Total protection	6		Protection against conditions on ships decks. Occasional immersion. Water must not enter
—	—	—	7	1 m	Permanent immersion up to 1 metre. Water must not enter
—	—	—	8	X	Permanent immersion to specified depth and/or pressure

Common ratings are IP11, IP21, IP22, IP23, IP44, IP54, IP55

humidity, C is altitude, D is water, E is dust, F is corrosive substances, G and H are mechanical stress and vibration, and so on to R (wind effects). The final character is a number which defines the degree of the effect. AA4, for example, defines a temperature range of −5°C to +40°C. Classifications can be combined to give an environment definition.

Utilisation covers the capability of people coming into contact with the equipment (e.g. BA5 skilled technician), electrical resistance of the body, and the proximity of earthed equipment (e.g. BC4m earthed metallic surroundings) and similar considerations.

8.8. Safety considerations

8.8.1. General aspects

The well known author Isaac Asimov postulated three laws of robotics which are, slightly modified:

(1) No robot shall, through action or inaction, allow harm to come to a human being.
(2) No robot shall, through action or inaction, allow harm to come to itself except where this conflicts with the first law.
(3) No robot shall disobey the legitimate orders of a human being except where these conflict with the first two laws.

These three simple laws can be considered to be the basic requirements for process control design if a word 'plant' is substituted for 'robot'. In essence, the priorities must be human safety first, plant protection second, and production a poor third.

A plethora of legislative might ensures that most employees are in a safer environment at work than at home. For people who travel to and from work by car or bike, the most dangerous part of the working day is that journey.

The majority of industrial accidents are not electrocutions or burns or spectacular petrochemical explosions but a series of relatively trivial cuts, bruises, sprains, etc., from slipping on oily floors, incorrect use of tools, poor housekeeping or short cuts around safe working procedures. This is not grounds for complacency, however. Most industrial plants have the capacity to maim or kill, and great care must be taken to ensure that a safe working environment is maintained.

Safety legislature mainly falls under the Health and Safety at Work Act (1974) which puts the responsibility for the safe use of equipment on the manufacturer (who must provide sufficient information for the equipment to be used safely), the user (who must ensure that a piece of equipment is safe by virtue of its application and location) and employees (who must be competent to use the equipment and follow safe working procedures). In the USA, the Occupational Safety and Health Act (OSHA) affords similar protection.

Electrical installations generally fall under the Institute of Electrical Engineers (IEE) wiring regulations, currently the fifteenth edition, and the Electricity (Factory Act) Special Regulations (1908 and 1944). There is also a wide range of legislation for special circumstances such as mines, quarries and petrochemical industries.

Hazards can be considered to fall under hazards during normal operation, hazards during fault conditions, and hazards whilst plant is under maintenance or repair.

Hazards during normal operation cover normal design precautions such as using correctly specified materials, correctly stressed pressure vessels, etc., and ensuring that people cannot come into contact with hazardous material, moving machinery or exposed electrically live equipment. Safe working procedures need to be laid down (and followed!) for potentially dangerous operations.

Fault conditions can introduce additional hazards. The failure of a temperature sensor, for example, could lead to overheating of a chemical reactor and a subsequent fire or explosion. Risks can be reduced by making plants fault tolerant; if a temperature sensor failed, for example, the pressure vessel could be designed so that it could contain the maximum conceivable temperature and pressure. Alternatively, techniques such as redundancy or majority voting can be used. Care must be taken to ensure that the failure of one element of a redundancy based system is detected, or a supposedly two-out-of-three system could be operated unbeknown as a single route system.

Redundancy based systems are also vulnerable to common mode failures, which affect all parallel paths. Typical examples are services such as water, instrument air and electrical supply. Often the operators should be considered as a potential common mode failure.

Maintenance activities are possibly the most hazardous times. Maintenance work, particularly fault rectification, is usually carried out in an atmosphere of haste and stress, both of which are conducive to dangerous short cuts and a potential overlooking of hazards. (The author speaks from personal experience of fault finding on a hydraulic system without having carried out the fundamental step of blowing down the accumulator.) All plants must have a formal written procedure for isolating plant and making it safe–both electrically dead and immobile. Similarly, safe working practices need to be defined for all potentially hazardous jobs. Needless to say, staff training is essential.

Usually, it is the ad hoc repair jobs that result in accidents rather than routine maintenance work. These jobs are less controlled, the work ill defined and the plant usually operational and electrically live. There is also a tendency to bypass safety interlocks to 'get the plant away'. Once out, interlocks tend to stay out. Three Mile Island, Flixborough and (from first reports) Chernobyl all originated, to some extent, from ill advised maintenance work. A cool, logical atmosphere is required for fault finding. Undue pressure from production management can all too easily lead to an accident, maybe weeks after the fault has been 'repaired'.

8.8.2. Explosive atmospheres

An explosive mixture is formed when combustible materials are mixed with air. Combustible vapours occur in many chemical and petrochemical processes and, less obviously, powders from coal dust and even such apparently harmless materials as flour and custard powder, can ignite explosively.

Precautions must obviously be taken to prevent ignition of potentially explosive mixtures. Such ignition can occur from electrical sparks, hot surfaces, mechanical sparks (e.g. formed by rubbing surfaces) and electrostatic discharges. Equipment used in hazardous areas must be designed to prevent the above sources of ignition. The legislation governing such installations is profuse and complex, and the descriptions below should only be taken as a guide to the techniques used.

There are three factors which determine the degree of hazard in any particular location. The first of these determines the probability of an explosive gas being present. In the UK, three zones are defined:

Division 0, where an explosive mixture is present for long periods or continuously under normal operation.

Division 1, where an explosive mixture is likely to occur in normal operation.

Division 2, where an explosive mixture is not likely to occur in normal operation but may occur in abnormal or fault conditions. If an explosive mixture does occur, it will only persist for a short period. (This implies adequate ventilation.)

In the USA divisions 0 and 1 are combined, and called division 1. By default areas not classified as divisions 0 to 2 are deemed non-hazardous.

The second consideration is the ease of ignition of the mixture. There are, unfortunately, several ways of grouping gases, with notable differences between Europe and the USA. Table 8.3 gives an *approximate* relationship between the different standards.

Table 8.3 *Approximate relationships for gas grouping*

Test gas	*IEC*	*BS 1259 Intrinsic safety*	*BS 229 Flameproof*	*American NEC 50°*	*German VDE 0171*
Ammonia		2a			
Propane	IIA	2c	II	D	1
Ethylene	IIB	2d	IIIa Ethylene IIIb coal gas	C	2
Hydrogen	IIC	2e	IV	B	3a
Acetylene	IIC	2f	IV	A	3b, c and n

IEC group 1 is intended for use in mines subject to fire damp (methane).

The final consideration is the ignition temperature of the explosive mixture. This is the temperature of a surface which will ignite the gas, and should not be confused with flash point. The latter is the temperature at which sufficient vapour is produced for the vapour to ignite when in contact with a naked flame. Flash point temperature is lower than ignition temperature. Six classes are defined from T1 (450°C) to T7 (85°C). These six classes are further subdivided in the USA.

All the above classifications are applied to the equipment which is being considered for use in a hazardous environment (i.e. it is the equipment which is really being classified, not the environment, but the subtle difference is more pedantic hairsplitting than a practical consideration).

Equipment intended for use in hazardous areas is generally termed 'explosion proof'. There are several techniques for achieving this (see BS 5501, parts 1 to 9, and BS 5345). The commonest are flameproof enclosures (permitted in divisions 1 and 2), pressurisation (again permitted in divisions 1 and 2) and intrinsic safety (permitted in practically all locations).

A flameproof enclosure is one designed in such a way that it can withstand an internal explosion, and potential flamepaths (e.g. joints) are of such a length and cross section that the flame from the explosion cannot propagate to the outside atmosphere. The surface temperature of the enclosure must, at all times and under all conditions, not exceed the specified temperature classification. Flameproof equipment is, of necessity, bulky and heavy, and particular care needs to be taken to ensure that the integrity is not affected by maintenance work. Flameproof enclosures are the only practical solution for electrical motors.

With pressurisation, the equipment is kept separate from the explosive atmosphere by a positive pressure differential maintained by a purging gas (e.g. nitrogen). In division 2 applications, loss of pressurisation should raise an alarm, whilst in division 1 an automatic shut down interlock on pressure loss is required. The interlocks, which will operate in a *non*-safe atmosphere, will probably use the next technique: intrinsic safety.

Intrinsically safe equipment is designed such that no possible normal or fault condition can result in the ignition of an explosive mixture. Somewhat simplified, this limits operating voltages to 30 V *and* operating currents to 100 mA. Intrinsically safe equipment is subdivided into 'ia' where safety is maintained with two simultaneous faults, and 'ib' where safety is maintained with one fault.

The principle advantages of intrinsically safe equipment are ease of use, simplicity and the ability (in some circumstances) to be maintained and adjusted whilst live.

Some portable equipment can be made inherently safe (e.g. meters and radios) but problems exist with, say, transducers. These operate at low voltages and currents, but their cabling could, under fault conditions and intercable shorts, introduce high voltages and currents into the hazardous area. Safety is maintained by the use of safety barriers at the transition between safe and hazardous areas. These have no effect on normal signals, but prevent fault voltages or currents entering the hazardous area, as in fig.8.30a.

Figure 8.30b shows a typical zener barrier. As long as the operating voltages are less than the zener voltage, the barrier has no effect. If the input voltage rises due to a fault, one zener will

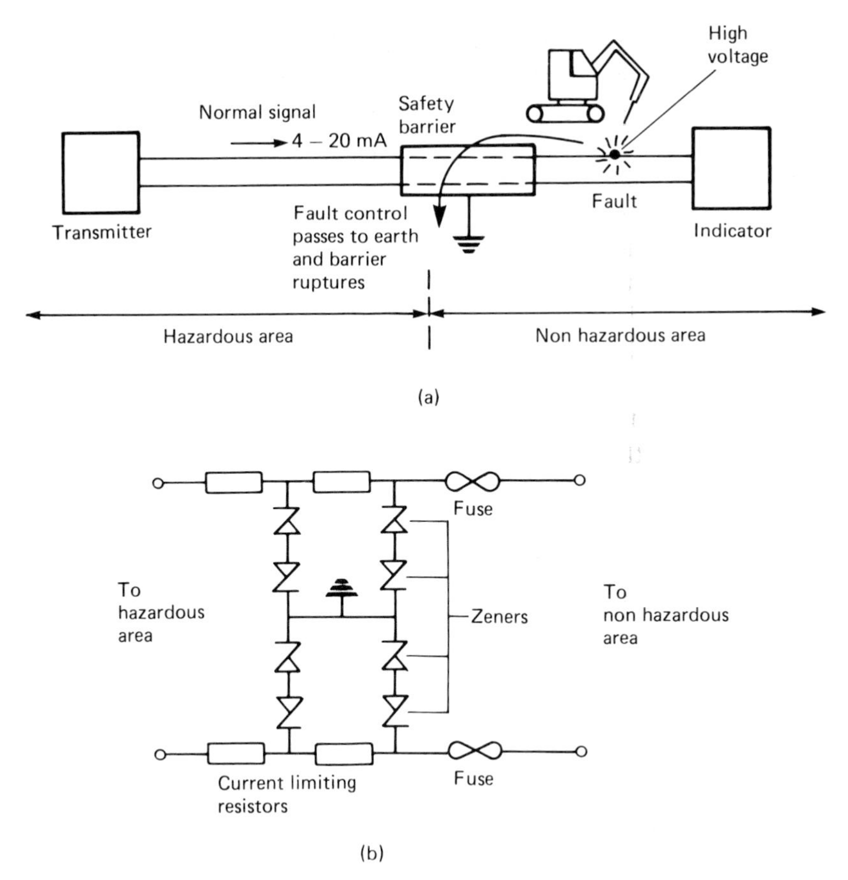

Fig.8.30 Zener safety barriers. (a) Principle of barrier protection. (b) Construction; note this is a certified device and cannot be 'home built'.

conduct, causing the fuse to blow. Barriers are certified, and consequently the fuses are not user replaceable.

The use of zener barriers allows great flexibility in design, and almost total freedom in the choice of 'safe area' equipment. Care should be taken to ensure that the 'hazardous area' equipment cannot store electrical energy, either capacitively or inductively, which could result in a spark in a fault condition. Cable parameters (self-capacitance and inductance) are theoretically important, but unlikely to be significant in practical installations.

Legislation governing the use of equipment in explosive atmospheres is complex, and places responsibility on both manufacturer and user. Compliance with the legislation is best demonstrated by certification by an independent recognised authority (such as BASEEFA in the UK). Electrical equipment for use in mines is covered by separate, and very specialised, legislation.

Pneumatic, hydraulic and fibre optic signals, of course, present no hazard in an explosive atmosphere, and are often chosen as an alternative to electrical instrumentation in difficult applications.

INDEX